# mODa 7 – Advances in Model-Oriented Design and Analysis

# Contributions to Statistics

V. Fedorov/W.G. Müller/I.N. Vuchkov (Eds.)
Model-Oriented Data Analysis
XII/248 pages, 1992

J. Antoch (Ed.)
Computational Aspects of Model Choice
VII/285 pages, 1993

W.G. Müller/H.P. Wynn/A.A. Zhigljavsky (Eds.)
Model-Oriented Data Analysis
XIII/287 pages, 1993

P.Mandl/M. Hušková (Eds.)
Asymptotic Statistics
X/474 pages, 1994

P. Dirschedl/R. Ostermann (Eds.)
Computational Statistics
VII/553 pages, 1994

C.P. Kitsos/W.G. Müller (Eds.)
MODA 4 – Advances in Model-Oriented Data Analysis
XIV/297 pages, 1995

H. Schmidli
Reduced Rank Regression
X/179 pages, 1995

W. Härdle/M.G. Schimek (Eds.)
Statistical Theory and Computational Aspects of Smoothing
VIII/265 pages, 1996

S. Klinke
Data Structures for Computational Statistics
VIII/284 pages, 1997

A.C. Atkinson/L. Pronzato/H.P. Wynn (Eds.)
MODA 5 – Advances in Model-Oriented Data Analysis and Experimental Design
XIV/300 pages, 1998

M. Moryson
Testing for Random Walk Coefficients in Regression and State Space Models
XV/317 pages, 1998

S. Biffignandi (Ed.)
Micro- and Macrodata of Firms
XII/776 pages, 1999

W. Härdle/Hua Liang/J. Gao
Partially Linear Models
X/203 pages, 2000

W.G. Müller
Collecting Spatial Data
2nd edition
XII/196 pages, 2001

A.C. Atkinson/P. Hackl/W.G. Müller (Eds.)
mODa 6 – Advances in Model-Oriented Design and Analysis
XVI/283 pages, 2001

C. Lauro/J. Antoch/V. Esposito Vinzi/G. Saporta (Eds.)
Multivariate Total Quality Control
XIII/236 pages, 2002

P.-A. Monney
A Mathematical Theory of Arguments for Statistical Evidence
XIII/154 pages, 2003

Y. Haitovsky/H.R. Lerche/Y. Ritov (Eds.)
Foundations of Statistical Inference
XII/230 pages, 2003

C. Dagum/G. Ferrari (Eds.)
Household Behaviour, Equivalence Scales Welfare and Poverty
XVI/296 pages, 2004

Alessandro Di Bucchianico ·
Henning Läuter · Henry P. Wynn
(Editors)

# mODa 7 – Advances in Model-Oriented Design and Analysis

Proceedings of the 7th International Workshop on Model-Oriented Design and Analysis held in Heeze, The Netherlands, June 14–18, 2004

With 17 Figures and 26 Tables

Physica-Verlag
A Springer-Verlag Company

**Series Editors**
Werner A. Müller
Martina Bihn

**Editors**
Dr. Alessandro Di Bucchianico
EURANDOM and
Eindhoven University of Technology
Department of Mathematics and
Computer Science
P.O. Box 513
5600 MB Eindhoven
The Netherlands
A.d.Bucchianico@TUE.nl

Prof. Dr. Henning Läuter
University of Potsdam
Institute of Mathematics
14415 Potsdam
laeuter@rz.uni-potsdam.de
Germany

Prof. Dr. Henry P. Wynn
London School of Economics
Houghton Street
London, WC2A 2AE, UK
h.wynn@lse.ac.uk
United Kingdom

ISSN 1431-1968
ISBN 3-7908-0213-1 Physica-Verlag Heidelberg New York

Library of Congress Control Number: 2004105253
Bibliographic information published by Die Deutsche Bibliothek
Die Deutsche Bibliothek lists this publication in the Deutsche Nationalbibliografie; detailed bibliographic data is available in the Internet at <http://dnb.ddb.de>.



Physica-Verlag is a part of Springer Science+Business Media

springeronline.com



Softcover design: Erich Kirchner, Heidelberg

SPIN 10994474     88/3130-5 4 3 2 1 0 – Printed on acid-free and non-aging paper

# Preface

This volume contains a substantial number of the papers presented at the mODa 7 conference. mODa stands for Model Oriented Data Analysis and previous conferences have been held in Wartburg (1987) (then in the GDR), St Kirik monastery, Bulgaria (1990), Petrodvorets, St Petersburg, Russia (1992), The island of Spetsos, Greece (1995), the Centre International de Rencontres Mathématiques, Marseilles, France (1998) and Puchberg/Schneeberg, Austria, (2001).

The purpose of these workshops has traditionally been to bring together scientists from the East and West interested in the optimal design of experiments and related topics and younger and senior researchers in the field. These traditions remain vital to the health of the series. During this period Europe has seen increasing unity and the organizers of and participants in mODa must take some satisfaction from the fact that the youthful ideals of the founders of the series are reflected in this transition.

The present conference and mODa 6 are supported by a European Union conference grant (contract HPCF-CT 2000 00045) whose funding emphasis is on younger participants. The company GlaxoSmithKline has very generously continued its support. We are very grateful for these substantial contributions.

mODa 7 has benefited from excellent administrative input from the staff of EURANDOM, particularly Lucienne Coolen. EURANDOM is a major European stochastics research institute, housed within, and partly supported by, the Eindhoven University of Technology, The Netherlands. The conference itself takes place at the conference centre Kapellerput in Heeze, not far from Eindhoven. Among the post-doctoral fellows and PhD students who helped and participated in mODa 7, we should thank Peter van de Ven for tireless work processing papers for this volume.

The mODa series, both the conferences and the volumes, have grown to a leadership position in experimental design and analysis. It is not surprising therefore that most of the main developments in the area have been represented over the years.

The most evident trend in the present volume is towards optimal design for nonlinear models and models with nonstandard assumptions. The papers by Atkinson, by Biedermann, Dette and Pepelyshev and by Uciński and Bogacka cover non-linear models arising from compartmental models in chemical kinetics given by first order differential equations. Trandafir and López-Fidalgo cover the familiar Michaelis-Menten model. The papers by Gauchi and Pázman and by Pázman and Pronzato cover special theoretical aspects of non-linear models. The locally optimal design method of Melas also falls into the non-linear category. In all this work the main feature is that the information matrix is parameter dependent. This is also a feature of the generalized linear models covered by Pinto and Ponce de Leon who also look at Bayesian solutions.

The area of biased coin designs, up-and-down models, urn models is stimulated by its application to dose-response experiments and clinical trials more generally. The papers by Baldi Antognini, by Biswas and Mandal, by Giovagnoli and by Tymofyeyev, Rosenberger and Hu are in this area. It is good to see other papers on the application of optimal design ideas in medicine. Fedorov and Leonov investigate how to use optimal design methods for control in the presence of forced baseline measurements and Rabie and Flournoy study the situation of double (contingent) responses such as toxicity and disease failure. Yin, Wang, Wang and Zheng use a sophisticated control/learning strategic for patient control during anesthesia.

The main stream of optimal design is well represented. Maximum Entropy Sampling relates optimal design to information theoretic formulations. The papers by Anstreicher and Lee and by Wynn are closely related with the first covering bounds and computational aspects and the second sketching the links to $D$-optimality. The paper by Pronzato, Thierry and Wolsztynski use entropy as a basis for estimation, rather than design. Four papers cover different aspects of optimal design for the standard linear model, the core of the field: Harman looks at efficiency, how close a given design is to the global optimum; Müller and Kitsos look at combined optimal design and simultaneous inference (confidence, tolerance); Rodríguez, Ortiz and Martínez study design for models with non-constant (heteroscedastic) error variance; Torsney and Mandal continue their work on algorithms.

Two other areas are represented by single papers. Basso, Salmaso, Evangelaras and Koukouvinos cover the difficult subject of experimental design combined with non-parametric testing. Vuchkov presents the important tolerance design method which is an aspect of robust engineering, for the special case of mixture experiments.

Eindhoven,
January 2004

*Alessandro Di Bucchianico*
*Henning Läuter*
*Henry P. Wynn*

# Contents

**A Masked Spectral Bound for Maximum-Entropy Sampling**
*K.M. Anstreicher, J. Lee* ........ 1
1 Introduction ........ 1
2 The Masked Spectral Bound ........ 2
3 The Minimization Method ........ 3
4 Computational Results ........ 5
5 Alternative Use of Oppenheim's Inequality ........ 9
6 Conclusion ........ 10
References ........ 10

**Some Bayesian Optimum Designs for Response Transformation in Nonlinear Models with Nonconstant Variance**
*A.C. Atkinson* ........ 13
1 Introduction ........ 13
2 Transformations and First-Order Decay ........ 14
3 Optimum Design for a Multivariate Response ........ 15
4 Parameter Sensitivities and Transforming Both Sides ........ 16
5 Two Consecutive First-Order Reactions ........ 17
6 Efficiencies and Bayesian Optimum Designs ........ 17
7 Discussion ........ 20
References ........ 21

**Extensions of the Ehrenfest Urn Designs for Comparing Two Treatments**
*A. Baldi Antognini* ........ 23
1 Introduction ........ 23
2 The Ehrenfest Urn Design ........ 25
3 Symmetric Ehrenfest Design for Achieving Balance ........ 26
4 Asymmetric Ehrenfest Design for a Generic Target ........ 26
5 Ehrenfest-Brillouin Design ........ 27
6 Some Convergence Properties ........ 29
References ........ 30

**Nonparametric Testing for Main Effects on Inequivalent Designs**
*D. Basso, L. Salmaso, H. Evangelaras, C. Koukouvinos* .............. 33
1 Introduction .............................................. 33
2 The IMP Test .............................................. 35
3 A Comparative Simulation Study .............................. 37
4 Conclusions ................................................ 39
References .................................................. 40

**Maximin Optimal Designs for a Compartmental Model**
*S. Biedermann, H. Dette, A. Pepelyshev* .......................... 41
1 Introduction .............................................. 41
2 Locally $D$-optimal Designs .................................. 43
3 Standardized Maximin $D$-optimal Designs ..................... 45
References .................................................. 48

**Optimal Adaptive Designs in Phase III Clinical Trials for Continuous Responses with Covariates**
*A. Biswas, S. Mandal* ............................................ 51
1 Introduction .............................................. 51
2 Optimal Designs for Continuous Distributions .................. 52
2.1 Response Distributions with one Unknown Parameter ....... 52
2.2 Distributions Having more than one Parameter ............. 53
3 Presence of Covariates ...................................... 54
4 Conclusions ................................................ 56
References .................................................. 58

**Optimal Designs for Regression Models with Forced Measurements at Baseline**
*V.V. Fedorov, S. Leonov* ........................................ 61
1 Introduction .............................................. 61
2 Model ...................................................... 62
3 Optimal Designs for Model (1) ................................ 63
3.1 Equivalence of $D$-optimal Designs when only $\gamma$ are Unknown . 63
3.2 Unknown $\gamma$ and Population Variance $\Lambda$ .................. 66
4 Case when Baseline and Placebo Responses Coincide ............ 67
4.1 Only Response Parameters $\gamma$ Unknown .................... 68
4.2 Unknown Parameters $\gamma$ and $\Lambda$ ............................ 68
References .................................................. 68

**Small Size Designs in Nonlinear Models Computed by Stochastic Optimization**
*J.-P. Gauchi, A. Pázman* .......................................... 71
1 Introduction .............................................. 71
2 Optimality Criteria Expressed as an Integral .................. 72
3 Density of the Estimator .................................... 73

4 The Penalty for the Boundary of $\Theta$ .... 74
5 Accelerated Method of Stochastic Optimization .... 75
6 Examples and Numerical Results .... 77
References .... 78

**Asymptotic Properties of Biased Coin Designs for Treatment Allocation**
*A. Giovagnoli* .... 81
1 Introduction .... 81
2 Markovian Experiments .... 82
3 Different Types of Biased Coin Designs .... 84
3.1 BCD's for Comparing Two Treatments .... 84
3.2 Biased Coin Design for Targeting a Quantile .... 85
4 Asymptotic Properties of the Adjustable Biased Coin Designs .... 86
5 Asymptotic Properties of Up-and-Down Designs .... 87
References .... 88

**Lower Bounds on Efficiency Ratios Based on $\Phi_p$-Optimal Designs**
*R. Harman* .... 89
1 Introduction .... 89
2 Bounds on the $E_k$-optimal Values Based on the Eigenvalues of $\Phi_p$-optimal Information Matrices .... 91
3 Bounds on the $E$-efficiency and the Minimal Efficiency Ratio of $\Phi_p$-optimal Designs .... 91
4 Example: the Minimal Efficiency Ratio of the $D$-optimal Design for the Model of Spring Balance Weighing .... 93
5 Appendix - Proofs .... 94
References .... 96

**On a Functional Approach to Locally Optimal Designs**
*V.B. Melas* .... 97
1 Introduction .... 97
2 Outline of the Problem .... 98
3 Basic Analytical Results .... 99
4 The Three-Parameter Logistic Distribution .... 101
References .... 104

**Optimal Design Criteria Based on Tolerance Regions**
*C.H. Müller, C.P. Kitsos* .... 107
1 Introduction .... 107
2 $\beta$-expectation Bayesian Tolerance Regions .... 109
3 Optimal Designs .... 112
References .... 114

**Simultaneous Choice of Design and Estimator in Nonlinear Regression with Parameterized Variance**
*A. Pázman, L. Pronzato* ..... 117
1 Introduction ..... 117
2 Randomized Designs and Uniform Strong Law of Large Numbers .. 118
3 Penalized Weighted LS and Two-Stage LS Estimation ..... 119
3.1 Penalized Weighted LS Estimation ..... 120
3.2 Two-stage LS Estimation ..... 121
4 Choosing the Design and the Estimator ..... 122
References ..... 124

**Minimum Entropy Estimation in Semi-Parametric Models: a Candidate for Adaptive Estimation?**
*L. Pronzato, É. Thierry, É. Wolsztynski* ..... 125
1 Introduction ..... 125
2 Minimizing Entropy ..... 126
3 Adaptive Estimation in the Location Model ..... 128
4 Adaptive Estimation in Nonlinear Regression ..... 130
References ..... 132

**Optimal Designs for Contingent Response Models**
*H.S. Rabie, N. Flournoy* ..... 133
1 Introduction ..... 133
2 The Contingent Response Model ..... 134
3 The Design Problem ..... 135
4 Locally $D$-optimal Designs ..... 136
4.1 Unequal Slopes $\beta_1 \neq \beta_2; \boldsymbol{\Theta} = (\alpha_1, \beta_1, \alpha_2, \beta_2); \boldsymbol{\theta} = (\alpha_2, \beta_2, \mu, r)$ 136
4.2 Equal Slopes $\beta_1 = \beta_2 = \beta$; $\boldsymbol{\Theta} = (\alpha_1, \beta, \alpha_2)$; $\boldsymbol{\theta} = (\alpha_2, \beta, \mu)$ ... 138
5 Locally $c$-optimal Designs ..... 138
5.1 Unequal Slopes $\beta_1 \neq \beta_2; \boldsymbol{\Theta} = (\alpha_1, \beta_1, \alpha_2, \beta_2); \boldsymbol{\theta} = (\alpha_2, \beta_2, \mu, r)$ ..... 139
5.2 Equal Slopes $\beta_1 = \beta_2 = \beta; \boldsymbol{\Theta} = (\alpha_2, \beta, \alpha_1); \boldsymbol{\theta} = (\alpha_2, \beta, \mu)$ .... 139
6 Conclusion ..... 140
References ..... 141

**Bayesian $D$-Optimal Designs for Generalized Linear Models with a Varying Dispersion Parameter**
*E. Rodrigues Pinto, A. Ponce de Leon* ..... 143
1 Introduction ..... 143
2 The Quasi-Likelihood ..... 144
3 The Extended Quasi-Likelihood ..... 145
4 The Model ..... 146
5 The Equivalence Theorem ..... 147
6 Cake Mix Example ..... 148
7 Final Considerations ..... 150

References .......... 151

**$L$–optimum Designs in Multi–factor Models with Heteroscedastic Errors**
*C. Rodríguez, I. Ortiz, I. Martínez* .......... 153
1 Introduction .......... 153
2 Product Models .......... 154
3 Additive Models .......... 155
3.1 Additive Models with Constant Term .......... 155
3.2 Orthogonal Additive Model .......... 158
4 Some Remarks .......... 159
References .......... 160

**Multiplicative Algorithms for Constructing Optimizing Distributions : Further Developments**
*B. Torsney, S. Mandal* .......... 163
1 Introduction .......... 163
2 Optimality Conditions .......... 164
3 Algorithms .......... 164
3.1 Properties of the Iteration (3) .......... 165
4 Optimal Distribution on Spaces .......... 166
5 Objective Choices .......... 168
6 Conclusions .......... 170
References .......... 171

**Locally Optimal Designs for an Extension of the Michaelis-Menten Model**
*C. Trandafir, J. López-Fidalgo* .......... 173
1 Introduction .......... 173
2 Theoretical Background .......... 174
2.1 The Nonlinear Regression Model .......... 174
2.2 The Design Criteria .......... 175
3 Extension of the Michaelis-Menten Model .......... 175
4 Compound Optimal Designs for 3 Parameters .......... 177
4.1 $c$-optimal Designs .......... 177
4.2 $c$-efficiencies of the $D$-optimal Design .......... 178
4.3 Compound Optimal Designs .......... 178
References .......... 180

**Asymptotic Properties of Urn Designs for Three-arm Clinical Trials**
*Y. Tymofyeyev, W.F. Rosenberger, F. Hu* .......... 183
1 Introduction .......... 183
2 Spectral Analysis of Generating Matrix .......... 185
3 Asymptotic Distribution of $\mathbf{Y_n}$ .......... 186
4 Asymptotic Distribution of $\mathbf{N}_n$ .......... 188

5 Discussion ... 189
References ... 190

**$T$-Optimum Designs for Multiresponse Dynamic Heteroscedastic Models**
*D. Uciński, B. Bogacka* ... 191
1 Introduction ... 191
2 $T$-optimality Criterion for Heteroscedastic Models ... 193
3 Numerical Construction of Optimum Designs ... 195
4 Conclusion ... 197
References ... 198

**Error Transmission in Mixture Experiments**
*I.N. Vuchkov* ... 201
1 Introduction ... 201
2 Mean and Variance Models for Mixture Experiments ... 202
3 Mean and Variance Models for Experiments with Mixture and Process Variables ... 204
4 Example ... 207
References ... 209

**Maximum Entropy Sampling and General Equivalence Theory**
*H.P. Wynn* ... 211
1 Introduction ... 211
1.1 Maximum Entropy Sampling ... 212
1.2 Limit of Bayes Case ... 213
2 Continuous Theory ... 214
2.1 Constrained Measures for the MES Case ... 215
3 General MES and $D$-optimality ... 216
References ... 218

**Towards Identification of Patient Responses to Anesthesia Infusion in Real Time**
*G. Yin, H. Wang, L.Y. Wang, H. Zheng* ... 219
1 Introduction ... 219
2 System ... 220
2.1 Basic Two-Step Setup for Real-Time Learning Algorithms ... 222
3 Recursive Algorithms ... 222
4 Convergence and Rates of Convergence ... 224
References ... 226

**List of Contributors** ... 229

**List of Referees** ... 233

**List of Figures** ........ 235

**List of Tables** ........ 237

**Index** ........ 239

# A Masked Spectral Bound for Maximum-Entropy Sampling

K.M. Anstreicher[1] and J. Lee[2]

[1] Tippie College of Business, University of Iowa, Iowa City, IA. U.S.A.
kurt-anstreicher@uiowa.edu

[2] IBM T.J. Watson Research Center, Yorktown Heights, NY. U.S.A.
jonlee@us.ibm.com

**Summary.** We introduce a new "masked spectral bound" for the maximum-entropy sampling problem. This bound is a continuous generalization of the very effective "spectral partition bound". Optimization of the masked spectral bound requires the minimization of a nonconvex, nondifferentiable objective over a semidefiniteness constraint. We describe a nonlinear affine scaling algorithm to approximately minimize the bound. Implementation of the procedure obtains excellent bounds at modest computational expense.

**Key words:** maximum-entropy sampling, experimental design, semidefinite programming, spectral partition bound.

## 1 Introduction

Let $C$ be a symmetric positive definite matrix of order $n$. Let $s$ be an integer satisfying $0 < s < n$. For subsets $S$ and $T$ of $N := \{1, 2, \ldots, n\}$, we let $C[S,T]$ denote the submatrix of $C$ having rows indexed by $S$ and columns indexed by $T$.

The *maximum-entropy sampling problem* is to calculate

$$\text{MESP}: \qquad z(C,s,n) := \max\{\ln\det C[S,S] \ : \ S \subset N,\ |S| = s\}.$$

This problem was introduced in Shewry and Wynn (1987) and first used in an applied monitoring design context in Caselton and Zidek (1984) (also see Guttorp, Le, Sampson, and Zidek (1992), Muramatsu (1998), Sebastiani and Wynn (2000), Wu and Zidek (1992)). From an algorithmic perspective, the problem has been studied extensively; see Anstreicher, Fampa, Lee, and Williams (1996, 1999), Hoffman, Lee, and Williams (2001), Ko, Lee, and Queyranne (1995), Lee (1998), Lee and Williams (2003) and the surveys Lee (2000, 2001). In a typical application, $C$ is a sample covariance matrix obtained from time-series observations of one variable at $n$ locations, and it is desired to choose $s$ locations from which to conduct subsequent data collection. The use of entropy as a selection criterion, together with the assumption that values at the $n$ locations are drawn from a multivariate normal distribution, then leads naturally to MESP. (Note that ln det $C[S,S]$ is, up to constants, the entropy of the Gaussian random variables having covariance matrix $C[S,S]$.)

Exact algorithms to compute a maximum-entropy design use the "branch-and-bound" framework. Besides lower-bounding heuristics used to find good candidate solutions, a key ingredient is the upper-bounding method. A fast method that can provide a reasonably sharp upper bound on $z(C,n,s)$ is critical to the success of such an approach. Much of the aforementioned algorithmic work concentrates on developing effective upper bounds. The present article continues in this direction.

The organization of the paper is as follows. In Section 1 we describe the masked spectral bound, which is derived using Oppenheim's inequality (see Oppenheim (1930)). The new bound is a generalization of the spectral partition bound of Hoffman, Lee, and Williams (2001), which itself is a generalization of both the eigenvalue bound of Ko, Lee, and Queyranne (1995) and the diagonal bound of Hoffman, Lee, and Williams (2001). Optimization of the masked spectral bound requires the minimization of a nonconvex, nondifferentiable objective over a semidefiniteness constraint. In Section 3 we consider the application of a nonlinear affine scaling algorithm to approximately optimize the masked spectral bound. In Section 4 we give computational results. We find that the performance of the masked spectral bound is superior to several known bounds, and the computational expense to obtain the bound is quite reasonable. Finally, in Section 5 we describe some alternative bounds based on Oppenheim's inequality.

NOTATION: $I$ is an identity matrix; $E$ is a matrix of all ones; $e$ is a vector of all ones; det is determinant; for a matrix $X$, $\mathrm{diag}(X)$ is the vector of diagonal entries in $X$; for a vector $x$, $\mathrm{Diag}(x)$ is the diagonal matrix such that $x = \mathrm{diag}(\mathrm{Diag}(x))$; for a matrix (or vector) $X$, we denote its transpose by $X'$; $X \succeq 0$ denotes that $X$ is symmetric and positive semidefinite; $\circ$ is Hadamard (that is, element-wise) product; $A \bullet B := \mathrm{tr}(AB')$; $\lambda_\ell(X)$ is the $\ell$th greatest eigenvalue of $X$. Other basics concerning matrix algebra can be found in Horn and Johnson (1985, 1991).

## 2 The Masked Spectral Bound

A *mask* is an $X \succeq 0$ having $\mathrm{diag}(X) = e$. We define the associated *masked spectral bound* as

$$\xi_C(X) := \textstyle\sum_{\ell=1}^{s} \ln\left(\lambda_\ell\left(C \circ X\right)\right).$$

Special cases include the *diagonal bound* $\xi_C(I)$, the *eigenvalue bound* $\xi_C(E)$, and the *spectral partition bound* $\xi_C(X)$, where $X$ is a block-diagonal matrix with diagonal blocks being matrices of all 1's. Obviously the spectral partition bound is a generalization of the diagonal and eigenvalue bounds, and the masked spectral bound is a further generalization.

Validity of the masked spectral bound ($\xi_C(X) \geq z(C,s,n)$) is based on (i) *Oppenheim's inequality* (see Oppenheim (1930))

$$\det A \leq \det A \circ B / \prod_{j=1}^{n} B_{jj} ,$$

where $A \succeq 0$ and $B \succeq 0$ (see (Horn and Johnson, 1985, Thm 7.8.6)), and (ii) the eigenvalue inequalities $\lambda_\ell(A) \geq \lambda_\ell(B)$, where $A \succeq 0$, and $B$ is a principal submatrix of $A$ (see (Horn and Johnson, 1985, Theorem 4.3.15)).

We would like to minimize $\xi_C(X)$ over all masks. While the set of masks is indeed a convex set, the function $\xi_C(\cdot)$ is not convex. So, we will have to be content with heuristics that seek a good local minimum of $\xi_C(\cdot)$.

By exploiting the identity

$$\ln \det C[S,S] = \ln \det C + \ln \det C^{-1}[N \setminus S, N \setminus S],$$

*any* bound for the complementary problem of choosing a maximum entropy set of $n-s$ points with respect to the covariance matrix $C^{-1}$ translates to a bound for the original problem (see Anstreicher, Fampa, Lee, and Williams (1996, 1999)).

There are other bounds that do not fit neatly into the present framework. The "linear integer programming bound" introduced in Lee and Williams (2003) is a strengthening of the spectral partition bound. Although quite effective, that bound is computationally very intensive. The "nonlinear-programming bound" of Anstreicher, Fampa, Lee, and Williams (1996, 1999) is based on a concave, continuous relaxation of MESP. That bound, while quite effective on some problems, is somewhat inconsistent and requires parameter tuning that is not well understood.

## 3 The Minimization Method

For $\tilde{X} \succeq 0$, let $u_\ell(C \circ \tilde{X})$ be an eigenvector, of Euclidean norm 1, associated with $\lambda_\ell(C \circ \tilde{X})$. Then, as long as $\lambda_s(C \circ \tilde{X}) > \lambda_{s+1}(C \circ \tilde{X})$, we have that the gradient of $\xi_C(\cdot)$ at $\tilde{X}$ is the matrix

$$\nabla_X \xi_C(\tilde{X}) = C \circ \sum_{\ell=1}^{s} \lambda_\ell(C \circ \tilde{X}) u_\ell(C \circ \tilde{X}) u_\ell(C \circ \tilde{X})' .$$

This can be derived using standard results concerning symmetric functions of eigenvalues (see Tsing, Fan, and Verriest (1994), for example). Note that we must define the $u_\ell$ properly when $\lambda_\ell(C \circ \tilde{X}) = \lambda_{\ell+1}(C \circ \tilde{X})$ for any $\ell = 1, \ldots, s-1$; in such situations, we just take care that the associated $\{u_\ell\}$ form an orthonormal basis for each of the eigenspaces corresponding to distinct eigenvalues.

When $\lambda_s(C \circ \tilde{X}) = \lambda_{s+1}(C \circ \tilde{X})$, $\xi_C(\cdot)$ is not differentiable at $\tilde{X}$. Thus the problem of finding an optimal mask corresponds to minimizing a nondifferentiable, nonconvex function over a semidefiniteness constraint. There is at present very little literature on problems of this type. A number of recent papers have considered the extension of methods for general nonlinear programming (NLP) to include semidefiniteness (or "Linear Matrix Inequality") constraints. Methodologies based on primal-dual algorithms, augmented Lagrangians, and sequential quadratic programming are described in Benson and Vanderbei (2003), Fares, Apkarian, and Noll (2001) and Fares, Noll, and Apkarian (2002), respectively. The difficulty with such methods in our context is that they are likely to fail numerically, due to the lack of differentiability of the objective.

An alternative approach to our problem would be to use a general method for the unconstrained minimization of a nondifferentiable function, for example the well-known Bundle-Trust (BT) algorithm (see Schramm and Zowe (1988)). To represent the problem of minimizing $\xi_C(\cdot)$ in the form required by such an algorithm, consider the function $\text{svec}(\cdot) : \mathbb{R}^{n\times n} \to \mathbb{R}^{n(n-1)/2}$ that takes the superdiagonal components of a matrix $X$ and "stacks" them into a vector $x$:

$$x = \text{svec}(X) = (X_{12}, X_{13}, \ldots, X_{1n}, X_{23}, X_{24}, \ldots, X_{2n}, \ldots X_{n-1,n})'.$$

Similarly let $X = \text{Smat}(x)$ be the symmetric matrix, with unit diagonal, such that $\text{svec}(\text{Smat}(x)) = x$. The problem of obtaining an optimal mask can then be written in the form

$$\text{MOP}: \qquad \min\ \left\{\xi_C(\text{Smat}(x)) - \rho \min\{\lambda_n(\text{Smat}(x)), 0\}\right\},$$

where $\rho$ is a sufficiently large penalty parameter.

The original objective $\xi_C(\cdot)$ could be exponentiated in MOP to extend the domain of definition to include indefinite matrices. However, for $\rho$ sufficiently large, the use of $\xi_C(\cdot)$ should present no difficulties unless $s$ is close to $n$, in which case the complementary problem could be considered.

From the standpoint of nondifferentiable optimization (NDO), the main difficulty with MOP is the number of variables, $m = n(n-1)/2$. For example, $n = 60$ gives $m = 1770$, and $n = 120$ gives $m = 7140$. In addition, algorithms for general NDO, such as the BT method, know nothing of the special structure that underlies MOP and instead treat the objective function to be minimized as a "black box."

One recently developed method for NDO (see Burke, Lewis, and Overton (2003)) was designed more specifically for minimizing nondifferentiable functions of the eigenvalues of a matrix, such as $\xi_C(\cdot)$. However, in the control applications for which this algorithm was designed the number of variables is typically small, although the matrices that are functions of the controls can be much larger. For a problem with $m$ variables, the algorithm of Burke, Lewis, and Overton (2003) samples the gradient at $> m$ points in a neighborhood of the current iterate, and then obtains a search direction via minimization of a convex quadratic function of $m$ nonnegative variables. For the dimensions $m$ arising in our application, this work per iteration could be prohibitively large.

Our approach to attempt to minimize $\xi_C(\cdot)$ is based on a heuristic adaptation of the well-known affine scaling algorithm for linear programming (LP).

For a given $\tilde{X} \succ 0$ with $\operatorname{diag}(\tilde{X}) = e$, let $G = \nabla_X \xi_C(\tilde{X})$, and consider the linear semidefinite programming problem

$$\text{SDP}: \qquad \min\{G \bullet X \ : \ \operatorname{diag}(X) = e,\ X \succeq 0\}.$$

The affine scaling direction for SDP (see Faybusovich (1994)) is based on minimizing the linear objective over the "Dikin ellipsoid" induced by the barrier $-\ln \det X$, intersected with the linear equality constraints. For the constraints $\operatorname{diag}(X) = e$, it is straightforward to show (see for example Muramatsu (1998)) that the affine scaling direction $D$ at $\tilde{X}$ is given by

$$D := \tilde{X}(G - \operatorname{Diag}(u))\tilde{X},$$

where $u = (\tilde{X} \circ \tilde{X})^{-1} \operatorname{diag}(\tilde{X} G \tilde{X})$. Given the direction $D$, we consider a step of the form

$$X := \tilde{X} - \alpha\beta^k D,$$

where $0 < \beta < 1$, and the initial step parameter $\alpha$ corresponds to a fixed fraction of either a "short step" or a "long step." The short step is based on the limit of the Dikin ellipsoid that is used to define $D$, while the long step is based on the limit of the true feasible region $X \succeq 0$; see (Muramatsu, 1998, Sec 2) for details. We attempt a step with $k = 0$, and we accept the resulting $X$ if $\xi_C(X) < \xi_C(\tilde{X})$. If not, we "backtrack" by incrementing $k$ a limited number of times in an attempt to decrease $\xi_C(\cdot)$. For the highest allowed $k$, we accept $X$ even if $\xi_C(X) > \xi_C(\tilde{X})$. In our implementation we use $\beta = 1/3$ and $k \leq 2$.

The above strategy for minimizing $\xi_C(\cdot)$ is clearly quite heuristic. For example, due to the nondifferentiability of $\xi_C(\cdot)$, we cannot insure descent on every iteration. However, the use of non-improving steps in algorithms for NDO (for example the well-known subgradient method) is quite common. Even in the case when $G$ is constant, the affine scaling method for SDP (either short-step or long-step) may fail to converge to a solution (see Müller (2001)). The affine scaling method for LP has been very successful in practice, and it is usually applied with very long steps even though it is known that theoretically the algorithm with such step-sizes could fail to converge (see Mascarenhas (1997)).

## 4 Computational Results

We have implemented the affine scaling heuristic for approximately minimizing $\xi_C(\cdot)$ in MATLAB. In Figure 1, we illustrate the performance of the algorithm on a problem with $n = 63$, $s = 31$. The data for this example, which has previously been used to evaluate other bounding schemes for $z(C, s, n)$, comes from an environmental monitoring application (see Guttorp, Le, Sampson, and Zidek (1992)). The algorithm is applied to the complementary problem, and is initialized at a matrix $X_1 = .975E + .025I$ (recall that $X = E$

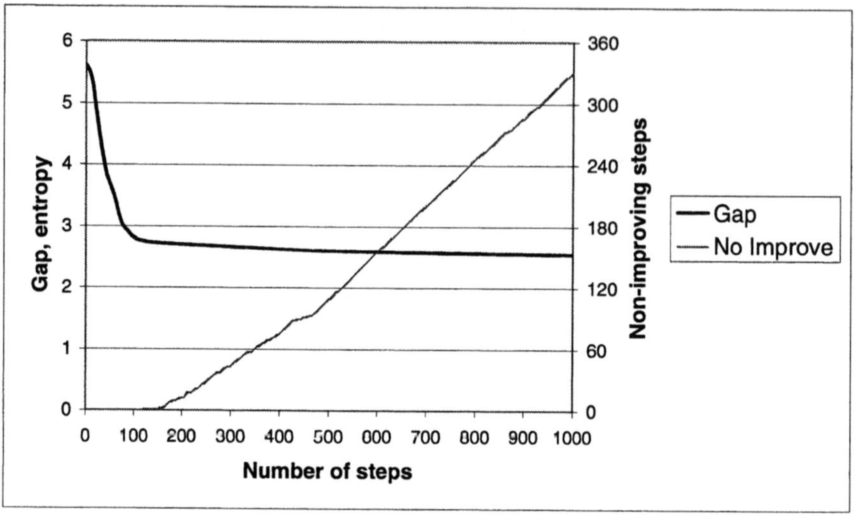


**Fig. 1.** Decrease in masked spectral bound for problem with $n = 63$, $s = 31$.

corresponds to the eigenvalue bound, and $X = I$ corresponds to the diagonal bound). The algorithm is run for 1000 iterations, using an initial step-length $\alpha$ on each iteration corresponding to 50% of a short step. In the figure, we give the gap between the masked spectral bound and the best solution found by a heuristic procedure (the heuristic value is approximately 0.0075 below the optimal value, previously computed using the algorithm of Anstreicher, Fampa, Lee, and Williams (1999)). The gap for the eigenvalue bound ($X_0 = E$) is approximately 5.71454. In the figure we also give the cumulative number of non-improving steps. The first 120 iterations are all improving, and decrease the gap to approximately 2.75362, a reduction of almost 52%. The second non-improving step is on iteration 152, after which non-improving steps become more and more frequent. The algorithm nevertheless continues to make gradual progress, eventually decreasing the gap to 2.54639, a reduction of about 55.4% compared to $X_0 = E$. This gap is slightly more than the best gap for the problem obtained in Hoffman, Lee, and Williams (2001). A substantially smaller gap was computed in Lee and Williams (2003). However, the better gaps obtained in Hoffman, Lee, and Williams (2001) and Lee and Williams (2003) are obtained using significantly more computational effort. On a 400 MHz Windows PC the first 120 iterations of the affine scaling method require about 14 seconds, and all 1000 iterations require a total of about 240 seconds. It is to be expected that non-improving iterations are more time-consuming due to backtracking of the steplength. It should also be noted that our implementation is not at all optimized for efficiency, since our primary goal is to evaluate the potential strength of the masked spectral bounds rather than how quickly the bounds can be obtained.

The performance illustrated in Figure 1 appears to be typical for our method. After an initial number of improving steps the algorithm performs non-improving steps more and more frequently, but still obtains gradual decrease in the bound. In our code the iteration sequence is terminated if 50 consecutive steps fail to produce an improvement in the best bound, but using a maximum of 1000 steps this early termination criterion is rarely satisfied.

To evaluate our methodology, we have attempted to minimize the masked spectral bound for a number of different test problems. In Figure 2 we give results based on the environmental monitoring data set with $n = 63$ from Guttorp, Le, Sampson, and Zidek (1992). For each $s = 3, 4, \ldots, 60$ we give the gaps between several different bounds and the highest entropy found using a heuristic procedure. In the figure 'Eig' and 'Diag' denote the eigenvalue and diagonal bounds, and 'OneBig' denotes the bound obtained using an $X$ having a single block of ones corresponding to the best solution found by the primal heuristic, and all remaining off-diagonal components zero. 'MS_Eig' and 'MS_OB' refer to the masked spectral bounds computed using an initial $X_0$ equal to the $X$ that gives the Eig and OneBig bounds, respectively. In both cases the iterative sequence for minimizing $\xi_C(\cdot)$ was initiated at $X_1 = .95X_0 + .05(I + E)$, and the algorithm was run for 1000 iterations. For all of the bounds except Eig we computed bounds based on the original and complementary problems, and report the better of the two. (The original bound was better for $s$ less than 20, 26, 16 and 25 for Diag, OneBig, MS_Eig and MS_OB, respectively.) It is clear from the figure that the masked spectral

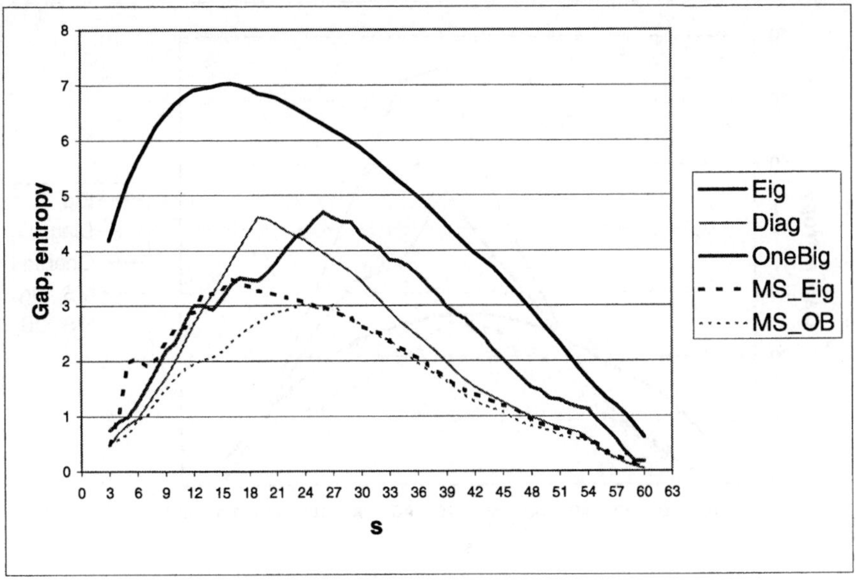


**Fig. 2.** Comparison of bounds on problems with $n = 63$.

bound performs very well on this data compared to the previously described Eig, Diag and OneBig bounds. The substantial reduction in the gap comparing MS_Eig to Eig and the excellent performance of MS_OB are both noteworthy.

We also considered problems based on another data set with $n = 124$. In Lee and Williams (2003) this data was used to evaluate a number of bounds for $z(C, s, n)$, including the Eig and Diag bounds and several variants of the NLP-based bounds from Anstreicher, Fampa, Lee, and Williams (1996, 1999). In Figure 3 we give the gaps between different bounds and a heuristic solution, for $s = 10, 20, \ldots, 120$. For all of the bounds considered in Lee and Williams (2003), the original problem gave a better bound than the complementary problem for $s \leq 70$. Consequently each bound considered here was based on the original problem for $s \leq 70$, and the complementary problem for $s \geq 80$. The MS_Eig and MS_OB bounds were computed using 500 iterations of the affine scaling procedure, requiring up to 800 seconds per instance on a 400 MHz Windows PC. Among all of the bounds considered in Lee and Williams (2003), Diag gives the best bound for $s \leq 40$, and Eig gives the best bound for $50 \leq s \leq 110$. For $s = 120$ a partition bound based on blocks of size two is slightly better than Diag. As shown in (Lee and Williams, 2003, Table 1.2) the performance of the different NLP bounds is quite poor on these problems. Considering the bound values from Lee and Williams (2003) together with Figure 3, we conclude that the MS_OB bound is superior to all previously computed bounds on these problems for $10 \leq s \leq 120$. Unfortunately the

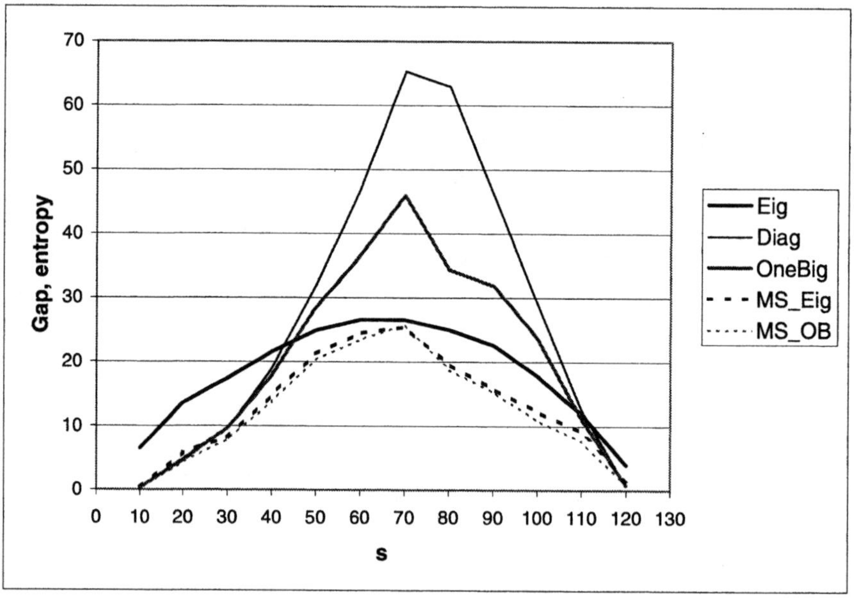


**Fig. 3.** Comparison of bounds on problems with $n = 124$.

gaps for intermediate values of $s$ (say $40 \le s \le 100$) are still quite high, probably too high to permit the computation of an optimal solution using branch-and-bound in a reasonable amount of time. It is important to note, however, that the optimal values for these problems are unknown, and a substantial fraction of the gap could be due to suboptimality of the lower bound provided by the heuristics.

## 5 Alternative Use of Oppenheim's Inequality

Not bothering with the logarithms in this section, the bound that we seek is

$$P_v: \qquad v := \min\left\{\prod_{\ell=1}^{s} \lambda_\ell(C \circ X) \;:\; X \succeq 0,\ \mathrm{diag}(X) = e\right\}.$$

However, by arguing a bit differently we can obtain an alternative bound

$$P_u: \qquad u := \min\left\{\prod_{\ell=1}^{s} \lambda_\ell(C \circ X) / \prod_{\ell=1}^{s} \mathrm{diag}_{[\ell]}(X) \;:\; X \succeq 0\right\},$$

where $\mathrm{diag}_{[\ell]}(X)$ denotes the $l$th least component of $\mathrm{diag}(X)$. Another variation is

$$P_w: \qquad w := \min\left\{\prod_{\ell=1}^{s} \lambda_\ell(C \circ \hat{X}) \;:\; X \succeq 0,\ \hat{X}_{ij} := X_{ij}/\sqrt{X_{ii}X_{jj}}\right\}.$$

All of these are upper bounds on $\max\{\det C[S,S] \;:\; |S| = s\}$. Even though we do not know algorithms to calculate these bounds exactly, it is potentially useful to know the relationship between them.

**Theorem 1.** $u \le v = w$ .

*Proof.* Certainly we have $u \le v$ since every $X$ that is feasible for $P_v$ is feasible for $P_u$ and has the same objective value. Also, we have $w \le v$ since every $X$ that is feasible for $P_v$ has $\hat{X} = X$ in $P_w$.

Next, we observe that for every $X$ that is feasible for $P_w$, $\hat{X}$ is just the Hadamard product of $X$ with $Y := y'y$, where $y_i := 1/\sqrt{X_{ii}}$. It is immediate that $\mathrm{diag}(\hat{X}) = e$ and that $Y \succeq 0$. Furthermore, by the Schur Product Theorem (see Horn and Johnson (1991)), we have that $\hat{X} \succeq 0$. Therefore $\hat{X}$ is feasible for $P_v$. Finally, the objective value of $\hat{X}$ in $P_v$ is the same as that of $X$ in $P_w$. Therefore, we have established that $v \le w$. □

Based on Theorem 1, it might be profitable to attempt to solve $P_u$ instead of $P_v$ to obtain a tighter bound for MESP. However, our limited experience indicates that the objective of $P_u$ is substantially more difficult to minimize than that of $P_v$, due to the presence of the additional nondifferentiable term in the denominator. A minimization framework that deals more carefully with nondifferentiability of the objective might overcome this added difficulty. On the other hand, we believe that it is likely, but cannot prove, that in the solution of $P_u$ many of the components of $\mathrm{diag}(X)$ are equal.

## 6 Conclusion

The spectral partition bound of Hoffman, Lee, and Williams (2001) and the related integer linear programming bound of Lee and Williams (2003) yield significant improvements over spectral bounds based on pre-defined combinatorial masks (for example the eigenvalue bound $\xi_C(E)$ or the diagonal bound $\xi_C(I)$). However, these bounds are potentially too computationally intensive to implement within branch-and-bound due to the effort required to perform the required combinatorial local search. In this paper we have shown, using the new masked spectral bound, that significant improvements over a pre-defined mask can also be achieved through methods of nonlinear programming. Moreover, the required optimization in this case does *not* suffer from the combinatorial explosion associated with local search. We therefore believe that the masked spectral bound is an excellent candidate to enhance the performance of branch-and-bound algorithms for MESP, for example the algorithm based on eigenvalue bounds described in Ko, Lee, and Queyranne (1995).

## References

K.M. Anstreicher, M. Fampa, J. Lee, and J. Williams. Continuous relaxations for constrained maximum-entropy sampling. In *Integer programming and combinatorial optimization (Vancouver, BC, 1996)*, volume 1084 of *Lecture Notes in Comp. Sci.*, pages 234–248. Springer, Berlin, 1996.

K.M. Anstreicher, M. Fampa, J. Lee, and J. Williams. Using continuous nonlinear relaxations to solve constrained maximum-entropy sampling problems. *Math. Prog., Ser A*, 85(2):221–240, 1999.

H.Y. Benson and R.J. Vanderbei. Solving problems with semidefinite and related constraints using interior-point methods for nonlinear programming. *Math. Prog, Ser. B*, 95:279–302, 2003.

J.V. Burke, A.S. Lewis, and M.L. Overton. A robust gradient sampling algorithm for nonsmooth, nonconvex optimization. Technical report, October 2003.

W.F. Caselton and J.V. Zidek. Optimal monitoring networks. *Stat. and Prob. Let.*, 2:129–178, 1984.

B. Fares, P. Apkarian, and D. Noll. An augmented Lagrangian method for a class of LMI-constrained problems in robust control theory. *International J. Control*, 74:348–360, 2001.

B. Fares, D. Noll, and P. Apkarian. Robust control via sequential semidefinite programming. *SIAM J. Control Optim.*, 40:1791–1820, 2002.

L. Faybusovich. Dikin's algorithm for matrix linear programming problems. *Lec. Notes in Control and Info. Sci.*, 197:237–247, 1994.

P. Guttorp, N.D. Le, P.D. Sampson, and J.V. Zidek. Using entropy in the redesign of an environmental monitoring network. Technical Report 116, Dept. of Stat., Univ. of British Columbia, 1992.

A. Hoffman, J. Lee, and J. Williams. New upper bounds for maximum-entropy sampling. In A.C. Atkinson, P. Hackl, and W.G. Müller, editors, *mODa 6 - Advances in Model-Oriented Design and Analysis (Puchberg/Schneeberg, June 25-29, 2001)*, pages 143–153. Springer, 2001.

R.A. Horn and C.R. Johnson. *Matrix Analysis*. Cambridge Univ. Press, Cambridge, 1985.

R.A. Horn and C.R. Johnson. *Topics in Matrix Analysis*. Cambridge Univ. Press, Cambridge, 1991.

C.W. Ko, J. Lee, and M. Queyranne. An exact algorithm for maximum entropy sampling. *Oper. Res.*, 43(4):684–691, 1995.

J. Lee. Constrained maximum-entropy sampling. *Oper. Res.*, 46(5):655–664, 1998.

J. Lee. Semidefinite programming in experimental design. In Wolkowicz H., R. Saigal, and L. Vandenberghe, editors, *Handbook of Semidefinite Programming*, volume 27 of *Int. Ser. in Oper. Res. and Manag. Sci.*, pages 528–532. Kluwer Acad. Publ., Boston, 2000.

J. Lee. Maximum-entropy sampling. In A.H. El-Shaarawi and W.W. Piegorsch, editors, *Encyclopedia of Environmetrics*, volume 3, pages 1229–1234. Wiley, 2001.

J. Lee and J. Williams. A linear integer programming bound for maximum-entropy sampling. *Math. Prog., Ser. B*, 94(2-3):247–256, 2003.

W.F. Mascarenhas. The affine scaling algorithm fails for $\lambda = .999$. *SIAM J. Optim.*, 7:34–46, 1997.

W.G. Müller. *Collecting Spatial Data.* Contrib. to Stat. Physica-Verlag, Heidelberg, revised edition, 2001. Optimum design of experiments for random fields.

M. Muramatsu. Affine scaling algorithm fails for semidefinite programming. *Math. Prog.*, 83:393–406, 1998.

A. Oppenheim. Inequalities connected with definite Hermitian forms. *J. London Math. Soc.*, 5:114–119, 1930.

H. Schramm and J. Zowe. A combination of the bundle approach and trust region concept. In J. Guddat, P. Kall, K. Lommatzsch, M. Vlach, and K. Zimmermann, editors, *Advances in Mathematical Optimization and Related Topics.* Akademie Verlag, Berlin, 1988.

P. Sebastiani and H.P. Wynn. Maximum entropy sampling and optimal Bayesian experimental design. *J. Royal Stat. Soc, Ser. B. Stat. Meth.*, 62(1):145–157, 2000.

M.C. Shewry and H.P. Wynn. Maximum entropy sampling. *J. of Appl. Stat.*, 46:165–170, 1987.

N.K. Tsing, M.K.H. Fan, and E.I. Verriest. On analyticity of functions involving eigenvalues. *Linear Alg. and its Appl.*, 207:159–180, 1994.

S. Wu and J.V. Zidek. An entropy based review of selected NADP/NTN network sites for 1983-86. *Atmos. Env.*, 26A:2089–2103, 1992.

# Some Bayesian Optimum Designs for Response Transformation in Nonlinear Models with Nonconstant Variance

A.C. Atkinson

London School of Economics, London WC2A 2AE, UK a.c.atkinson@lse.ac.uk

**Summary.** Often the responses from mechanistic models have to be transformed to achieve error distributions that are symmetric and have constant variance. The paper develops methods for the design of experiments when the value of the power transformation $\lambda$ is not precisely known. Because of the nature of the relationship between response and the mechanistic model, it is necessary to transform both sides of the model. Expressions are given for the parameter sensitivities in the transformed model and examples given of optimum designs for particular values of $\lambda$, together with the efficiency of these designs as $\lambda$ varies. When the uncertainty about $\lambda$ is expressed by a prior distribution, Bayesian $D$-optimum designs can be found. It is shown, for one example, that these designs can have high efficiency for a range of values of $\lambda$.

**Key words:** Bayesian $D$-optimality, Box-Cox transformation, Chemical kinetics, $D$-optimum design, Mechanistic model, Parameter sensitivities, Transform both sides.

## 1 Introduction

This paper is concerned with the design of experiments for the mechanistic models which arise in pharmacokinetics and chemical kinetics. Usually the variance of the measurements of concentrations to which these models are fitted increases with the mean of the observations. Atkinson (2003) describes several examples. The variance can be stabilised, and the error distribution made symmetrical, by transforming the response. But, with the mechanistic models that are considered here, it is necessary to transform both sides of the model. The optimum design will then depend on the transformation that is appropriate as well as on the model.

The transformation is indexed by the parameter $\lambda$. Calculations and plots in Atkinson (2003) indicate that the optimum designs often change rapidly with $\lambda$. The example, for two consecutive second-order chemical reactions, in Atkinson (2004) shows that this dependence of the designs on $\lambda$ can be complicated. In practice the exact value of $\lambda$ will often not be known. The purposes of this paper are to calculate the efficiency of the designs for specific $\lambda$ as the true value of the transformation parameter varies and to find Bayesian designs that are optimum for a set of values of $\lambda$.

The paper starts in Section 2 with the simple, but illuminating, example of first-order decay. The next section reviews optimum design theory for multivariate responses. In Section 4 the transformation model is described together with parameter sensitivities. The main numerical example of two consecutive first-order reactions is introduced in Section 5. The main theoretical results are in Section 6, where efficiencies are defined and Bayesian $D$-optimum designs for $\lambda$ are introduced. Both are calculated for $D$-optimum designs for two consecutive first-order reactions. The paper concludes with a brief discussion in Section 7.

## 2 Transformations and First-Order Decay

The experimental design problem considered is that of selecting a set of times $t_i$ at which readings $y_i$ should be taken with weights $w_i$ in order to obtain good estimates of the parameter vector $\psi$ in the nonlinear mechanistic model

$$y = \eta(t, \psi). \tag{1}$$

This becomes a statistical problem when error is introduced.

Let the expected value of Y be $\mathrm{E}(Y) = \mu$ and the variance be $\mathrm{Var}\, Y$. If the variance of the random variable $Y$ increases as a power of the expected value so that

$$\mathrm{Var}\, Y \propto \{E(Y)\}^{2(1-\lambda)}, \tag{2}$$

Taylor expansion shows that power transformation of the response provides a random variable with variance approximately independent of the mean. For transformation of just the response $y$ in a regression model, the normalized power transformation of Box and Cox (1964) is

$$z(\lambda) = \begin{cases} (y^\lambda - 1)/(\lambda \dot{y}^{\lambda-1}) & \lambda \neq 0 \\ \dot{y} \log y & \lambda = 0 \end{cases}, \tag{3}$$

where $\dot{y} = \exp\left(\sum_i \log y_i/n\right)$, the geometric mean of the observations. When $\lambda = 1$, there is no transformation. The model to be fitted is (1) with response $z(\lambda)$ and additive errors of constant variance.

However, when, for example, $\eta(t, \psi)$ is a mechanistic model based on chemical kinetics, the relationship between the response and the concentrations of the other reactants needs to be preserved after transformation. This is achieved by transformation of both sides of the model, as described in Chapter 4 of Carroll and Ruppert (1988). For fixed $\lambda \neq 0$, estimation of the parameters $\psi$ after transformation does not depend on whether the response is $z(\lambda)$ or the non-normalized $y^\lambda$. Simplification of the model and the introduction of observational error on this transformed scale leads to the statistical model

$$y^\lambda = \{\eta(t, \psi)\}^\lambda + \varepsilon. \tag{4}$$

Atkinson (2003) assumed that the value of $\lambda$ was known and compared designs for different $\lambda$. Designs for estimation of $\lambda$ when the response of a regression model is to be transformed are described by Atkinson and Cook (1996).

A simple example of the effect of assumptions about the mean-variance relationship on experimental design comes from the model for first-order decay in which the concentration of chemical $A$ at time $t$ is given by the nonlinear function

$$[A] = \eta_A(t, \theta) = e^{-\theta t} \qquad (\theta, t \geq 0), \tag{5}$$

if it is assumed that the initial concentration of $A$ is 1. If we put $\lambda = 1$ in (4) we have the usual model with additive errors of constant variance. Box and Lucas (1959) show that, for this error model, the locally $D$-optimum design consists of taking all measurements at a time $t^* = 1/\theta$.

However, if the log transformation is appropriate to obtain errors of constant variance and $[A]$ is measured, taking logarithms of both sides of (5), combined with additive errors, yields the statistical model

$$\log y_i = \log\{\eta_A(t_i, \theta)\} + \varepsilon_i = -\theta t_i + \varepsilon_i. \tag{6}$$

The optimum design now puts all observations at the maximum possible time, when the concentration is as small as possible, an implausible experimental strategy.

Analysis of two sets of pharmacokinetic data in Atkinson (2003) yields values of $\lambda$ of 0.35 and 0.1. In comparison, Horwitz's rule for analytical chemistry based on the analysis of thousands of data sets (Lischer (1999)) yields a value around 0.14 for $\lambda$. Clearly transformation of such data is often required.

## 3 Optimum Design for a Multivariate Response

The $i$th reading comes from running the experiment for a time $t_i$ and taking a measurement of one or more responses. As in Box and Lucas (1959), the measurements are taken at a single time point for each run. Here only continuous designs are found, in which the design $\xi$ is a continuous measure specifying both a set of $k$ distinct points in a design region $\mathcal{T}$ and the proportions, $w_i$, of observations taken at these points. The times $t_i$ are the points of support of the design $\xi$ and $w_i$ the design weights. Locally $D$-optimum designs maximize the logarithm of the determinant of the information matrix, $\log|M(\xi, \psi)|$, in which a best guess $\psi^o$ is taken for the values of the parameters.

If the concentration of more than one chemical is measured, there will be a model for each expected response giving a matrix $F_i$ of parameter sensitivities for the $i$th response, $i = 1, \ldots, m$,

$$F_i = \{f_{ij}(t_u, \psi)\} = \left\{\frac{\partial \eta_i(t_u, \psi)}{\partial \psi_j}\right\},$$

where $u = 1, \ldots, k$ denote the design points and $j = 1, \ldots, p$ denote the parameters. In the absence of transformation the observations follow the model

$$y_{iu} = \eta_i(t_u, \psi) + \varepsilon_{iu},$$

with

$$\mathrm{E}(\varepsilon_{iu}) = 0, \qquad \mathrm{E}(\varepsilon_{iu}\varepsilon_{\ell v}) = \begin{cases} 0 & \text{if } u \neq v \\ \sigma_{i\ell} & \text{if } u = v \end{cases},$$

when the variance-covariance matrix of the responses is

$$\Sigma = \{\sigma_{i\ell}\}_{i,\ell=1,\ldots,m}.$$

For normally distributed errors the information matrix is given by

$$M(\xi, \psi) = \sum_{i=1}^{m}\sum_{\ell=1}^{m} \sigma^{i\ell} F_i^T W F_\ell, \tag{7}$$

where $\Sigma^{-1} = \{\sigma^{i\ell}\}_{i,\ell=1,\ldots,m}$ and $W = \mathrm{diag}\{w_1, \ldots, w_k\}$.

# 4 Parameter Sensitivities and Transforming Both Sides

The parameter sensitivities for the model in Section 2 with $\lambda = 1$ will be written

$$f_j^1(t, \psi) = \frac{\partial \eta(t, \psi)}{\partial \psi_j}. \tag{8}$$

The parameter sensitivities in the transformed model (4) are then

$$f_j^\lambda(t, \psi) = \frac{\partial\{\eta(t, \psi)\}^\lambda}{\partial \psi_j} = \lambda\{\eta(t, \psi)\}^{\lambda-1} \frac{\partial \eta(t, \psi)}{\partial \psi_j} = \lambda\{\eta(t, \psi)\}^{\lambda-1} f_j^1(t, \psi). \tag{9}$$

For fixed $\lambda$, multiplication by $\lambda$ in (9) does not change the optimum design, so the sensitivities have the easily calculated form

$$f_j^\lambda(t, \psi) = \{\eta(t, \psi)\}^{\lambda-1} f_j^1(t, \psi) = f_j^1(t, \psi)/\{\eta(t, \psi)\}^{1-\lambda}. \tag{10}$$

If $\lambda < 1$, the variance of the observations increases with the value of $\eta(t, \psi)$. Thus transformation of both sides for such values of $\lambda$ will increase the relative value of the sensitivities for times where the response is small. We can expect that designs for $\lambda < 1$ will include observations at lower concentrations than those when no transformation is needed.

## 5 Two Consecutive First-Order Reactions

Section 6 of Atkinson (2003) shows how the locally optimum design for $\theta$ in the exponential decay model of Section 2 changes as a function of $\lambda$. In this section locally $D$-optimum designs are found for the model for two consecutive first-order reactions introduced by Box and Lucas (1959).

The two reactions can be written

$$A \overset{\theta_1}{\rightarrow} B \overset{\theta_2}{\rightarrow} C. \tag{11}$$

When both reactions are first order, an explicit algebraic solution can be found for the concentrations of the reactants as a function of time. If the initial concentration of $A$ is one and that of $B$ and $C$ are zero, then $\eta_A(t,\theta)$ follows the exponential decay (5) with $\theta = \theta_1$. The other concentrations are given by

$$\begin{aligned} \eta_B(t,\theta) &= \frac{\theta_1}{\theta_1 - \theta_2}\left(e^{-\theta_2 t} - e^{-\theta_1 t}\right) \\ \eta_C(t,\theta) &= 1 - \eta_A(t,\theta) - \eta_B(t,\theta) \end{aligned}. \tag{12}$$

The parameter sensitivities $f^1_{Bj}(t,\theta)$ and $f^1_{Cj}(t,\theta) = -f^1_{Aj}(t,\theta) - f^1_{Bj}(t,\theta)$ are readily found by differentiation of the concentrations in (12). The parameter sensitivities for the transformed model are then found from (10).

The locally optimum designs when the concentrations of all three reactants are measured are plotted in Figure 1 for $\lambda$ from 0.01 to one. In the calculations $\theta_1 = 0.7$, $\theta_2 = 0.2$ and the three measurements were taken as uncorrelated with equal variances. For $\lambda$ close to one, the design has two approximately equally weighted design points. However, for $\lambda$ below 0.64 there is only one design point, except for a region around 0.25 where there are two unequally weighted points. Below 0.12 there is a single design point at $t = 20$, the edge of the design region.

## 6 Efficiencies and Bayesian Optimum Designs

Since the optimum designs in Figure 1 vary so much with $\lambda$ it is likely that a design for the wrong $\lambda$ will be inefficient. In this section efficiencies are calculated for designs as $\lambda$ varies and Bayesian optimum designs introduced that are intended to have high efficiency over a set of values of $\lambda$.

Let the optimum design for a specified $\lambda_0$ be $\xi_0^*$ and for some other $\lambda$ be $\xi_\lambda^*$. The value of the information matrix depends not only on $\xi$ and $\psi$ but also on the parameter $\lambda$. When the value of the transformation parameter is $\lambda$ the information matrix for the design $\xi_0^*$ can be written as $M(\xi_0^*,\psi,\lambda)$. Then the efficiency of the design $\xi_0^*$ for some $\lambda$ is the ratio of determinants

$$\mathit{Eff}(\xi_0^*,\lambda) = \{|M(\xi_0^*,\psi,\lambda)|/|M(\xi_\lambda^*,\psi,\lambda)|\}^{1/p}, \tag{13}$$

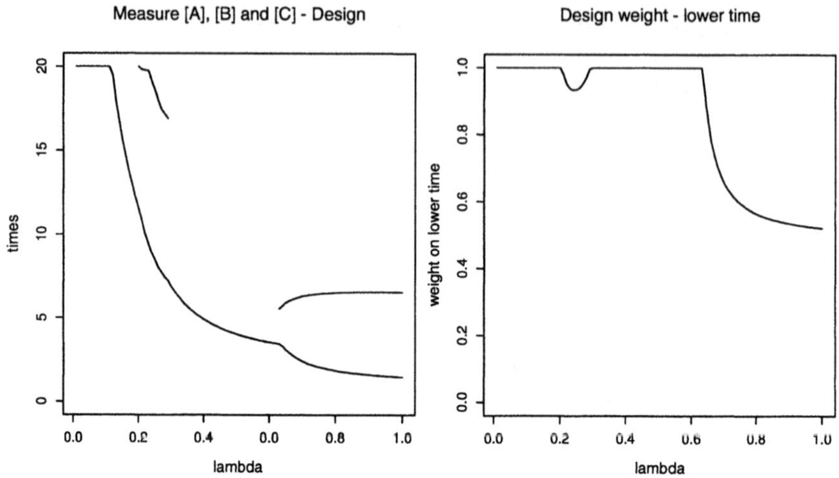


**Fig. 1.** Two consecutive first-order reactions: locally-optimum designs when [A], [B] and [C] are measured. Left panel, times; right-panel, design weights.

where $p$ is the number of parameters in the model. The dotted lines in Figures 2 and 3 are the efficiencies over a range of values of $\lambda$ of the $D$-optimum designs for five values of $\lambda_0$: 0.2, 0.4, 0.6, 0.8 and 1. Each is, of course, 100% when $\lambda = \lambda_0$. What is particularly noticeable is that all designs except that for $\lambda_0 = 0.2$ are similarly inefficient for low values of $\lambda$. The design for $\lambda_0 = 0.2$ is itself inefficient for high values of $\lambda$.
Bayesian optimum designs are found by introducing a prior distribution with weight $q_i$ for the $i$th value of $\lambda_0$. The resulting compound design criterion, when the prior is concentrated on $n(q)$ points, maximizes

$$\Phi(\xi, \psi, \lambda_0) = \sum_{i=1}^{n(q)} q_i \log |M(\xi, \psi, \lambda_i)|. \tag{14}$$

Comparison with (13) shows that $\xi_B^*$, the Bayesian $D$-optimum design, is also the design that maximizes a weighted product of the efficiencies of the designs for the elements of $\lambda_0$. The effect of the logarithm in (14) is to remove the effect of the determinants $|M(\xi_\lambda^*, \psi, \lambda)|$ since these are constants at the $n(q)$ values where optimization takes place.

Three Bayesian optimum designs are listed in Table 1, each for a five-point prior for $\lambda$ with all $q_i = 0.2$. The first design has the values of $\lambda$ uniformly distributed over 0.2, 0.4, 0.6, 0.8 and 1. It is a two-point design, the efficiency of which is shown by the continuous line in Figure 2. It has good efficiency for high values of $\lambda$, but its efficiency falls off rapidly below $\lambda = 0.3$. The second design in Table 1 is also a two-point design. The prior values of $\lambda$ concentrate on a lower range of values than before, 0.1 to 0.5. Plots not given here show that the resulting design is efficient over this range, but of

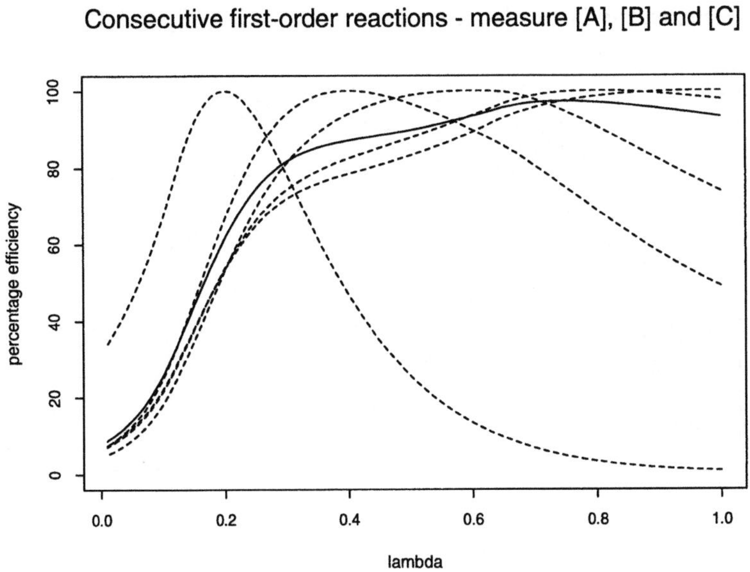


**Fig. 2.** Two consecutive first-order reactions when [A], [B] and [C] are measured. Continuous line, efficiency of Bayesian optimum design for $\lambda_0 = (0.2, 0.4, 0.6, 0.8, 1.0)^T$; dashed lines, efficiencies of the optimum designs for individual members of $\lambda_0$.

**Table 1.** Bayesian optimum designs for three five-point priors on $\lambda$.

| Prior $\lambda_0$ | $t, w$ | Design | | |
|---|---|---|---|---|
| 0.2, 0.4, 0.6, 0.8, 1.0 | $t_i$ | 2.043 | 6.883 | |
| | $w_i$ | 0.452 | 0.548 | |
| 0.1, 0.2, 0.3, 0.4, 0.5 | $t_i$ | 5.139 | 17.911 | |
| | $w_i$ | 0.591 | 0.409 | |
| 0.1, 0.2, 0.3, 0.9, 1.0 | $t_i$ | 1.633 | 8.228 | 20.000 |
| | $w_i$ | 0.280 | 0.591 | 0.129 |

decreasing efficiency for higher values of $\lambda$. An efficient design for a wider range of values of $\lambda$ is in the last row of Table 1. It is for a prior concentrated at the ends of the region and the resulting design has three points of support at low medium and high values of time. As the plot of the efficiency in Figure 3 shows, the design has good efficiency for values of $\lambda$ as low as 0.2 and a steady efficiency between 75 and 80% for most values above this. By suitable choice of the prior, a design has been found that is robust to the choice of $\lambda$.

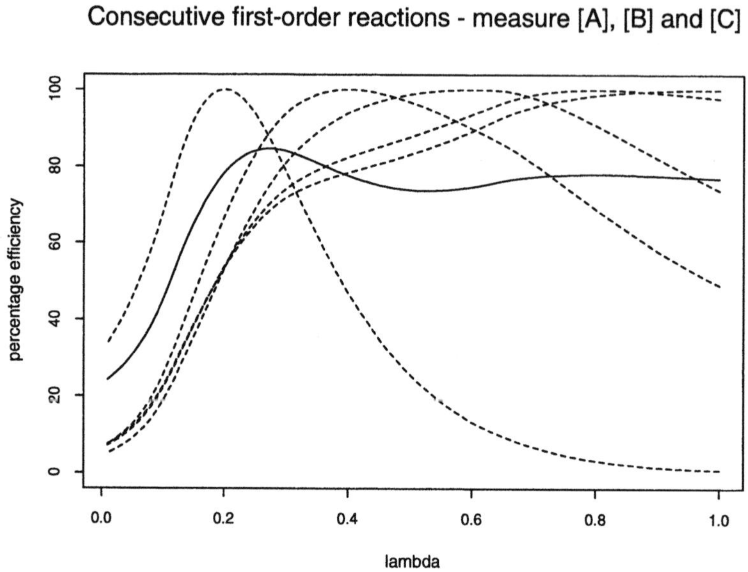


**Fig. 3.** Two consecutive first-order reactions when [A], [B] and [C] are measured. Continuous line, efficiency of Bayesian optimum design for $\lambda_0 = (0.1, 0.2, 0.3, 0.9, 1.0)^T$; dashed lines, efficiencies of the optimum designs for individual values 0.2, 0.4, 0.6, 0.8, 1.0 for $\lambda$.

# 7 Discussion

The plot of the design in Figure 1 shows, at least for this example, how dependent the locally $D$-optimum design is on the assumed value of $\lambda$. The examples in Atkinson (2003, 2004) for consecutive second-order reactions show even more complicated changes of design with $\lambda$, particularly if several responses are measured. The plots of efficiencies in Figures 2 and 3 reveal that designs for a single value of $\lambda$ are inefficient. However, the Bayesian $D$-optimum designs introduced in this paper illustrate how, by a suitable choice of prior, designs can be found that have good efficiency for a range of values of $\lambda$.

Transformations were introduced in this paper in order to provide statistical models in which the observations had constant variance. An alternative (Bogacka and Wright (2004)) is to use weighted least squares with weights proportional to $\mathrm{E}(Y)^{-(1-\lambda)}$. The resulting parameter sensitivities and so designs, are identical to those of Section 4. An important difference between such methods and the transformation studied here is in the model implied for fitting the data. In the transformation model, the original observations will have skewed distributions, which become symmetrical, with constant error variance, after the appropriate transformation. However, weighting and models with structured variances lead to symmetrical distributions of error of non-constant variance in the original scale of the observations.

Several extensions of this work are possible. One important direction is to the design of experiments for industrial processes where the dynamics of the model are described by sets of differential equations. Any correlation of errors is ignored, for example as in Bauer, Bock, Körkel, and Schlöder (2000). So a second extension is to design when the errors are correlated. Recent papers describing algorithms for optimum designs in the presence of correlated errors include Müller and Pázman (2003) and Uciński and Atkinson (2004).

## References


A. C. Atkinson. Horwitz's rule, transforming both sides and the design of experiments for mechanistic models. *Applied Statistics*, 52:261–278, 2003.

A. C. Atkinson. Transforming both sides and optimum experimental design for a nonlinear model arising from second-order chemical kinetics. *Tatra Mountains Mathematical Publications*, 2004. (In press).

A. C. Atkinson and R. D Cook. Designing for a response transformation parameter. *Journal of the Royal Statistical Society B*, 59:111–124, 1996.

I. Bauer, H. G. Bock, S. Körkel, and J. P. Schlöder. Numerical methods for optimum experimental design in DAE systems. *Journal of Computational and Applied Mathematics*, 120:1–25, 2000.

B. Bogacka and F. Wright. A non-linear design problem in a chemical kinetic model with non-constant error variance. 2004. (Submitted).

G. E. P. Box and D. R. Cox. An analysis of transformations (with discussion). *Journal of the Royal Statistical Society, Series B*, 26:211–246, 1964.

G. E. P. Box and H. L. Lucas. Design of experiments in nonlinear situations. *Biometrika*, 46:77–90, 1959.

R. J. Carroll and D. Ruppert. *Transformation and Weighting in Regression.* Chapman and Hall, London, 1988.

P. Lischer. Good statistical practice in analytical chemistry. In B. Grigelionis, editor, *Probability Theory and Mathematical Statistics*, pages 1–12. VSP, Dordrecht, 1999.

W. G. Müller and A. Pázman. Measures for designs in experiments with correlated errors. *Biometrika*, 90:423–434, 2003.

D. Uciński and A. C. Atkinson. Experimental design for processes over time. *Studies in Nonlinear Dynamics and Econometrics*, 2004. (In press).

# Extensions of the Ehrenfest Urn Designs for Comparing Two Treatments

A. Baldi Antognini

University of Bologna, via Belle Arti 41, 40126, Bologna, Italy
baldi@stat.unibo.it

**Summary.** In this paper we discuss sequential urn designs for comparing two treatments. The Ehrenfest urn design suggested by Chen is shown to be a special case of the Adjustable Biased Coin Design proposed by Baldi Antognini and Giovagnoli. Some generalizations of Chen's design are proposed that converge to a general target proportion of allocations $\pi^*$. A further extension is suggested, the Ehrenfest - Brillouin design, that converges to the target faster.

**Key words:** sequential urn design, Ehrenfest urn design.

## 1 Introduction

Assume we want to carry out a sequential experiment to compare the efficacy of several treatments. In general, some form of randomization is required to protect the experiment against various types of bias. Urns are a classical tool as a randomization device. For the problem of the balanced allocation of two treatments, Wei (1977) was among the first authors to suggest a sequential procedure based on a generalized Friedman urn, and recently Chen (2000) has suggested one based on the Ehrenfest urn. Urn randomization mechanisms have also found widespread application in adaptive, that is response driven, experimental design: see Rosenberger (1996), Bandyopadhyay and Biswas (2000), Rosenberger (2002) and Bai, Hu, and Shen (2002).

Assume there are $v$ urns $U_1, \ldots, U_v$, standing for the $v$ treatments, containing a total number of $w$ balls. Eligible subjects arrive sequentially and for each subject a ball is chosen at random from all the $w$ balls of all the urns. If the chosen ball is in urn $U_i$, then treatment $T_i$ is allocated to the corresponding subject ($i = 1, 2, \ldots, v$). Furthermore the selected ball is removed from the urn to which it belonged and a given number of balls is placed in each of the other urns, according to a well-defined rule. These operations are repeated at each trial and this causes the sequence of the allocations. At step $k$ the random vector $\delta(k) = (\delta_{k,1}; \delta_{k,2}; \ldots; \delta_{k,v})$, where $\delta_{k,i} = 1$ if the $k$th unit is allocated to $T_i$, 0 otherwise, will indicate the allocation and $\mathbf{W}(k) = (W_1(k), \ldots, W_v(k))$ will denote the corresponding occupancy vector, where $W_i(k)$ is the number of balls in $U_i$ after $k$ steps and

$\mathbf{W}(0) = \mathbf{w}_0$. Let $\Im_n = \sigma\{\delta(1), W(1), \ldots, \delta(n), W(n)\}$ be the $\sigma$-algebra representing the natural history. An urn design consists of a sequence of random vectors $\delta(1), \ldots, \delta(n), \ldots$ such that for all $n \in \mathbb{N}$ we have that $\sum_{i=1}^{v} \delta_{n,i} = 1$, the random vector $\delta(n+1)$ is $\Im_n$-measurable and

$$P(\delta_{n+1,i} = 1|\Im_n) = \frac{W_i(n)}{W_1(n) + \ldots + W_v(n)} \text{ for all } i = 1, \ldots, v, \tag{1}$$

together with a replacement rule $P(\mathbf{W}(n+1) = \mathbf{x}|\Im_n)$.

In this paper we shall focus on $v = 2$ treatments, thus it is enough to consider the one-dimensional process $\{W_1(n)\}_{n \in \mathbb{N}}$. For simplicity we shall put $\delta_{k,1} = \delta_k$ and $W_1(n) = W_n$. After $k$ steps, $N_1(k) = \sum_{i=1}^{k} \delta_i$ and $N_2(k) = k - N_1(k)$ will denote the number of observations that are allocated to $T_1$ and to $T_2$, respectively, and $\pi_k = k^{-1} \sum_{i=1}^{k} \delta_i$ will be the proportion of allocations to $T_1$. In many instances there is a given proportion $\pi^*$ (to $T_1$) which can be regarded as optimal by the experimenter. For example, assuming a linear model and letting $\sigma_1$ and $\sigma_2$ be the standard deviations corresponding to the responses under the two treatments, the proportion that minimizes the variance of the estimated difference is $\pi^* = \sigma_1/(\sigma_1 + \sigma_2)$ (the so-called Neyman allocation). The balanced design (that is $\pi^* = 1/2$) is a special case.

By martingale theory it can be shown that (Theorem 4 in Baldi Antognini and Giovagnoli (2003))

**Theorem 1.** *For any urn design given by (1)*

$$\lim_{n \to \infty} \left[ \frac{N_1(n)}{n} - \frac{1}{n} \sum_{i=1}^{n} \frac{W_1(i-1)}{W_1(i-1) + W_2(i-1)} \right] = 0 \ a.s. \tag{2}$$

The replacement rule will consist in returning the ball that has been drawn to one of the urns, so that the total number $w$ of balls is kept constant; thus (1) simplifies to

$$P(\delta_{k+1} = 1) = \frac{W_k}{w} \text{ for all } k \in \mathbb{N}. \tag{3}$$

We shall assume all the designs to be non-adaptive. Furthermore the replacement rules will be such that

$$P(W_{n+1} = x|\Im_n) = P(W_{n+1} = x|W_n, \delta_n). \tag{4}$$

**Lemma 1.** *Under condition (4), the process $\{W_n\}_{n \in \mathbb{N}}$ is a homogeneous Markov chain on $\{0, 1, \ldots, w\}$.*

This Lemma follows from the Fundamental Lemma in Giovagnoli (2004) by letting $\delta_k = Y_k$ and $W_k = X_k$.

This paper starts with a discussion, in Section 2, of sequential urn designs for comparing two treatments when balance is desirable. The Ehrenfest urn design ($ED$) suggested in Chen (2000) is shown to be a special case of the Adjustable Biased Coin Design ($ABCD$) proposed in Baldi Antognini and Giovagnoli (2004), which is also defined in Giovagnoli (2004). In Section 3 a generalization of the $ED$, $Sym - ED$, is proposed, based on an aperiodic symmetric extension of the Ehrenfest model; the expected imbalance is shown to be zero at each step. Moving from balance to a general target proportion of allocations $\pi^*$, in Section 4 we extend the Ehrenfest process in an aperiodic asymmetric way ($Asym - ED$) which allows us to define sequential randomized designs that converge to any $\pi^*$. Finally, a further extension of $Asym - ED$ is suggested, the Ehrenfest - Brillouin design $EBD$, based on the recently introduced Ehrenfest - Brillouin process (Aoki (2002), Viarengo (2002)). This is an ergodic Markov chain depending on a set of parameters the choice of which modifies the speed of convergence of functions of interest to their ergodic mean. In Section 6 we show that we can find an Ehrenfest - Brillouin design that converges to the target faster than the corresponding $Asym - ED$.

## 2 The Ehrenfest Urn Design

The Ehrenfest urn process is well-known. Assume initially there are $w$ balls equally divided between two urns, $\mathbf{W}(0) = (w/2; w/2)$. At each step one ball is chosen at random and replaced in the other urn. By Lemma 2, $\{W_k, k \geq 1; W_0 = w/2\}$ is a random walk on $\{0, 1, \ldots, w\}$ with transition probabilities

$$P(i;j) = P(W_{n+1} = j | W_n = i) = \begin{cases} \frac{w-i}{w} & j = i+1 \\ \frac{i}{w} & j = i-1 \\ 0 & \text{otherwise} \end{cases}, \quad i,j = 0, 1, \ldots, w.$$

This chain has period 2: its stationary distribution is Bin$(w; 1/2)$. This replacement rule is used by Chen for his urn design, which we denote by $ED(w)$. Observe that it does not depend on $\delta_n$. At each step $k$ we have

$$W_k = \frac{w}{2} + k - 2N_1(k) \tag{5}$$

and from (3)

$$P(\delta_{k+1} = 1 | W_k) = \frac{1}{2} - \frac{2N_1(k) - k}{w}. \tag{6}$$

Chen has studied the balance and unpredictability of design $ED(w)$ both asymptotically and for small samples. Some of his results are a special case of Baldi Antognini and Giovagnoli (2004) since it is straightforward to see that

**Theorem 2.** *The Ehrenfest urn design $ED(w)$ is a special case of the ABCD defined in Baldi Antognini and Giovagnoli (2004), obtained by letting $F(x) = 2^{-1} - x/w$ for $-w/2 \leq x \leq w/2$.*

## 3 Symmetric Ehrenfest Design for Achieving Balance

First of all we extend the Ehrenfest urn design $ED(w)$ to a general initial occupancy $\mathbf{W}(0) = (w_1; w_2)$, $w_1, w_2 \in \mathbb{R}^+$ and $w_1 + w_2 = w$. At each step one ball is drawn at random and assigned to either urn by means of a random mechanism. One possibility, which we refer to as $Sym - ED(w_1, w_2; p)$, is to place it in the opposite urn with given probability $p$, where $0 \leq p \leq 1$. Let $C_k$ equal 1 if at step $k$ the ball is placed into the other urn, 0 otherwise; $\{C_n\}_{n\in\mathbb{N}}$ is a sequence of Bernoulli random variables i.i.d., $B(p)$, independent of the design sequence $\{\delta_n\}_{n\in\mathbb{N}}$ and of the urn occupancy $\{W_n\}_{n\in\mathbb{N}}$. For $Sym - ED(w_1, w_2; p)$, (5) becomes

$$W_k = w_1 - \sum_{i=1}^{k} C_i \left(1 - 2\delta_i\right) \text{ for all } k \in \mathbb{N}. \tag{7}$$

The chain $\{W_k, k \geq 1;\ W_0 = w_1\}$ is now aperiodic with transition probabilities

$$P(i;j) = \begin{cases} p\frac{w-i}{w} & j = i+1 \\ 1-p & j = i \\ p\frac{i}{w} & j = i-1 \\ 0 & \text{otherwise} \end{cases} \qquad i,j = 0,1,\ldots,w. \tag{8}$$

Again, it can be shown that the stationary distribution is $\text{Bin}(w; \frac{1}{2})$. Because of aperiodicity:

$$\lim_{n\longrightarrow\infty} P(W_n = j) = \binom{w}{j}\left(\frac{1}{2}\right)^w, \quad j = 0,1,\ldots,w. \tag{9}$$

**Theorem 3.** *For all $w_1, w_2 \in \mathbb{R}^+$, $Sym - ED(w_1, w_2; p)$ is asymptotically balanced. In particular $ED(w/2, w/2; 1/2)$ coincides with $ED(w)$.*

The reason for the asymptotic balance is that, although starting from an asymmetric number of balls in the urns, this procedure deals with the two treatments symmetrically.

## 4 Asymmetric Ehrenfest Design for a Generic Target

Another way of modifying $ED$, which we will denote as $Asym{-}ED(w_1, w_2; p)$, consists in placing the ball that has been drawn in $U_1$ with probability $p$

$(0 \le p \le 1)$ independently of the urn from where it comes. Let $\tilde{C}_k$ equal 1 if at step $k$ the ball is placed in $U_1$, 0 otherwise; $\left\{\tilde{C}_n\right\}_{n\in\mathbb{N}}$ is an *i.i.d.* sequence of Bernoulli random variables $B(p)$, independent of $\{\delta_n\}_{n\in\mathbb{N}}$ and of $\{W_n\}_{n\in\mathbb{N}}$. It is easy to see that

$$W_1(k) = w_1 - N_1(k) + \sum\nolimits_{i=1}^{k} \tilde{C}_i \text{ for all } k \in \mathbb{N}. \tag{10}$$

The Markov chain $\{W_k, k \ge 1; W_0 = w_1\}$ is aperiodic with transition probabilities

$$P(i;j) = \begin{cases} p\dfrac{w-i}{w} & j = i+1 \\ p\dfrac{i}{w} + (1-p)\dfrac{w-i}{w} & j = i \\ (1-p)\dfrac{i}{w} & j = i-1 \\ 0 & \text{otherwise} \end{cases} \qquad i,j = 0,1,\ldots,w. \tag{11}$$

and stationary distribution Bin $(w;p)$, so that

$$\lim_{n\longrightarrow\infty} P(W_n = j) = \binom{w}{j} p^j (1-p)^{w-j}, \quad j = 0,1,\ldots,w.$$

Observe that the $Sym - ED(w_1, w_2; p)$ defined by (8) is not a special case of $Asym - ED(w_1, w_2; p)$ given by (11). From (10) we obtain for all $k$

$$\frac{N_1(k)}{k} = \frac{w_1}{k} - \frac{W_k}{k} + \frac{1}{k}\sum_{i=1}^{k} \tilde{C}_i \tag{12}$$

and it can be seen that

**Theorem 4.** *Under* $Asym - ED(w_1, w_2; p)$

$$\lim_{n\to\infty} \frac{N_1(n)}{n} = p \quad a.s.$$

*and* $N_1(n)/n$ *is asymptotically normal.*

Given a target proportion $\pi^* \in [0;1]$ of allocations to treatment $T_1$, the Ehrenfest design $Asym - ED(w_1, w_2; \pi^*)$ converges to it for any choice of $w_1$ and $w_2$.

## 5 Ehrenfest-Brillouin Design

The $Asym - ED(w_1, w_2; p)$ (11) can be further generalized: at each step the probability of returning the ball to either urn can be made to depend on both the number of balls in the urn and the values of two real parameters $\alpha_1$ and $\alpha_2$. The probability of drawing the ball from either urn is the same as before, namely (3), whereas the replacement in urn $U_1$ is driven by the process defined by Garibaldi and Penco (2000), which they call the *Ehrenfest-Brillouin process*, namely

$$P\left(\tilde{C}_{k+1}=1\right)=\frac{\alpha_1+W_k-\delta_{k+1}}{\alpha+w-1} \text{ for all } k\in\mathbb{N},$$

where $\alpha=\alpha_1+\alpha_2$. The idea, which has been utilized by Aoki (2002) in an economic context, consists in assuming the existence of another pair of hypothetical urns $\tilde{U}_1$ and $\tilde{U}_2$, where at each stage $\tilde{U}_i$ contains the same number of balls as $U_i$ plus $\alpha_i$ $(i=1,2)$, and replacement in $U_i$ has the same probability as drawing from $\tilde{U}_i$. The design generated by this procedure will be denoted $EBD(w_1,w_2;\alpha_1,\alpha_2)$. Relation (10) still holds, however the sequence $\left\{\tilde{C}_n\right\}_{n\in\mathbb{N}}$ is no longer i.i.d., nor is it independent of $\{\delta_n\}_{n\in\mathbb{N}}$ and $\{W_n\}_{n\in\mathbb{N}}$. Parameters $\alpha_1$ and $\alpha_2$ can be interpreted as a form of 'correlation' between adding the ball to the urn and the number of balls already in the urn at that stage. The process $\{W_k,k\geq 1;W_0=w_1\}$ generated by (13) satisfies the Markov property, because of (4). It is an aperiodic random walk on $\{0,1,\ldots,w\}$ with transition probabilities for $i,j=0,1,\ldots,w$

$$P(i;j)=\begin{cases}\frac{\alpha_1+i}{\alpha+w-1}\frac{w-i}{w} & j=i+1\\ \frac{\alpha_1+i-1}{\alpha+w-1}\frac{i}{w}+\frac{\alpha_2+w-i-1}{\alpha+w-1}\frac{w-i}{w} & j=i\\ \frac{\alpha_2+w-i}{\alpha+w-1}\frac{i}{w} & j=i-1\\ 0 & \text{otherwise.}\end{cases} \tag{13}$$

The stationary distribution is shown in Viarengo (2002) to be the generalized Pòlya distribution Polya$(w;\alpha_1,\alpha_2)$:

$$\lim_{n\to\infty}P(W_n=k)=\binom{w}{k}\frac{\alpha_1^{[k]}\alpha_2^{[w-k]}}{\alpha^{[w]}},\quad k=0,1,\ldots,w, \tag{14}$$

where $y^{[x]}=y(y+1)\ldots(y+x-1)$. As is well-known, the Bose-Einstein, Fermi-Dirac and Maxwell-Boltzmann "statistics" are all special cases.

1. When $\alpha_1,\alpha_2$ vary, the compositions of urns $\tilde{U}_1$ and $\tilde{U}_2$, and thus the probabilities of returns to urns $U_1$ and $U_2$, vary as well. This dependence decreases as $\alpha$ increases and tends to $\infty$.
2. If $\alpha_1,\alpha_2<0$, the constraint $W_n\leq|\alpha_1|$ and $w\leq|\alpha|$ must be satisfied. For $\alpha=-w$ there is maximum negative correlation between the composition of urn $U_i$ and the probability of returning the ball to it , whereas as $\alpha\to-\infty$ the correlation tends to 0.
3. If $|\alpha_1|\to\infty$ and $\alpha_1/\alpha=p$ the probabilities tend to be independent both of the urn composition and of where the ball comes from, which gives $Asym-ED(w_1,w_2;p)$ as a special case.

## 6 Some Convergence Properties

A suitable choice of the parameters $\alpha_1$ and $\alpha_2$ makes the Ehrenfest-Brillouin design $EBD(w_1, w_2; \alpha_1, \alpha_2)$ converge to any specified target proportion. Applying Theorem 1 we can shown that

**Theorem 5.** *For any urn design (3), let $\xi$ be the stationary distribution of the Markov process $\{W_n\}_{n\in\mathbb{N}}$ and $E_\xi$ the ergodic mean, then by the SLLN we have*

$$\lim_{n\to\infty} \frac{N_1(n)}{n} = \frac{E_\xi}{w_1 + w_2} \quad a.s. \text{ for all } w_1, w_2 \in \mathbb{R}^+. \tag{15}$$

**Corollary 1.** *Given a target $\pi^* \in [0,1]$ and a design $EBD(w_1, w_2; \alpha_1, \alpha_2)$ such that $\alpha_1/\alpha = \pi^*$ we have*

$$\lim_{n\to\infty} \frac{N_1(n)}{n} = \pi^* \quad a.s.$$

It follows from (2) and (15) that the convergence of an Ehrenfest urn design towards a target is given by the convergence of the time average of the Markov chain $\{W_n\}_{n\in\mathbb{N}}$ to the ergodic mean $E_\xi$. Thus an approximate measure of the speed of convergence of the design is given by the convergence of $\{W_n\}_{n\in\mathbb{N}}$ to the limit variable $W^* \sim \xi$.

The following result shows that the speed of convergence of $Asym - ED(w_1, w_2; p)$ towards the ergodic mean $wp$ is geometric of order $w^{-1}$, independent of $p$; however under the Ehrenfest - Brillouin sequential procedure $EBD(w_1, w_2; \alpha_1, \alpha_2)$ the convergence to $w(\alpha_1/\alpha)$ is still geometric, but the order depends on the values of parameters $\alpha_1$ and $\alpha_2$.

**Theorem 6.** *Under $EBD(w_1, w_2; \alpha_1, \alpha_2)$ we have*

$$E\left[W_n - w\frac{\alpha_1}{\alpha}\right] = E\left[W_0 - w\frac{\alpha_1}{\alpha}\right]\left(1 - \frac{\alpha}{w(w+\alpha-1)}\right)^n.$$

*In particular for $Asym - ED(w_1, w_2; p)$ we have*

$$E[W_n - wp] = E[W_0 - wp]\left(1 - \frac{1}{w}\right)^n.$$

For any target $\pi^*$ and any $w$, choosing negative $\alpha_1$ and $\alpha_2$ makes the Ehrenfest-Brillouin design converge to $\pi^*$ more rapidly than the corresponding asymmetric Ehrenfest design.

**Theorem 7.** *For any target $\pi^* \in [0,1]$ and design $EBD(w_1, w_2; \alpha_1, \alpha_2)$ given by (13 ), if $\alpha_1/\alpha = \pi^*$, with $\alpha < -w$, then*

$$\lim_{n\to\infty} \frac{N_1(n)}{n} = \pi^* \quad a.s.$$

*Moreover, the convergence is faster than that of $Asym - ED(w_1, w_2; \pi^*)$.*

When $\pi^*$ is known it would seem natural to start from an initial urn composition such that $w_1/w = \pi^*$. By choosing $\alpha_1 = -w_1$ and $\alpha_2 = -w_2$ in $EBD(w_1, w_2; \alpha_1, \alpha_2)$ (maximum negative correlation) the speed of convergence is maximal since the Ehrenfest - Brillouin process does not evolve; it becomes a sampling-with-replacement scheme because of (13). In this case the procedure $EBD(w_1, w_2; -w_1, -w_2)$ coincides with a completely randomized design in which at each step treatment $T_1$ is assigned with probability $\pi^*$. Needless to say, the smaller is $w$, the more "deterministic" the design is, and therefore the faster the convergence is, as the transition probabilities (13) show. There is a trade-off between the degree of randomization and the rate of convergence which, in a case of balance, has been explored in Baldi Antognini and Giovagnoli (2004).

*Acknowledgement.* The author was supported by the 2003 Research Project of the Department of Statistical Sciences, University of Bologna: "Asymptotic properties of stochastic processes: methodological tools and biomedical physical and engineering applications". The author wishes to thank Professors Alessandra Giovagnoli and Derek Boothman for helpful advice.

## References

M. Aoki. *Modeling aggregate behavior and fluctuations in economics: stochastic views of interacting agents.* Cambridge University Press, 2002.

Z.D. Bai, F. Hu, and L. Shen. An adaptive design for multi-arm clinical trials. *Journal of Multivariate Analysis*, 81:1–18, 2002.

A. Baldi Antognini and A. Giovagnoli. On the large sample optimality of sequential designs for comparing two treatments. *Submitted*, 2003.

A. Baldi Antognini and A. Giovagnoli. A new "biased coin design" for the sequential allocation of two treatments. *J. R. Statist. Soc. A*, 53:1–14, 2004.

U. Bandyopadhyay and A. Biswas. A class of adaptive designs. *Sequential Analysis*, 19:45–62, 2000.

Y. Chen. Which design is better? Ehrenfest urn versus biased coin. *Advances in Applied Probability*, 32:738–49, 2000.

U. Garibaldi and M.A. Penco. Ehrenfest's urn model generalized: an exact approach for market participation models. *Statistica Applicata*, 12:249–72, 2000.

A. Giovagnoli. Markovian properties of some biased coin designs. In A. Di Bucchianico, H. Läuter, and H.P. Wynn, editors, *mODa 7—Advances in Model-Oriented Design and Analysis (Heeze, June 14-18, 2004)*, Contributions to Statistics, pages 81–88, Heidelberg, 2004. Physica-Verlag.

W.F. Rosenberger. New directions in adaptive design. *Statistical Science*, 11: 137–149, 1996.

W.F. Rosenberger. Randomized urn models and sequential design. *Sequential Analysis*, 21:1–28, 2002.

P. Viarengo. *Birth and death chain: theory and applications based on interacting agents.* PhD thesis, University of Bologna, Italy, 2002.

L.J. Wei. A class of designs for sequential clinical trials. *Journal of the American Statistical Association*, 72:382–386, 1977.

# Nonparametric Testing for Main Effects on Inequivalent Designs

D. Basso[1], L. Salmaso[2], H. Evangelaras[3], and C. Koukouvinos[4]

[1] Department of Statistics, University of Padova, Str. C. Battisti 241, 35121 Padova, Italy dario@stat.unipd.it
[2] Department of Management and Engineering, University of Padova, Str. Stradella San Nicola 3, 36100 Vicenza, Italy salmaso@stat.unipd.it
[3] Department of Mathematics, National Technical University of Athens, Zografou 15773, Athens, Greece harris11@central.ntua.gr
[4] Department of Mathematics, National Technical University of Athens, Zografou 15773, Athens, Greece ckoukouv@math.ntua.gr

**Summary.** A class of inequivalent orthogonal designs is considered and the aim is to provide a "permutation" test for active effects. Rather than permuting responses of a single observed design matrix (or, equivalently, permuting rows of the design matrix keeping the responses fixed), matrices are exchanged in order to obtain the permutation distribution. Such test called IMPT (Inequivalent Matrices Permutation Test) behaves well both under the null hypothesis and in power, even with heavy-tailed distributions. A comparative simulation study with Lenth's and the parametric $F$-test is also presented, showing a better performance of the IMPT with respect to both other tests in the case of Cauchy errors distribution and a similar performance with the $F$-test with normal error distribution.

**Key words:** factorial design, permutation test, Lenth's method.

## 1 Introduction

Screening designs very often represent the first approach to experimental situations in which many explanatory factors are available and we are interested in establishing which ones are important. The two-level factorial designs represent an instrument which is easy to use and interpret. In screening experiments sometimes it may be difficult to verify in practice the normality of the responses both for replicated and for unreplicated designs. When considering unreplicated designs it is not possible to obtain an estimate of the variance of the errors, therefore the usual inferential techniques, aimed at identifying the significantly active factors, are unsuitable. Various solutions to this problem have been proposed in the literature; some, such as the *Normal plot*, prove not to be very objective; others presuppose assumptions, such as normality of errors, especially when little or nothing is known about the phenomenon being analyzed. Hamada and Balakrishnan (1998) propose a review of existing tests conceived for unreplicated two-level factorial designs. Among the tests considered, we quote those of Daniel (1959), Holms and Berrettoni (1969),

Zahn (1975), Box and Meyer (1986) and Lenth (1989). The behaviour of the tests, once they have been recalibrated to have the same performances under $H_0$, are compared in the presence of a growing number of active effects. The study makes use of simulations which also use non-normal errors. Leaving aside the description of procedures (on this topic see Hamada and Balakrishnan (1998)), our intention is to dwell upon the conclusions. One of the best tests in the review Hamada and Balakrishnan (1998) is Lenth's test. He considered a robust estimator of the contrasts (here denoted by $c_i, i = 1, \dots, k$) standard error $\tau$, which he termed the *pseudo standard error*, defined as $PSE = 1.5\,\text{median}_{\{|c_i|<2.5s_0\}}|c_i|$, where $s_0 = 1.5\,\text{median}|c_i|$. That is, PSE is a trimmed median which attempts to remove contrasts corresponding to active effects. Active effects can be then identified using the margin error $ME = t_{1-\alpha;df}\,PSE$, where $\alpha$ is the desired type I error and $df = (k-1)/3$. The degrees of freedom $(k-1)/3$ comc from an approximation of $PSE^2$ by a scaled $\chi^2$ distribution. For details, see Lenth (1989). Note that Lenth's test is a general procedure, which can be applied both to replicated and unreplicated designs.

One of the desirable properties for a testing procedure is the robustness of the test in the conclusions, that is to say the ability of the test to identify active effects even without any assumption on the distribution of the experimental errors (Pesarin and Salmaso, 2002). In this paper we mainly concentrate on this problem and we propose a new nonparametric testing procedure for a class of orthogonal replicated designs presented in Evangelaras, Georgiou, and Koukouvinos (2004). Note that the proposed testing procedure can also be applied to unreplicated designs. Loughin and Noble (1997) propose a nonparametric permutation test which orders the effects observed as an absolute value and compares them to the maximum of the effects obtained from the permutations of the response, which is residualized, at each step, with respect to the previously tested effects. The aim of the residualization of the response is to obtain permutation estimates of the effects after removing those already tested. In this way the possibility of comparing the observed effect with permutation estimates equal to the effects tested previously is excluded.

The test proposed here is called the IMPT (*Inequivalent Matrices Permutation Test*) and obtains permutation estimates of the effects based on the exchanging of inequivalent design matrices instead of permuting the response. The inequivalent matrices considered are those proposed by Evangelaras, Georgiou, and Koukouvinos (2004), which are orthogonal arrays.

An orthogonal array $OA(n, k, r, t)$ is a $n \times k$ array with entries from a set of $r$ distinct symbols arranged so that for any collection of $t$ columns of the array each of the $r^t$ row vectors appears equally often. In application to factorial designs, each column corresponds to a factor, the symbols are the factors levels and each row represents a combination of the factor levels (run). Thus, we call $n$ the number of runs in the OA, $k$ the number of factors, $r$ the number of levels for each factor and $t$ the strength of the array. So, every $OA(n, k, r, t)$ defines a $n$-run factorial design for $k$ $r$-level factors. Orthogonal arrays of strength 2 are often called *orthogonal main effect plans*, since the estimators of the $k$ main effects are uncorrelated. A wide class of orthogonal arrays with $r = 2$ levels and strength (at least) two, can be obtained from Hadamard matrices. An $n$-dimensional Hadamard matrix is an $n$ by $n$ matrix of $1's$ and $-1's$ with $HH^T = H^TH = nI_n$. This relation means that any two rows or columns of the Hadamard matrix are orthogonal. Two Hadamard matrices $H_1$ and $H_2$ are called equivalent (or $H$-equivalent) if one can be obtained from the other by a sequence of row negations, row permutations, column negations and column permutations. In these matrices, $n$ is necessarily 2 or a multiple of 4. A Hadamard matrix is said to be *normalized* if its first row and column consist of 1's. If not we can normalize the Hadamard matrix by multiplying rows and columns by -1 where is needed. When a Hadamard matrix is normalized, then all the columns but the first have an equal number of +1's and −1's. If we remove the first column of 1's from a normalized Hadamard matrix of order $n$, we obtain a saturated OA$(n, n-1, 2, 2)$. With respect to factorial designs, we can therefore use such an orthogonal array to study the influence of $n-1$ experimental factors on the response using $n$ experimental runs (or factor level combinations). In this paper, we have applied the IMP test to the 63 non-isomorphic OA$(24, 5, 2, 2)$ that are embedded into Hadamard matrices of order 24. Two such orthogonal arrays are said to be isomorphic if one can be obtained from the other by permutation of rows, columns and sign changes in columns.

## 2 The IMP Test

Following Loughin and Noble's set up, we have devised a permutation test for effects in two-level factorial designs and as a practical case we apply this test to the class of inequivalent designs for $n = 24$ and $k = 5$. The reference model is the usual linear model:

$$y_i = \mu + \sum_{j=1}^{k} \beta_j x_{ij} + \varepsilon_i, \tag{1}$$

where $y_i$ $(i = 1, \ldots, n)$ is the $i$th experimental response, $\mu$ is the overall mean, $\beta_j (j = 1, \ldots, k)$ is the effect of the $j$th factor, $x_{ij}$ $(i = 1, \ldots, n; j = 1, \ldots, k)$ is the $j$th element of the $i$th row of the design matrix, and $\varepsilon_i$ $i = 1, \ldots, n$ is the $i$th experimental error, which is assumed to be uncorrelated with the other errors and to have equal variance. Permuting the random response vector is the same as permuting the rows of the design matrix $\mathbf{X}$, holding the response vector fixed. In this way we can interpret the vector of permuted responses $\mathbf{y}^*$ as the projection of the observed response vector $\mathbf{y}$ onto a space generated by the columns of $\mathbf{X}^*$, the matrix obtained from the design matrix through permutation of the rows. Since two matrices $\mathbf{X}$ and $\mathbf{X}^*$ are said to be equivalent if the second can be obtained from the first only through permutations of rows and/or columns, the result is that the vector $\mathbf{y}^*$ is the projection of $\mathbf{y}$ onto a space equivalent to the one generated by the columns of $\mathbf{X}$. Thus the permutation estimates $\hat{\beta}_1^*, \ldots, \hat{\beta}_k^*$ represent the coefficients which minimize the distance between $\mathbf{y}$ and its projection in the subspace generated by $\mathbf{X}^*$, equivalent to the one generated by $\mathbf{X}$. The least squares estimator of the vector of effects is:

$$\hat{\beta} = (\mathbf{X}'\mathbf{X})^{-1}\mathbf{X}'\mathbf{y}. \tag{2}$$

We can therefore also consider the effect estimators as projections onto the space of the columns of matrix $\mathbf{X}$, and the the estimators of the effects obtained from the permutations as projections onto the space generated by the columns of $\mathbf{X}^*$. The basic idea behind permutation tests is that if an effect is null, then its projection in any subspace is also null (unless there is a random error). This concept does not change if we use inequivalent subspaces to the one generated by the design matrix, instead of equivalent ones. The test we propose is based on this important consideration.

As in Loughin and Noble's test, the main idea is that if the maximum of the estimated effects is non significant, then all other effects will also be non significant. The proposed procedure is *stepwise*, that is we test the null hypothesis on the effects based on their decreasing ordering. Given the arbitrariness of the "high" and "low" levels, we order the observed effects according to their absolute value.

As the observed design matrix $\mathbf{X}$ we have chosen to consider one of the 63 inequivalent matrices with $n = 24$ observations and $k = 5$ effects (for the sake of simplicity we consider the first matrix to be the one which generated the responses) presented by Evangelaras, Georgiou, and Koukouvinos (2004) and we have chosen to assign each column of $\mathbf{X}$ to a factor (named $A$, $B$, $C$, $D$, $E$, or, equivalently, 1, 2, 3, 4, 5). Once the experiment has been performed according to a design matrix $\mathbf{X}$, with $n = 24$ and $k = 5$, the algorithm of the IMP test is as follows:

1. Obtain the estimates in a decreasing order $|\hat{\beta}_{(1)}^{obs}| \geq |\hat{\beta}_{(2)}^{obs}| \geq |\hat{\beta}_{(3)}^{obs}| \geq |\hat{\beta}_{(4)}^{obs}| \geq |\hat{\beta}_{(5)}^{obs}|$, where "obs" stands for "observed". This means that the estimates $\hat{\beta}_{(j)}^{obs}, j = 1, \ldots, 5$ are the ones corresponding to the design matrix $\mathbf{X}$.
2. Put $\tilde{\mathbf{y}} = \mathbf{y} - \sum_{\mathrm{j} \neq \mathrm{s}} \hat{\beta}_{(\mathrm{j})} \mathbf{x}_{\mathrm{j}}$.
3. For $q = 2, \ldots, 63$ estimate $\hat{\beta}_1^q, \hat{\beta}_2^q, \hat{\beta}_3^q, \hat{\beta}_4^q, \hat{\beta}_5^q$ from $\tilde{\mathbf{y}}$ with $\mathbf{X}^q$.
4. Put $\hat{\beta}_{(1)}^q = \max_j \{|\hat{\beta}_j^q|\}$.
5. Calculate the $p$-value for $\hat{\beta}_{(s)}^{obs} : p_{(s)} = \frac{\#[\hat{\beta}_{(1)}^q \geq |\hat{\beta}_{(s)}^{obs}|]}{63}$.

Repeat steps 2), 3), 4) and 5) for $s = 1, \ldots, 5$.

Since the 63 projection matrices are inequivalent, at the $s$th step ($s = 1, \ldots, k$) of the algorithm we residualise the response vector compared to the estimates of the $k-1$ effects different from the one being tested ($\hat{\beta}_{(s)}$), that is we consider:

$$\tilde{\mathbf{y}} = \mathbf{y} - \sum_{j \neq s}^{k} \hat{\beta}_{(j)} \mathbf{x_j}, \tag{3}$$

where the $\hat{\beta}_j, (j = 1, \ldots, k; j \neq s)$ represents the estimate of the $j$th ordered observed effect different from the one considered and $\mathbf{x}_j, s$ $(j = 1, \ldots, k; j \neq s)$ are the columns of the design matrix from which the effect of the corresponding factor was estimated. The use of residualization ensures that the tested effect is the maximum of the considered effects. We can compare the $s$th observed effect with the maximum of the effects estimated with $\tilde{\mathbf{y}}$ from the remaining 62 inequivalent matrices $\mathbf{X}^q$, $q = 2, \ldots, 63$.

The use of the residualization also eliminates all the estimates of permutation effects equal to those observed, thus not inflating $p$-values. Loughin and Noble warn against what they define as the *masking effect*: for example it may happen that when there are $h \leq k$ effects actually under $H_1$, the estimates of the permutation effects are on average higher, leading us to accept $H_0$ for the first $h-1$ ordered effects. They avoid this problem by suggesting a step-up procedure, that is by testing $H_0$ on the observed effects in increasing order and residualizing the response compared with the estimates of the effects previously tested. A further solution is to residualize the response with respect to all the observed effects except the one being tested. In this way we obtain permutation estimates (projections) which are not inflated by all the other effects.

## 3 A Comparative Simulation Study

We now present a comparative simulation study in order to show the behaviour of the IMP test under the null hypothesis (Table 1) and with respect to power. Two distributions are considered for the experimental errors: the standard normal and the standard Cauchy. We have chosen the normal, as often occurs in literature, and the Cauchy to get an idea of the robustness of the test when errors are heavy-tailed. A power comparison of the IMP test with Lenth's test and the parametric $F$-test is also presented in Table 2. These results were obtained considering 1000 iterations, and represent, for each effect, the proportion of rejecting the null hypothesis at different nominal significance levels $\alpha$'s.

The significance levels which can actually be achieved have steps of 1/63 because there are 63 inequivalent matrices. As can be seen, the test is conservative. However, the significance levels observed in the simulations show

**Table 1.** Behaviour of the IMP test under $H_0$ with standard Normal and Cauchy errors.

| | N(0,1) | | | | | Ca(0,1) | | | | |
|---|---|---|---|---|---|---|---|---|---|---|
| | True Effects | | | | | | | | | |
| $\alpha$ | $A=0$ | $B=0$ | $C=0$ | $D=0$ | $E=0$ | $A=0$ | $B=0$ | $C=0$ | $D=0$ | $E=0$ |
| 0.0158 | 0.030 | 0.036 | 0.019 | 0.038 | 0.036 | 0.023 | 0.020 | 0.006 | 0.015 | 0.018 |
| 0.0317 | 0.035 | 0.049 | 0.046 | 0.044 | 0.043 | 0.025 | 0.026 | 0.018 | 0.021 | 0.021 |
| 0.0476 | 0.043 | 0.056 | 0.062 | 0.051 | 0.047 | 0.028 | 0.029 | 0.024 | 0.023 | 0.024 |
| 0.0634 | 0.047 | 0.061 | 0.069 | 0.055 | 0.050 | 0.032 | 0.029 | 0.027 | 0.025 | 0.025 |
| 0.0793 | 0.057 | 0.068 | 0.075 | 0.060 | 0.056 | 0.038 | 0.031 | 0.033 | 0.032 | 0.031 |
| 0.0952 | 0.063 | 0.075 | 0.080 | 0.062 | 0.067 | 0.044 | 0.034 | 0.036 | 0.035 | 0.032 |
| 0.1111 | 0.067 | 0.081 | 0.088 | 0.065 | 0.064 | 0.049 | 0.043 | 0.041 | 0.040 | 0.039 |

a homogenous trend at any nominal significance level under $H_0$. The power simulations show the effectiveness of the test , in fact from the first nominal levels the active effects are detected. In particular note that even with Cauchy errors the IMP test works well, both under $H_0$ and under $H_1$. In particular, the inactive effects under $H_1$ maintain the same significance levels observed under $H_0$. The absolute value of the effects has been fixed at maximum levels of respectively 2 for Cauchy and 1 for Normal to guarantee their comparability. In fact the result is that if $X \sim N(0,1)$ and $Y \sim Ca(0,1)$, $P[|X| \leq 1] \simeq 0.692$ and $P[|Y| \leq 2] \simeq 0.674$. For a comparison involving Lenth's test and the $F$-test we considered the same observed design matrix (the $1^{st}$ of the 63 inequivalent matrices) and applied Lenth's test algorithm to $k = 5$ factors. To stand the most favourable way we considered as active each effect exceeding the $ME$, obtained with a $t_{1-\alpha;23/3}$ distribution (as suggested by the author). As can be seen in Table 2, the IMP test is more powerful than Lenth's one at any significance levels, and tends to be less conservative for the inactive effects when standard normal errors are considered. Also when errors follow a standard Cauchy distribution the IMP test shows a better behaviour: the active effects are recognized but while Lenth's test significance levels are very close to (or grater than) the nominal levels for the inactive effects, IMP test is more conservative and robust as well. As regards the comparison with the $F$-test we can underline that under the normality assumption for error components, the power behaviour of the IMP test is very close to the behaviour of the $F$-test, while under the Cauchy error distribution, as it can be expected, the achieved significance levels of the $F$-test are very inflated, especially at low nominal significance levels.

**Table 2.** A comparison between IMP test, Lenth's test and the parametric $F$-test under $H_1$ with standard Normal and Cauchy errors.

| | N(0,1) | | | | | Ca(0,1) | | | | |
|---|---|---|---|---|---|---|---|---|---|---|
| | True Effects | | | | | | | | | |
| | $A=1$ | $B=0.75$ | $C=0.5$ | $D=0$ | $E=0$ | $A=2$ | $B=1.5$ | $C=1$ | $D=0$ | $E=0$ |
| $\alpha$ | **IMP test** | | | | | | | | | |
| 0.0158 | 0.982 | 0.860 | 0.189 | 0.030 | 0.026 | 0.471 | 0.342 | 0.068 | 0.015 | 0.017 |
| 0.0317 | 0.984 | 0.892 | 0.559 | 0.036 | 0.030 | 0.483 | 0.362 | 0.205 | 0.021 | 0.020 |
| 0.0476 | 0.986 | 0.900 | 0.594 | 0.037 | 0.033 | 0.491 | 0.374 | 0.221 | 0.023 | 0.023 |
| 0.0634 | 0.989 | 0.906 | 0.611 | 0.041 | 0.039 | 0.504 | 0.380 | 0.239 | 0.025 | 0.028 |
| 0.0793 | 0.989 | 0.914 | 0.639 | 0.043 | 0.041 | 0.516 | 0.388 | 0.251 | 0.032 | 0.030 |
| 0.0952 | 0.990 | 0.919 | 0.644 | 0.046 | 0.047 | 0.525 | 0.401 | 0.265 | 0.035 | 0.032 |
| 0.1111 | 0.993 | 0.932 | 0.651 | 0.048 | 0.050 | 0.535 | 0.417 | 0.270 | 0.040 | 0.038 |
| $\alpha$ | **Lenth's test** | | | | | | | | | |
| 0.0158 | 0.663 | 0.444 | 0.238 | 0.001 | 0.000 | 0.282 | 0.185 | 0.092 | 0.037 | 0.051 |
| 0.0317 | 0.802 | 0.572 | 0.324 | 0.002 | 0.002 | 0.373 | 0.244 | 0.139 | 0.050 | 0.069 |
| 0.0476 | 0.874 | 0.666 | 0.397 | 0.004 | 0.002 | 0.432 | 0.292 | 0.167 | 0.062 | 0.089 |
| 0.0634 | 0.913 | 0.731 | 0.451 | 0.005 | 0.004 | 0.465 | 0.325 | 0.197 | 0.075 | 0.102 |
| 0.0793 | 0.937 | 0.785 | 0.504 | 0.007 | 0.004 | 0.511 | 0.359 | 0.228 | 0.088 | 0.112 |
| 0.0952 | 0.950 | 0.835 | 0.552 | 0.008 | 0.004 | 0.542 | 0.384 | 0.255 | 0.097 | 0.127 |
| 0.1111 | 0.968 | 0.867 | 0.591 | 0.009 | 0.005 | 0.569 | 0.422 | 0.286 | 0.108 | 0.138 |
| $\alpha$ | **$F$ test** | | | | | | | | | |
| 0.0158 | 0.983 | 0.830 | 0.464 | 0.017 | 0.012 | 0.367 | 0.257 | 0.118 | 0.003 | 0.004 |
| 0.0317 | 0.995 | 0.903 | 0.576 | 0.029 | 0.033 | 0.427 | 0.316 | 0.172 | 0.010 | 0.011 |
| 0.0476 | 0.996 | 0.936 | 0.598 | 0.042 | 0.046 | 0.459 | 0.353 | 0.208 | 0.017 | 0.021 |
| 0.0634 | 0.998 | 0.957 | 0.699 | 0.056 | 0.059 | 0.490 | 0.391 | 0.235 | 0.025 | 0.033 |
| 0.0793 | 0.998 | 0.968 | 0.736 | 0.070 | 0.072 | 0.528 | 0.415 | 0.266 | 0.037 | 0.053 |
| 0.0952 | 1.000 | 0.973 | 0.764 | 0.085 | 0.087 | 0.547 | 0.438 | 0.299 | 0.060 | 0.059 |
| 0.1111 | 1.000 | 0.977 | 0.782 | 0.101 | 0.101 | 0.570 | 0.465 | 0.325 | 0.070 | 0.073 |

# 4 Conclusions

The idea of exchanging matrices instead of permuting the vector of responses (or, equivalently, of permuting the design matrix rows while keeping the response vector fixed) is introduced. This extends the usual approach to permutation techniques. The permutation of the response is a particular case of the exchanging of inequivalent matrices. One could consider a test that combines both techniques, in order to reach the desired nominal levels. The test, still in progress, promises good future developments. The choice was to use the projection of designs 24 into 5, but it can be applied in general to factorial designs when an appropriate number of inequivalent matrices is available. In the case when interactions are not confounded in the inequivalent matrices, tests for the interaction effects can be provided. Furthermore the IMP test may also be applied to unreplicated factorials based on Hadamard matrices.

In this paper we presented the IMP test only applied to 24 runs designs to test only 5 main effects, but as pointed out before the "permutation" procedure is very general and can be applied to other cases.

## References

G.E.P. Box and R.D. Meyer. An analysis for unreplicated fractional factorials. *Technometrics*, 28(1):11–18, 1986.

C. Daniel. Use of half-normal plots in interpreting factorial two-level experiments. *Technometrics*, 1:311–341, 1959.

H. Evangelaras, S. Georgiou, and C. Koukouvinos. Evaluation of inequivalent projections of Hadamard matrices of order 24. *Metrika*, 2004. (to appear).

M. Hamada and N. Balakrishnan. Analyzing unreplicated factorial experiments: a review with some new proposals. *Statist. Sinica*, 8(1):1–41, 1998. With comments by Claudio Benski, Perry D. Haaland, Russell V. Lenth and a rejoinder by the authors.

A.G. Holms and J.N. Berrettoni. Chain-pooling ANOVA for two-level factorial replication free experiments. *Technometrics*, 11:725–746, 1969.

R.V. Lenth. Quick and easy analysis of unreplicated factorials. *Technometrics*, 31(4):469–473, 1989.

T.M. Loughin and W. Noble. A permutation test for effects in an unreplicated factorial design. *Technometrics*, 39(2):180–190, 1997.

F. Pesarin and L. Salmaso. Exact permutation tests for unreplicated factorials. *Appl. Stoch. Models Bus. Ind.*, 18(3):287–299, 2002. 2nd International Symposium on Business and Industrial Statistics (Yokohama, 2001).

D.A. Zahn. Modifications of and revised critical values for half-normal plot. *Technometrics*, 17:189–200, 1975.

# Maximin Optimal Designs for a Compartmental Model

S. Biedermann[1], H. Dette[2], and A. Pepelyshev[3]

[1] Ruhr-Universität Bochum, Fakultät für Mathematik, 44780 Bochum, Germany
stefanie.biedermann@ruhr-uni-bochum.de

[2] Ruhr-Universität Bochum, Fakultät für Mathematik, 44780 Bochum, Germany
holger.dette@ruhr-uni-bochum.de

[3] St.-Petersburg State University, Department of Mathematics, Univ. av. 28, St.-Petersburg, 198504, Russia andrey@ap7236.spb.edu

**Summary.** For an important example from the class of compartmental models we determine optimal designs, which are robust against misspecifications of the unknown model parameters. We propose a maximin approach based on $D$-efficiencies and provide designs that are optimal with respect to the particular choice of various parameter regions.

**Key words:** Compartmental model, robust optimal design, maximin $D$-optimality, local optimality.

## 1 Introduction

We consider a compartmental model, which is commonly applied in the chemical sciences and pharmacokinetics as far as two-step reactions are concerned (see, *e.g.*, Box and Lucas (1959), Gibaldi and Perrier (1982) or Atkinson (2003)). Suppose a substance $A$ decomposes to form substance $B$ which then in turn decomposes to form substance $C$. Assuming that the reactions are irreversible and first order, the concentrations $[A]$, $[B]$ and $[C]$ of products $A$, $B$ and $C$, respectively, as functions of the time $x \geq 0$ satisfy the kinetic differential equations

$$\frac{d[A]}{dx} = -\theta_1[A], \quad \frac{d[B]}{dx} = \theta_1[A] - \theta_2[B], \quad \frac{d[C]}{dx} = \theta_2[B], \tag{1}$$

where $\theta_1 > \theta_2 > 0$ denote unknown constants measuring the specific rates of the first and second decomposition, respectively. Solving system (1) with initial conditions $[A] = 1$ and $[B] = [C] = 0$, we obtain the amount of intermediate product $B$ after time $x \geq 0$ has elapsed as

$$\eta(x,\theta) = \frac{\theta_1}{\theta_1 - \theta_2}\left(e^{-\theta_2 x} - e^{-\theta_1 x}\right). \tag{2}$$

In a statistical setup, we thus model observations $Y$ on $[B]$ at different times $x$ as a regression model $Y = \eta(x,\theta) + \varepsilon$, where the observation errors $\varepsilon$ are independent and from an exponential family with zero expectation and constant variance $\sigma^2$. The Fisher information $I(x, \theta)$ for estimating the parameter $\theta = (\theta_1, \theta_2)^T$ at time $x$ is then given by the expression $I(x, \theta) = f(x, \theta) f^T(x, \theta)$, where $f(x, \theta) = (f_1(x, \theta), f_2(x, \theta))^T$ is the gradient of the regression function $\eta(x, \theta)$ with respect to $\theta$. An (approximate) design $\xi$ is a probability measure with finite support on the non-negative real axis $\mathbb{R}_0^+$, that is the observations are taken at the support points of the measure proportional to the corresponding masses. The Fisher information matrix of the design $\xi$ is given by

$$M(\xi, \theta) = \int_0^\infty I(x, \theta)\, d\xi(x), \tag{3}$$

and an optimal design maximizes a real-valued function of the Fisher information matrix, which is usually referred to as an optimality criterion (see, *e.g.*, Silvey (1980)). A typical example is $D$-optimality where the determinant of the Fisher information is maximized with respect to the design $\xi$.

Since an appropriate choice of the experimental conditions can improve the quality of the statistical inference substantially, much effort has been devoted to the problem of finding optimal designs for the compartmental model (2) (see, *e.g.*, Atkinson (2003), Atkinson et al. (1993), Box and Lucas (1959), Dette and O'Brien (1999), Hamilton and Watts (1985) or Hill and Hunter (1974)). The Fisher information and thus the optimal designs with respect to the common optimality criteria depend on the unknown parameter. Following Chernoff (1953), several authors assume that an initial guess of $\theta$ is available and determine so-called locally optimal designs (see Box and Lucas (1959) or Hill and Hunter (1974)). However, misspecifications of the parameter for a locally optimal design can lead to poor results in the subsequent data analysis. A more robust alternative is to assume sufficient knowledge of $\theta$ to specify a prior distribution for this parameter and to average the respective optimality criteria over the plausible values of $\theta$ defined by the prior. This leads to so-called Bayesian optimality criteria (see, *e.g.*, Atkinson et al. (1993) or Pronzato and Walter (1985)). As an alternative for the construction of robust designs, we propose a maximin approach based on $D$-efficiencies, which only requires the specification of a certain range for the unknown parameter. We feel that this is a more realistic scenario since practitioners will often have difficulties to specify a prior distribution for the unknown parameter $\theta$. Most of the literature on maximin design discusses models with only one nonlinear parameter or minimally supported designs (see, *e.g.*, Imhof (2001)). Model (2), however, contains two nonlinear parameters and therefore standard techniques are not readily applicable. In section 2, we provide some insight in the structure of locally $D$-optimal designs for the compartmental model (2). In section 3, we define the standardized maximin $D$-optimality criterion, determine various designs optimal with respect to the maximin criterion and investigate the properties of these optimal designs. It turns out that in many cases the standardized maximin $D$-optimal designs are supported on three points so that they can also be used for model checking.

## 2 Locally $D$-optimal Designs

We start our investigation of the locally $D$-optimal designs by presenting a result relating $D$-optimal designs with respect to several parameter combinations, thus yielding a substantial simplification of the optimization problem.

**Lemma 1.** *Let $x^*(\theta)$ denote the vector of support points of the locally D-optimal design with respect to the parameter $\theta$. Then*

$$x^*(\gamma\theta) = \frac{1}{\gamma} x^*(\theta)$$

*for any $\gamma > 0$. Moreover, the locally D-optimal designs with respect to the parameters $\theta$ and $\gamma\theta$ have the same weights at the corresponding support points.*

*Proof.* The assertion of Lemma 1 is a simple consequence of the homogeneity of the linear differential equation formulation (1) and applies quite generally. □

By Lemma 1 it is sufficient to compute locally $D$-optimal designs with $\theta_1 = 1$, since the optimal designs with respect to any other parameter can easily be obtained from these designs by rescaling.

Numerical calculations suggest that the locally $D$-optimal designs in the compartmental model (2) are supported on exactly two points. Furthermore, minimally supported $D$-optimal designs are equally weighted, which is a standard result in design theory (see, *e.g.*, Silvey (1980), Lemma 5.1.3). Some locally $D$-optimal designs for several representative situations with respect to the choice of $\theta$ are listed in Table 1. As an interesting result from Table 1

**Table 1.** The support points $x_1^*, x_2^*$ of the locally $D$-optimal designs $\xi^* = \{x_1^*, x_2^*; 1/2, 1/2\}$ with respect to the parameter $\theta = (1, \theta_2)^T$.

| $\theta_2$ | $x_1^*$ | $x_2^*$ | $\theta_2$ | $x_1^*$ | $x_2^*$ | $\theta_2$ | $x_1^*$ | $x_2^*$ |
|---|---|---|---|---|---|---|---|---|
| 0.1 | 0.9283 | 11.0171 | 0.4 | 0.8186 | 3.9018 | 0.7 | 0.7164 | 2.8599 |
| 0.2 | 0.8907 | 6.1603 | 0.5 | 0.7825 | 3.4353 | 0.8 | 0.6868 | 2.6634 |
| 0.3 | 0.8554 | 4.6515 | 0.6 | 0.7483 | 3.1076 | 0.9 | 0.6594 | 2.5020 |

we obtain that a large distance between $\theta_1$ and $\theta_2$ results in relatively large support points of the locally $D$-optimal designs $\xi^*$. In the following lemma, we show a result on the asymptotic behavior of the design $\xi^*$ when $\theta_2$ tends to its upper boundary.

**Lemma 2.** *If $\theta_2$ tends to 1 $(= \theta_1)$ from below, the locally D-optimal design $\xi^* = \xi^*(\theta_2)$ converges weakly to the design with support points $x_1^* = (3 - \sqrt{3})/2$, $x_2^* = (3 + \sqrt{3})/2$ and equal weights, which is D-optimal for the linear regression model*

$$Y = a_1 \left( x - \frac{x^2}{2} \right) e^{-x} + a_2 \frac{x^2}{2} e^{-x} + \varepsilon.$$

*Proof.* Let $\theta_2 = 1 - z$. By a Taylor expansion of the regression gradient at the point $z = 0$ we obtain

$$f_1(x, \theta) = \left( x - \frac{x^2}{2} \right) e^{-x} + o(z), \quad f_2(x, \theta) = -\frac{x^2}{2} e^{-x} + o(z),$$

and the assertion follows. □

So far, we have considered design regions for the time $x$, which are unbounded, that is the time $x$ is allowed to vary from zero to infinity. However, in many situations there exists a boundary, $x_{max}$ say, on the maximal amount of time that is feasible for the particular experiment. In this case, Lemma 1 has to be modified appropriately, that is if $\xi^* = (x_i^*; w_i^*)$ is the locally $D$-optimal design with respect to the parameter $\theta$ on the design space $[0, x_{max}]$ with support points $x_i^*$ and corresponding weights $w_i^*$, then $\xi_\gamma^* = (\frac{1}{\gamma} x_i^*; w_i^*)$ is locally $D$-optimal with respect to the parameter $\gamma\theta$ on the design space $[0, \frac{1}{\gamma} x_{max}]$. Table 2 gives several examples of locally $D$-optimal designs for the compartmental model (2) with a restricted design space. The efficiencies (eff) of the restricted designs $\xi_{x_{max}}^*$ are given by the square root of the ratio of their particular criterion value and the criterion value of the locally $D$-optimal design $\xi^*$ with respect to an unrestricted design interval. Observing

**Table 2.** Locally $D$-optimal designs with respect to several restricted design spaces $[0, x_{max}]$ and different values of $\theta$.

| $\theta = (1, 0.5)^T$ | | | | $\theta = (1, 0.1)^T$ | | | | $\theta = (1, 0.05)^T$ | | | |
|---|---|---|---|---|---|---|---|---|---|---|---|
| $x_{max}$ | $x_1^*$ | $x_2^*$ | eff | $x_{max}$ | $x_1^*$ | $x_2^*$ | eff | $x_{max}$ | $x_1^*$ | $x_2^*$ | eff |
| $\infty$ | 0.783 | 3.435 | 1 | $\infty$ | 0.928 | 11.017 | 1 | $\infty$ | 0.958 | 21.004 | 1 |
| 3 | 0.758 | 3 | 0.979 | 10 | 0.928 | 10 | 0.994 | 19 | 0.958 | 19 | 0.995 |
| 2.5 | 0.713 | 2.5 | 0.891 | 8 | 0.932 | 8 | 0.944 | 15 | 0.958 | 15 | 0.945 |
| 2 | 0.646 | 2 | 0.728 | 6 | 0.936 | 6 | 0.820 | 11 | 0.958 | 11 | 0.824 |
| 1.5 | 0.548 | 1.5 | 0.495 | 4 | 0.914 | 4 | 0.589 | 7 | 0.963 | 7 | 0.603 |
| 1 | 0.410 | 1 | 0.240 | 2 | 0.717 | 2 | 0.224 | 3 | 0.875 | 3 | 0.234 |
| 0.5 | 0.228 | 0.5 | 0.049 | 1 | 0.436 | 1 | 0.053 | 1 | 0.441 | 1 | 0.027 |
| 0.25 | 0.120 | 0.25 | 0.008 | 0.5 | 0.236 | 0.5 | 0.009 | 0.5 | 0.236 | 0.5 | 0.005 |

the entries of Table 2, we notice that if the largest support point of the unrestricted $D$-optimal design $\xi^*$ is larger than $x_{max}$, then the boundary itself is in the support of $\xi^*_{x_{max}}$. Moreover, the efficiencies of the restricted optimal designs decrease considerably with decreasing $x_{max}$. As a consequence, we suggest using the unrestricted design whenever possible, or trying to choose the boundary $x_{max}$ as large as possible from the experimental circumstances.

## 3 Standardized Maximin $D$-optimal Designs

Following Müller (1995) or Imhof (2001) we propose as robust designs for the compartmental model (2) designs, which maximize the minimal $D$-efficiency calculated over a certain range for the parameter $\theta$, thus protecting the experiment against the worst case scenario. That means that we maximize the expression

$$\Psi_{-\infty}(\xi) = \inf_{\theta \in \Theta} \left\{ \left( \frac{\det M(\xi, \theta)}{\det M(\xi_\theta^*, \theta)} \right)^{1/2} \right\}, \tag{4}$$

where, throughout this article, $\xi_\theta^*$ denotes the locally $D$-optimal design with respect to $\theta$. A design maximizing $\Psi_{-\infty}(\xi)$ is called standardized maximin $D$-optimal (with respect to $\Theta$). The parameter space $\Theta$ of possible values of $\theta$ has to be specified in advance by the experimenter. An advantage of this approach compared to a Bayesian setup is that it is not required to specify a prior distribution for the unknown parameter $\theta$, which is not possible in all circumstances. The only "prior knowledge" needed to use the standardized maximin $D$-criterion is an approximate range $\Theta$ for the parameter $\theta$.

A powerful tool for checking maximin optimality of a design is an equivalence theorem, which can be found in Dette, Haines, and Imhof (2003).

**Theorem 1.** *A design $\xi^*$ is standardized maximin $D$-optimal with respect to $\Theta$ if and only if there exists a prior $\pi^*$ supported on the set*

$$\mathcal{N}(\xi^*) = \left\{ \theta \in \Theta \mid \Psi_{-\infty}(\xi^*) = \left( \frac{\det M(\xi, \theta)}{\det M(\xi_\theta^*, \theta)} \right)^{1/2} \right\}$$

*such that the inequality*

$$d(\xi^*, x) = \int_{\mathcal{N}(\xi^*)} f^T(x, \theta) M^{-1}(\xi^*, \theta) f(x, \theta)\, d\pi^*(\theta) \le 2 \tag{5}$$

*holds for all $x$ within the design space. Moreover, there is equality in (5) for all support points of the design $\xi^*$.*

Following Dette, Haines, and Imhof (2003) we call the prior $\pi^*$ least favourable distribution. The designs given in the following discussion were calculated numerically using Theorem 1 whereby their optimality was carefully checked. These calculations are quite complicated, since the least favourable distribution is two-dimensional. A MATLAB programme is available from the third author.

There are two obvious ways in which a parameter space $\Theta$ can be specified. On the one hand, it might be sensible to assume a one-dimensional closed interval as an admissible range for each parameter. Since the inequality $\theta_1 > \theta_2 > 0$ always holds, that means that $\Theta$ is given by a closed rectangle in the positive half plane, that is $\Theta = \Theta(z_1, z_2, z_3, z_4) := [z_1, z_2] \times [z_3, z_4],\ z_i > 0,\ z_1 > z_4$. On the other hand, the fact that $\theta_1 > \theta_2$ can result in a triangular region $\Theta$, that is $\Theta = \Theta(z_1, z_2) := \{(z_1, z_2) | 0 < z_1 \le \theta_2 < \theta_1 \le z_2\}$.

The following lemma applies Lemma 1 to simplify the maximin optimization problem for $\Theta$ from the above-mentioned cases considerably.

**Lemma 3.** *Denote by $x_i^* = x_i^*(z_1, \ldots, z_j)$ the support points of the standardized maximin D-optimal design with respect to $\Theta(z_1, \ldots, z_j)$, $j = 4$ or $j = 2$. For any $\gamma > 0$ the design with the same weights and support points $\frac{1}{\gamma}x_i^*$ is $\Psi_{-\infty}$-optimal with respect to the parameter space $\gamma\Theta = \Theta(\gamma z_1, \ldots, \gamma z_j)$.*

*If $x_i$ denote the support points of the $\Psi_{-\infty}$-optimal design with respect to $\Theta$ on the restricted design space $[0, x_{max}]$, then the design with the same weights and support points $\frac{1}{\gamma}x_i$ is $\Psi_{-\infty}$-optimal with respect to $\gamma\Theta$ on $[0, \frac{1}{\gamma}x_{max}]$.*

*Proof.* The standardized maximin $D$-optimal design with respect to $\gamma\Theta$ can be obtained by maximizing

$$\min_{\theta\in\gamma\Theta} \frac{\det \int_0^{x_{max}} I(x,\theta)\, d\xi(x)}{\det \int_0^{x_{max}} I(x,\theta)\, d\xi_\theta^*(x)} = \min_{\theta\in\Theta} \frac{\det \int_0^{x_{max}} I(x,\gamma\theta)\, d\xi(x)}{\det \int_0^{x_{max}} I(x,\gamma\theta)\, d\xi_\theta^*(x)}$$

$$= \min_{\theta\in\Theta} \frac{\frac{1}{\gamma^4}\det \int_0^{x_{max}} I(\gamma x,\theta)\, d\xi(x)}{\frac{1}{\gamma^4}\det \int_0^{x_{max}} I(\gamma x,\theta)\, d\xi_\theta^*(x)} = \min_{\theta\in\Theta} \frac{\det \int_0^{\frac{x_{max}}{\gamma}} I(x,\theta)\, d\tilde{\xi}(x)}{\det \int_0^{\frac{x_{max}}{\gamma}} I(x,\theta)\, d\xi_{\gamma\theta}^*(x)},$$

where the design $\tilde{\xi}$ is derived from $\xi$ by the relation $\tilde{\xi}(\{x\}) = \xi(\{\gamma x\})$. The above equalities are obtained by direct calculations on the entries of $I(x, \theta)$, whereas the last transformation in the denominator is an application of Lemma 1. □

In Tables 3 and 4 we present some $\Psi_{-\infty}$-optimal designs corresponding to rectangular parameter regions. The term min eff$_D$ stands for the criterion value of the optimal design. The intervals $[z_1, z_2]$, $[z_3, z_4]$ in Table 3 are cho-

**Table 3.** Standardized maximin $D$-optimal designs $\xi^*$ for model (2) with unrestricted design space with respect to various rectangular parameter spaces.

| $z_1$ | $z_2$ | $z_3$ | $z_4$ | $x_1$ | $x_2$ | min eff$_D$ | $z_1$ | $z_2$ | $z_3$ | $z_4$ | $x_1$ | $x_2$ | min eff$_D$ |
|---|---|---|---|---|---|---|---|---|---|---|---|---|---|
| 0.7 | 0.8 | 0.3 | 0.4 | 1.06 | 4.78 | 0.983 | 0.9 | 1.1 | 0.2 | 0.5 | 0.86 | 4.40 | 0.891 |
| 0.9 | 1.0 | 0.3 | 0.4 | 0.88 | 4.33 | 0.986 | 2.2 | 2.8 | 0.2 | 0.5 | 0.37 | 3.47 | 0.885 |
| 0.9 | 1.1 | 0.3 | 0.4 | 0.84 | 4.25 | 0.978 | 2.2 | 2.8 | 0.2 | 0.7 | 0.37 | 2.94 | 0.809 |
| 0.9 | 1.1 | 0.3 | 0.5 | 0.83 | 3.95 | 0.954 | 2.0 | 3.0 | 0.2 | 0.7 | 0.37 | 2.93 | 0.787 |

sen relatively small. In this case, the $\Psi_{-\infty}$-optimal designs are supported on two points and equally weighted. For the examples in Table 4 we choose

**Table 4.** Standardized maximin $D$-optimal designs $\xi^*$ for model (2) with unrestricted design space with respect to various rectangular parameter spaces.

| $z_1$ | $z_2$ | $z_3$ | $z_4$ | $x_1$ | $x_2$ | $x_3$ | $w_1$ | $w_2$ | $w_3$ | min eff$_D$ |
|---|---|---|---|---|---|---|---|---|---|---|
| 2 | 3 | 0.2 | 0.8 | 0.37 | 2.28 | 4.69 | 0.50 | 0.35 | 0.15 | 0.755 |
| 2 | 3 | 0.2 | 0.9 | 0.35 | 1.97 | 4.80 | 0.49 | 0.32 | 0.19 | 0.736 |
| 2 | 3 | 0.2 | 1 | 0.35 | 1.63 | 5.14 | 0.50 | 0.29 | 0.21 | 0.727 |
| 2 | 3 | 0.1 | 1 | 0.38 | 1.85 | 7.88 | 0.54 | 0.25 | 0.20 | 0.661 |

larger parameter regions $\Theta$, thus obtaining optimal designs supported on three points. Results for triangular parameter regions are depicted in Table 5.

**Table 5.** Standardized maximin $D$-optimal designs for model (2) with unrestricted design space with respect to various triangular parameter regions.

| $z_1$ | $z_2$ | $x_1$ | $x_2$ | $x_3$ | $w_1$ | $w_2$ | $w_3$ | min eff$_D$ |
|---|---|---|---|---|---|---|---|---|
| 1 | 2 | 0.44 | 1.64 | | 0.50 | 0.50 | | 0.822 |
| 0.5 | 1 | 0.88 | 3.28 | | 0.50 | 0.50 | | 0.822 |
| 0.4 | 1 | 0.79 | 2.43 | 5.76 | 0.38 | 0.39 | 0.22 | 0.761 |
| 1 | 3 | 0.25 | 0.84 | 2.18 | 0.35 | 0.41 | 0.24 | 0.740 |
| 0.3 | 1 | 0.83 | 2.61 | 7.15 | 0.36 | 0.41 | 0.22 | 0.728 |

In our numerical study, we observe that standardized maximin $D$-optimal designs with respect to rectangular or triangular parameter spaces have at most three support points. A third support point appears if a "large" $\Theta$ is chosen, since these $\Psi_{-\infty}$-optimal designs must be "good" despite a high level of uncertainty about the position of the parameters. In some cases a relatively small number of different stages for the experiment can reduce costs considerably. Moreover, three design points still allow model checking to some extent.

We, finally, deal with the problem of finding $\Psi_{-\infty}$-optimal designs on a restricted design space. The results which are similar for rectangular and triangular parameter regions $\Theta$ are given in Tables 6 and 7. Unlike our observations concerning locally $D$-optimal designs, we find that the largest support point of the restricted maximin optimal designs is not always given by the boundary value. Another noticeable aspect is that in both the rectangular and the triangular case the number of support points of the restricted design decreases with decreasing the boundary, whereas the $D$-efficiencies increase.

**Table 6.** Standardized maximin $D$-optimal designs for model (2) with respect to various rectangular parameter regions on restricted design spaces.

| $\Theta = [0.8, 1.2] \times [0.2, 0.5]$ | | | | $\Theta = [2, 3] \times [0.2, 1]$ | | | | | | | |
|---|---|---|---|---|---|---|---|---|---|---|---|
| $x_{max}$ | $x_1$ | $x_2$ | min eff$_D$ | $x_{max}$ | $x_1$ | $x_2$ | $x_3$ | $w_1$ | $w_2$ | $w_3$ | min eff$_D$ |
| $\infty$ | 0.86 | 4.41 | 0.854 | $\infty$ | 0.36 | 1.64 | 5.29 | 0.50 | 0.29 | 0.21 | 0.727 |
| 4 | 0.79 | 3.89 | 0.943 | 4 | 0.35 | 1.69 | 4 | 0.51 | 0.28 | 0.21 | 0.741 |
| 3 | 0.79 | 3 | 0.984 | 3 | 0.35 | 1.85 | 3 | 0.49 | 0.31 | 0.20 | 0.785 |
| 2 | 0.67 | 2 | 0.992 | 2 | 0.33 | 1.91 | | 0.5 | 0.5 | | 0.909 |
| 1 | 0.42 | 1 | 0.998 | 1 | 0.31 | 1 | | 0.5 | 0.5 | | 0.986 |
| 0.5 | 0.23 | 0.5 | 1.000 | 0.5 | 0.20 | 0.5 | | 0.5 | 0.5 | | 0.996 |

**Table 7.** Standardized maximin $D$-optimal designs for model (2) with respect to various triangular parameter regions on restricted design spaces.

| $z_1 = 0.5,\ z_2 = 1$ | | | | $z_1 = 0.3,\ z_2 = 1$ | | | | | | | |
|---|---|---|---|---|---|---|---|---|---|---|---|
| $x_{max}$ | $x_1$ | $x_2$ | min eff$_D$ | $x_{max}$ | $x_1$ | $x_2$ | $x_3$ | $w_1$ | $w_2$ | $w_3$ | min eff |
| $\infty$ | 0.88 | 3.28 | 0.82 | $\infty$ | 0.83 | 2.62 | 7.12 | 0.36 | 0.41 | 0.22 | 0.728 |
| 3 | 0.78 | 2.91 | 0.93 | 5 | 0.75 | 2.25 | 5 | 0.33 | 0.41 | 0.26 | 0.759 |
| 2 | 0.68 | 2 | 0.98 | 3 | 0.81 | 3 | | 0.5 | 0.5 | | 0.904 |
| 1 | 0.42 | 1 | 0.99 | 2 | 0.73 | 2 | | 0.5 | 0.5 | | 0.963 |
| 0.5 | 0.23 | 0.5 | 0.99 | 1 | 0.44 | 1 | | 0.5 | 0.5 | | 0.979 |

*Acknowledgement.* The support of the Deutsche Forschungsgemeinschaft (SFB 475, Komplexitätsreduktion in multivariaten Datenstrukturen, Teilprojekt A2, Sachbeihilfe De 502/18-1) is gratefully acknowledged.

## References

A.C. Atkinson. Horwitz's rule, transforming both sides and design of experiments for mechanistic models. *Applied Statistics*, 52:261–278, 2003.

A.C. Atkinson, K. Chaloner, A.M. Herzberg, and J. Juritz. Optimum experimental designs for properties of a compartmental model. *Biometrics*, 49: 325–337, 1993.

G.E.P. Box and H. Lucas. Design of experiments in non-linear situations. *Biometrika*, 46:77–90, 1959.

H. Chernoff. Locally optimal designs for estimating parameters. *Annals of Mathematical Statistics*, 24:586–602, 1953.

H. Dette, L. Haines, and L. Imhof. Maximin and Bayesian optimal designs for regression models. Preprint available at: http://www.ruhr-uni-bochum.de/mathematik3/preprint.htm, 2003.

H. Dette and T.E. O'Brien. Optimality criteria for regression models based on predicted variance. *Biometrika*, 86:93–106, 1999.

M. Gibaldi and D. Perrier. *Pharmacokinetics.* Dekker, New York, second edition, 1982.

D.C. Hamilton and D.G. Watts. A quadratic design criterion for precise estimation in nonlinear regression models. *Technometrics*, 27:241–250, 1985.

W.J. Hill and W.G. Hunter. Design of experiments for subsets of parameters. *Technometrics*, 16:425–434, 1974.

L. Imhof. Maximin designs for exponential growth models and heteroscedastic polynomial models. *Annals of Statistics*, 29(2):561–576, 2001.

C. H. Müller. Maximin efficient designs for estimating nonlinear aspects in linear models. *Journal of Statistical Planning and Inference*, 44:117–132, 1995.

L. Pronzato and E. Walter. Robust experiment design via stochastic approximation. *Mathematical Biosciences*, 75:103–120, 1985.

S.D. Silvey. *Optimal Design.* Chapman and Hall, London, 1980.

# Optimal Adaptive Designs in Phase III Clinical Trials for Continuous Responses with Covariates

A. Biswas[1] and S. Mandal[2]

[1] Applied Statistics Unit, Indian Statistical Institute, 203 B. T. Road, Kolkata – 700 108, India. atanu@isical.ac.in
[2] Department of Statistics, University of Manitoba, Winnipeg, MB, R3T 2N2, Canada. saumen_mandal@umanitoba.ca

**Summary.** Some optimal adaptive allocation design was given by Rosenberger, Stallard, Ivanova, Harper, and Ricks (2002) for phase III clinical trials involving two treatments having binary responses, but no covariate. We extend that idea to introduce an optimal adaptive allocation design for two treatments having some continuous responses. Moreover, we allow covariates in our model. Exact and limiting proportion of allocation for the proposed design are numerically evaluated.

**Key words:** adaptive design, exact proportion of allocation, limiting proportion of allocation, treatment difference.

## 1 Introduction

Adaptive allocation designs are sometimes used in phase III clinical trials with sequential entrance of patients with two or more competing treatments. The objective of such adaptive allocation designs is to allocate a larger number of patients to the better treatment in course of the trial. Several adaptive designs are available in literature for this purpose, most of which are for binary treatment responses (see Zelen (1969) for the play-the-winner rule; Wei and Durham (1978) for the randomized play-the-winner rule, Wei (1979) for the randomized Pòlya urn design, Durham, Flournoy, and Li (1998) for success driven design, Ivanova (2003) for the drop-the-loser rule). There is a limited amount of work on continuous treatment responses (see Bandyopadhyay and Biswas (2001)). Some real-life applications of adaptive designs are available from the works of Iglewicz (1983) who reported one unpublished application of adaptive design to a lung cancer trial by Marvin Zelen; Bartlett, Roloff, Cornell, Andrews, Dillon, and Zwischenberger (1985), Rout, Rocke, Levin, Gouws, and Reddy (1993), Tamura, Faries, Andersen, and Heiligenstein (1994), Biswas and Dewanji (2003). Following the work of Mukhopadhyay (1976), Rosenberger, Stallard, Ivanova, Harper, and Ricks (2002) provided an optimal allocation design for binary responses. The question naturally arises how to generalize that approach for more general response patterns. For example, we may need to construct optimal adaptive designs for continuous treatment responses, for responses involving more than one parameter and responses involving covariates influencing the responses.

The present paper is organized as follows. In Section 2, we discuss the optimal designs for continuous responses. In particular, we discuss simple exponential responses and normal responses. In Section 3, we generalize the normal response case in the presence of covariates. Section 4 ends with some discussions and indicates some possible directions of future study.

## 2 Optimal Designs for Continuous Distributions

### 2.1 Response Distributions with one Unknown Parameter

We illustrate the technique for exponentially distributed responses. Let the response by treatment $A$ and the response by treatment $B$ follow an exponential distributions with expectations $\lambda_A$ and $\lambda_B$, respectively. Suppose a total of $n$ patients are to be treated and $n_A$ and $n_B$ patients are treated by treatments $A$ and $B$, where $n_A + n_B = n$. Let $n_A/n_B = R$. Suppose $\widehat{\lambda}_A$ and $\widehat{\lambda}_B$ are the estimators of $\lambda_A$ and $\lambda_B$ using the data. We fix the variance of $\widehat{\lambda}_A - \widehat{\lambda}_B$ to some known quantity $K$, and obtain the allocation procedure accordingly. That is

$$\mathrm{Var}\,(\widehat{\lambda}_A - \widehat{\lambda}_B) \simeq \frac{\lambda_A^2}{n_A} + \frac{\lambda_B^2}{n_B} = \frac{(1+R)(\lambda_A^2 + R\lambda_B^2)}{nR} = K. \tag{1}$$

We assume a set up where a larger response is better and a response greater than a threshold $c$ is considered as a 'good' response. We then decide to minimize the expected total number of 'bad' responses, that is the expected total number of responses less than $c$. The choice of $c$ is of course important and it is the experimenter's task to choose $c$ at the outset. Thus, we minimize

$$n_A[1 - \exp(-c/\lambda_A)] + n_B[1 - \exp(-c/\lambda_B)] \tag{2}$$

subject to (1). Note that (2) is minimized at

$$R = \frac{\lambda_A}{\lambda_B}\sqrt{\frac{1 - \exp(-c/\lambda_B)}{1 - \exp(-c/\lambda_A)}}.$$

Consequently, we suggest the optimal adaptive allocation procedure as the conditional probability that the $(i+1)$st patient is treated by treatment $A$, given all the data up to the $i$th patient, as

$$p_{i+1} = \frac{\widehat{\lambda}_{Ai}\sqrt{1 - \exp(-c/\widehat{\lambda}_{Bi})}}{\widehat{\lambda}_{Ai}\sqrt{1 - \exp(-c/\widehat{\lambda}_{Bi})} + \widehat{\lambda}_{Bi}\sqrt{1 - \exp(-c/\widehat{\lambda}_{Ai})}},$$

where $\widehat{\lambda}_{Ai}$ and $\widehat{\lambda}_{Bi}$ are the estimators of $\lambda_A$ and $\lambda_B$ based on the data up to the first $i$ patients. Thus, the design is response-adaptive as it takes all the accumulated data up to that time point through the estimates $\widehat{\lambda}_{Ai}$ and $\widehat{\lambda}_{Bi}$. To implement, we carry out a randomized 50:50 allocation for the first $2m$ patients (to get some initial estimates of $\lambda_A$ and $\lambda_B$), and carry out the optimal allocation from the $(2m+1)$st patient onwards. We carry out a detailed simulation study to illustrate the exact proportions of allocation with standard deviations and also report the limiting proportion of allocation. These are reported in Table 1.

## 2.2 Distributions Having more than one Parameter

Suppose the responses from treatments A and B follow $N(\mu_A, \sigma_A^2)$ and $N(\mu_B, \sigma_B^2)$, where all the parameters are unknown. The treatment difference is measured by $\theta = \mu_A - \mu_B$. Based on $n_A$ and $n_B$ samples on $A$ and $B$, the estimator of this treatment difference is $\widehat{\theta}_n = \widehat{\mu}_{An} - \widehat{\mu}_{Bn}$, and we fix the variance of $\widehat{\theta}_n$ to a prefixed quantity $K$, that is

$$\mathrm{Var}\,(\widehat{\theta}_n) \simeq \frac{\sigma_A^2}{n_A} + \frac{\sigma_B^2}{n_B} = \frac{(1+R)(\sigma_A^2 + R\sigma_B^2)}{nR} = K. \tag{3}$$

Writing $\Phi(\cdot)$ as the standard normal cumulative distribution function, we minimize the expected number of responses less than $c$, that is

$$n_A \Phi\left(\frac{c-\mu_A}{\sigma_A}\right) + n_B \Phi\left(\frac{c-\mu_B}{\sigma_B}\right)$$

subject to (3). The minimum value is attained at

$$R = \sqrt{\frac{\Phi\left(\frac{c-\mu_B}{\sigma_B}\right)}{\Phi\left(\frac{c-\mu_A}{\sigma_A}\right)} \frac{\sigma_A}{\sigma_B}}.$$

Consequently, the conditional probability of allocating the $(i+1)$st patient to treatment A is

$$p_{i+1} = \frac{\sqrt{\Phi\left(\frac{c-\widehat{\mu}_{Bi}}{\widehat{\sigma}_{Bi}}\right)}\widehat{\sigma}_{Ai}}{\sqrt{\Phi\left(\frac{c-\widehat{\mu}_{Bi}}{\widehat{\sigma}_{Bi}}\right)}\widehat{\sigma}_{Ai} + \sqrt{\Phi\left(\frac{c-\widehat{\mu}_{Ai}}{\widehat{\sigma}_{Ai}}\right)}\widehat{\sigma}_{Bi}},$$

where $\widehat{\sigma}_{Ai}$ and $\widehat{\sigma}_{Bi}$ are the estimates of $\sigma_A$ and $\sigma_B$ based on the first $i$ responses. Detailed simulation study provides the proportion of allocations and also the standard deviations. A part of that is reported in Table 2. The optimal allocation proportion is also reported.

## 3 Presence of Covariates

Suppose a covariate vector $x$ influence the responses. We assume that the responses from treatment $A$ follow $N(\mu_A + x^T\beta, \sigma^2)$ and the responses from treatment $B$ follow $N(\mu_B + x^T\beta, \sigma^2)$. For simplicity, we have assumed common variance and also we assumed that the covariate vector influence the response in the same way for both the treatments. Let $Y_i$ be the response of the $i$th patient and let $\delta_i$ be an indicator which takes the values 1 or 0 according as the treatment $A$ or $B$ is applied to the $i$th patient. We write

$$\bar{Y}_{Ai} = \frac{\sum_{j=1}^{i} \delta_j Y_j}{\sum_{j=1}^{i} \delta_j}, \quad \bar{Y}_{Bi} = \frac{\sum_{j=1}^{i} (1-\delta_j) Y_j}{\sum_{j=1}^{i} (1-\delta_j)},$$

$$n_{Ai} = \sum_{j=1}^{i} \delta_j, \; n_{Bi} = \sum_{j=1}^{i} (1-\delta_j),$$

$$S_{xx}^{(i)} = \sum_{j=1}^{i} \delta_j (x_j - \bar{x}_{Ai})(x_j - \bar{x}_{Ai})^T + \sum_{j=1}^{i} \bar{\delta}_j (x_j - \bar{x}_{Bi})(x_j - \bar{x}_{Bi})^T,$$

$$S_{xy}^{(i)} = \sum_{j=1}^{i} Y_j x_j - n_{Ai}\bar{Y}_{Ai}\bar{x}_{Ai} - n_{Bi}\bar{Y}_{Bi}\bar{x}_{Bi},$$

$$\bar{\delta}_j = 1 - \delta_j,$$

$$\bar{x}_{Ai} = \frac{\sum_{j=1}^{i} \delta_j x_j}{\sum_{j=1}^{i} \delta_j}, \; \bar{x}_{Bi} = \frac{\sum_{j=1}^{i} (1-\delta_j) x_j}{\sum_{j=1}^{i} (1-\delta_j)}.$$

The normal equations are:

$$\sum_{i=1}^{n} \begin{pmatrix} \delta_i \\ 1-\delta_i \\ x_i \end{pmatrix} \begin{pmatrix} \delta_i & 1-\delta_i & x_i^T \end{pmatrix} \begin{pmatrix} \mu_A \\ \mu_B \\ \beta \end{pmatrix} = \sum_{i=1}^{n} Y_i \begin{pmatrix} \delta_i \\ 1-\delta_i \\ x_i \end{pmatrix},$$

which implies that

$$\begin{pmatrix} n_{An} & 0 & \sum_{i=1}^{n} \delta_i x_i^T \\ 0 & n_{Bn} & \sum_{i=1}^{n} \bar{\delta}_i x_i^T \\ \sum_{i=1}^{n} \delta_i x_i & \sum_{i=1}^{n} \bar{\delta}_i x_i & \sum_{i=1}^{n} x_i x_i^T \end{pmatrix} \begin{pmatrix} \mu_A \\ \mu_B \\ \beta \end{pmatrix} = \begin{pmatrix} n_{An}\bar{Y}_{An} \\ n_{Bn}\bar{Y}_{Bn} \\ \sum_{i=1}^{n} Y_i x_i \end{pmatrix}$$

Consequently, after some routine steps, we get

$$\widehat{\mu}_{An} - \widehat{\mu}_{Bn} = \bar{Y}_{An} - \bar{Y}_{Bn} - (\bar{x}_{An} - \bar{x}_{Bn})^T \widehat{\beta}_n,$$

with

$$\widehat{\beta}_n = {S_{xx}^{(n)}}^{-1} S_{xy}^{(n)}.$$

Let $X = (x_1, \cdots, x_n)$ be the total covariate history. Then

$$\text{Var}\,(\widehat{\mu}_{An} - \widehat{\mu}_{Bn}|X) \simeq \frac{\sigma^2}{n_A} + \frac{\sigma^2}{n_B} + \sigma^2 (\bar{x}_{An} - \bar{x}_{Bn})^T {S_{xx}^{(n)}}^{-1} (\bar{x}_{An} - \bar{x}_{Bn}),$$

and

$$\text{E}(\widehat{\mu}_{An} - \widehat{\mu}_{Bn}|X) = \mu_A - \mu_B.$$

At this stage, for simplicity and mathematical convenience, we assume that the $x_i$'s are independently and identically distributed as $N_p(\mu_x, \Sigma)$. Thus, given $(\delta_1, \cdots, \delta_i)$, we have

$$\bar{x}_{Ai} \sim N_p(\mu_x, \Sigma/n_{Ai}),\ \ \bar{x}_{Bi} \sim N_p(\mu_x, \Sigma/n_{Bi}),$$
$$S_{xx}^{(i)} \sim \text{Wishart}_p(i-2, \Sigma),$$

where all these statistics are independent. As

$$\left(\frac{n_{Ai} n_{Bi}}{i}\right) (\bar{x}_{Ai} - \bar{x}_{Bi})^T \left(\frac{S_{xx}^{(i)}}{i-2}\right)^{-1} (\bar{x}_{Ai} - \bar{x}_{Bi}) \sim T_p^2,$$

where $T^2$ denotes the Hotelling's $T^2$-distribution, we have

$$(\bar{x}_{An} - \bar{x}_{Bn})^T {S_{xx}^{(n)}}^{-1} (\bar{x}_{An} - \bar{x}_{Bn}) \sim \left(\frac{i}{n_{Ai} n_{Bi}}\right) \frac{p}{i-p-1} F(p, i-p-1),$$

where $F(\cdot,\cdot)$ indicates an $F$-distribution. Hence,

$$E((\bar{x}_{An} - \bar{x}_{Bn})^T {S_{xx}^{(n)}}^{-1} (\bar{x}_{An} - \bar{x}_{Bn})) = \left(\frac{n}{n_A n_B}\right) \frac{p}{n-p-1} \frac{n-p-1}{n-p-3}.$$

Consequently, the unconditional variance of $\widehat{\mu}_{An} - \widehat{\mu}_{Bn}$ is

$$\sigma^2 \left[\frac{1}{n_A} + \frac{1}{n_B} + \frac{n}{n_A n_B} \frac{p}{n-p-3}\right] = \frac{\sigma^2 (R+1)^2 (n-3)}{nR(n-p-3)}. \tag{4}$$

We fix the value of (4) as $K$. We minimize the expected number of responses less than $c$ when all the patients have covariate vector $x_0$, a fixed vector. The minimum is attained at

$$R = \frac{\Phi\left(\frac{c - \mu_B - x_0^T \beta}{\sigma}\right)}{\Phi\left(\frac{c - \mu_A - x_0^T \beta}{\sigma}\right)}.$$

Consequently, the conditional allocation probability for the $(i+1)$st patient to treatment $A$ is

$$p_{i+1} = \frac{\Phi\left(\frac{c-\widehat{\mu}_{Bi}-x_{i+1}^T\widehat{\beta}_i}{\widehat{\sigma}_i}\right)}{\Phi\left(\frac{c-\widehat{\mu}_{Bi}-x_{i+1}^T\widehat{\beta}_i}{\widehat{\sigma}_i}\right)+\Phi\left(\frac{c-\widehat{\mu}_{Ai}-x_{i+1}^T\widehat{\beta}_i}{\widehat{\sigma}_i}\right)}.$$

Detailed simulation shows the exact proportions of allocations along with the standard deviations (SD's). A part of the computation is reported in Table 3 in Section 4.

# 4 Conclusions

One could maximize some other suitable function subject to a fixed variance. For example, one can maximize the expected total response instead of dichotomizing the responses by setting a threshold $c$. Some such criteria are under study and the results will be reported in a separate study.

Note that the formulation of the present design is based on keeping the variance fixed at a minimum level. Here subjects are treated to obtain the target allocation proportion, no question of minimizing variance as in the usual optimal designs.

From the computations of the exact and limiting proportions of allocation in Tables 1-3, it is clear that the proposed method works well in the sense that a larger number of patients are allocated to the better treatment when treatment difference is present, rather than the case of equivalence of treatments. Moreover, as the treatment difference increases, the allocation gets more skewed in favour of the better treatment. From the computations, it is also clear that the standard deviations of the exact proportions of allocations decreases with the increase in treatment difference and with the increase in sample size. The exact proportion approaches to the limiting proportion as the sample size increases, a much expected scenario.
At present we are working the detailed property of the design. For example, the proposed methodology can be extended to obtain optimal design for $t$ $(t > 2)$ treatments at hand. Moreover, a detailed inference following up the design and comparison of this design with some possible competitors (by fixing level and power at a fixed alternative or by fixing the sample size) is required to understand the usefulness of the design in practice. We are currently working on that. Again, instead of minimizing expressions like (2), one could maximize the expected total of the responses. The details will be reported in a separate communication.

One could generalize the procedure to consider more than one constraints. At each stage, with $t$ treatments and two constraints, for example, we can formulate the problem of maximizing a criterion function $\varphi(p)$ subject to the following constraints:

**Table 1.** Exact proportion of allocations (and the SD's) and limiting proportion of allocations for exponential responses. Here $m = 5$, $c = 2$, $\lambda_2 = 1$ and 1000 simulations are taken.

| | $n = 50$ | | $n = 100$ | | Limiting |
|---|---|---|---|---|---|
| $\lambda_1$ | Proportion | SD | Proportion | SD | Proportion |
| 1 | 0.5000 | 0.0974 | 0.5000 | 0.0854 | 0.5000 |
| 1.5 | 0.4122 | 0.0932 | 0.4027 | 0.0759 | 0.3809 |
| 2 | 0.3579 | 0.0853 | 0.3405 | 0.0693 | 0.2995 |
| 2.5 | 0.3254 | 0.0852 | 0.2993 | 0.0669 | 0.2420 |
| 3 | 0.2958 | 0.0783 | 0.2617 | 0.0617 | 0.2000 |
| 3.5 | 0.2779 | 0.0754 | 0.2423 | 0.0594 | 0.1686 |
| 4 | 0.2616 | 0.0714 | 0.2249 | 0.0560 | 0.1443 |

**Table 2.** Exact proportion of allocations (and the SD's) and limiting proportion of allocations for normal responses. Here $m = 5$, $c = 0$, $\mu_2 = 1$, $\sigma_1 = 1$, $\sigma_2 = 1.2$ and 1000 simulations are taken.

| | $n = 50$ | | $n = 100$ | | Limiting |
|---|---|---|---|---|---|
| $\lambda_1$ | Proportion | SD | Proportion | SD | Proportion |
| 0 | 0.4609 | 0.0895 | 0.4605 | 0.0726 | 0.4545 |
| 0.5 | 0.5189 | 0.0883 | 0.5200 | 0.0707 | 0.5148 |
| 1 | 0.5924 | 0.0889 | 0.5966 | 0.0695 | 0.5967 |
| 1.5 | 0.6705 | 0.0831 | 0.6842 | 0.0641 | 0.6951 |
| 2 | 0.7491 | 0.0736 | 0.7789 | 0.0579 | 0.7962 |

**Table 3.** Exact proportion of allocations (and the SD's) and limiting proportion of allocations for normal responses. Here $m = 5$, $c = 0$, $\mu_2 = 1$, common $\sigma = 1$, only one covariate $x \sim N(1, 1)$, $\beta = 1$ and 1000 simulations are taken.

| | $n = 50$ | | $n = 100$ | |
|---|---|---|---|---|
| $\lambda_1$ | Proportion | SD | Proportion | SD |
| 0 | 0.5000 | 0.0704 | 0.5000 | 0.0729 |
| 0.5 | 0.5573 | 0.0730 | 0.5576 | 0.0702 |
| 1 | 0.6385 | 0.0698 | 0.6365 | 0.0687 |
| 1.5 | 0.7306 | 0.0658 | 0.7296 | 0.0689 |
| 2 | 0.8175 | 0.0603 | 0.8166 | 0.0604 |

$$g_1(p) = c_1 \quad \text{and} \quad g_2(p) = c_2,$$

where $p = (p_1, \cdots, p_t)^T$ with $p_j \geq 0$ and $\sum_j p_j = 1$, for constants $c_1$ and $c_2$. We are currently working on the detailed formulation and solution of such constrained optimality problems.

The present approach can be used to provide some optimal adaptive designs for binary responses with covariates. Some logistic response model can be assumed for both the treatments such as

$$\operatorname{logit} P(Y_i = 1|\delta_i = 1) = \alpha + \beta + \gamma^T x_i,$$

and

$$\operatorname{logit} P(Y_i = 1|\delta_i = 0) = \beta + \gamma^T x_i.$$

We can find the asymptotic variance of $\widehat{\alpha}_n$ and fix it to some prefixed quantity and minimize the expected number of failures subject to this. The details are under study.

*Acknowledgement.* The authors wish to thank the referee for some valuable comments which led to some improvements over an earlier version of the manuscript.

## References

U. Bandyopadhyay and A. Biswas. Adaptive designs for normal responses with prognostics factors. *Biometrika*, 88:409–419, 2001.

R. H. Bartlett, D. W. Roloff, R. G. Cornell, A. F. Andrews, P. W. Dillon, and J. B. Zwischenberger. Extracorporeal circulation in neonatal respiratory failure: A prospective randomized trial. *Pediatrics*, 76:479–487, 1985.

A. Biswas and A. Dewanji. A randomized longitudinal play-the-winner design for repeated binary data. *Australian and New Zealand Journal of Statistics*, 2003.

S.D. Durham, N. Flournoy, and W. Li. Sequential designs for maximizing the probability of a favorable response. *Canadian Journal of Statistics*, 3: 479–495, 1998.

B. Iglewicz. Alternative designs: sequential, multi-stage, decision theory and adaptive designs. In M.E. Buyse, J. Staquet, and R.J. Sylvester, editors, *Cancer Clinical Trials: Methods and Practice*, pages 312–334. Oxford University Press, 1983.

A. Ivanova. A play-the-winner-type urn design with reduced variability. *Metrika*, 58:1–13, 2003.

N. Mukhopadhyay. Sequential estimation of a linear function of means of three normal populations. *J. Amer. Statist. Assoc.*, 71:149–153, 1976.

W.F. Rosenberger, N. Stallard, A. Ivanova, C.N. Harper, and M.L. Ricks. Optimal adaptive designs for binary response trials. *Biometrics*, 57:909–913, 2002.

C.C. Rout, D.A. Rocke, J. Levin, E. Gouws, and D. Reddy. A reevaluation of the role of crystalloid preload in the prevention of hypotension associated with spinal anesthesia for elective cesarean section. *Anesthesiology*, 79: 262–269, 1993.

R.N. Tamura, D.E. Faries, J.S. Andersen, and J.H. Heiligenstein. A case study of an adaptive clinical trial in the treatment of out-patients with depressive disorder. *J. Amer. Statist. Assoc.*, 89:768–776, 1994.

L.J. Wei. The generalized Polya's urn for sequential medical trials. *Ann. Statist.*, 7:291–296, 1979.

L.J. Wei and S. Durham. The randomized play-the-winner rule in medical trials. *J. Am. Statist. Assoc.*, 73:838–843, 1978.

M. Zelen. Play-the-winner rule and the controlled clinical trial. *J. Am. Statist. Assoc.*, 64:131–146, 1969.

# Optimal Designs for Regression Models with Forced Measurements at Baseline

V.V. Fedorov[1] and S. Leonov[2]

[1] GlaxoSmithKline, 1250 So Collegeville Rd, PO Box 5089, Collegeville, PA 19426-0989, U.S.A. Valeri.V.Fedorov@gsk.com
[2] GlaxoSmithKline, 1250 So Collegeville Rd, PO Box 5089, Collegeville, PA 19426-0989, U.S.A. Sergei.2.Leonov@gsk.com

**Summary.** We discuss the construction of $D$-optimal designs for regression models with forced measurements. Such models may be used in dose response studies where a baseline measurement is taken for all patients, that is forced in the design.

**Key words:** Locally optimal designs, Random effects, Baseline, Placebo effect, Multiple measurements per patient.

## 1 Introduction

In clinical trials, often measurements are taken at baseline, before administering a dose of the drug or placebo. It is important to distinguish between baseline measurements and measurements at placebo because of "placebo effect", when patients expect benefit from the treatment in blinded studies; see Piantadosi (1997), Ch. 4.2; Roberts, Kewman, Mercier, and Hovell (1993). In this paper we consider optimal experimental designs for the case when two measurements are taken for each patient, one measurement, $y_1$, at baseline and the second one, $y_2$, after administering a dose $x$ of the drug.

When baseline measurement is considered as a covariate, then one of the modelling alternatives is $y_2 = g(\theta, x, y_1) + \varepsilon$, where $g$ is a response function, $x$ is a control variable (dose), $\theta$ are unknown parameters, and $\varepsilon$ are random errors. Unfortunately the above model contains uncontrolled (observed) variable $y_1$ in "response curve" portion which makes experimental design rather difficult. Assuming that the probability distribution function of $y_1$ is known, Atkinson and Fedorov (1988) addressed this problem by integrating out variable $y_1$ from optimality criteria.

We propose to view $y_1$ and $y_2$ as components of a two-dimensional vector of observations and to build models and optimal designs in this context. In the more general case, the baseline measurement may be followed by several observations, like in cross-over trials, and analyzed within the same framework.

## 2 Model

Introduce a mean response function for patient $j$,

$$\mathrm{E}Y_j = \mathrm{E}(y_{1j}, y_{2j}|x_j, \gamma_j)^T = \eta(x_j, \gamma_j) = [\eta_b(\gamma_j), \eta_1(x_j, \gamma_j)]^T,$$

where $\gamma_j$ are response parameters of patient $j$, with mean $\gamma$ and population variance-covariance matrix $\Lambda = (\lambda_{kl})$, $\lambda_{kl} = \mathrm{Cov}(\gamma_k, \gamma_\ell)$. The main attention in the paper is given to the case of linear response,

$$\eta(x, \gamma) = \begin{bmatrix} \gamma_0 \\ f^T(x)\gamma_{(1)} \end{bmatrix}, \; \gamma_{(1)} = (\gamma_1, \ldots, \gamma_m)^T, \; f(x) = [f_1(x), \ldots, f_m(x)]^T. \tag{1}$$

However, the results and numerical algorithms are directly extended to more general models. Note that the case of $\Lambda \to 0$ corresponds to fixed effects model setting. The variance-covariance matrix of measurement errors is denoted by $\Sigma = \Sigma(x_j, \gamma_j) = \mathrm{Var}(Y_j|x_j, \gamma_j)$. Larger elements $\Sigma_{12}$ and $\lambda_{0\ell}, \ell = 1, \ldots, m$, correspond to the stronger link between baseline information and an observed treatment effect.

Let a $(p \times 1)$ vector $\theta$ comprise all unknown parameters of the model including $\gamma$, elements of population covariance matrix $\Lambda$ and error variance $\Sigma$. If a vector of observations $Y$ is normally distributed with density $p(x, \theta)$, mean $\eta(x, \theta)$ and variance matrix $S(x, \theta)$, then the Fisher information matrix of a single multi-response observation is given by

$$\mu_{\alpha\beta}(x, \theta) = \mathrm{E}\left[\frac{\partial \log p(x, \theta)}{\partial \theta_\alpha} \frac{\partial \log p(x, \theta)}{\partial \theta_\beta}\right] = \frac{\partial \eta^T(x, \theta)}{\partial \theta_\alpha} S^{-1}(x, \theta) \frac{\partial \eta(x, \theta)}{\partial \theta_\beta}$$
$$+ \frac{1}{2} \operatorname{tr} \left[S^{-1}(x, \theta) \frac{\partial S(x, \theta)}{\partial \theta_\alpha} S^{-1}(x, \theta) \frac{\partial S(x, \theta)}{\partial \theta_\beta}\right], \; \alpha, \beta = 1, \ldots, p, \tag{2}$$

see Muirhead (1982), Ch. 1. Therefore, to construct optimal designs one needs an explicit expression for the matrix $S(x, \theta)$. For Model (1),

$$E_{Y_j, \gamma_j}(Y_j|x_j) = F^T(x_j)\gamma, \text{ with } F(x) = \begin{bmatrix} 1 & 0 \\ \mathbf{0}_{m,1} & f(x) \end{bmatrix}, \tag{3}$$

where $\mathbf{0}_{a,b}$ is an $(a \times b)$-matrix of zeros. In case of $m_\gamma$ response parameters and $k$ observations per patient, $F$ is an $(m_\gamma \times k)$-matrix; see Fedorov, Gagnon, and Leonov (2002) for examples. It follows from (3) that

$$S(x, \theta) = \begin{pmatrix} S_{11} & S_{12} \\ S_{12} & S_{22} \end{pmatrix} = F^T(x)\Lambda F(x) + \Sigma. \tag{4}$$

In the sequel, $S(x, \theta)$ is called a variance matrix, not to be confused with population variance matrix $\Lambda$ and error variance matrix $\Sigma$. For the linear model (1) matrix $S$ does not depend on values of response parameters $\gamma$.

Throughout the paper, we consider (locally) optimal designs in continuous, or approximate, setting, that is solutions to the following optimization problem:

$$\xi^* = \arg\min_{\xi} \Psi[M^{-1}(\xi, \theta)],$$

where $\Psi$ is an optimality criterion; $\xi = \{x_i, w_i\}$ is a continuous (normalized) design, weights $w_i$ satisfy $0 \leq w_i \leq 1$, $\sum_i w_i = 1$; $x_i \in \mathbf{X}$, where $\mathbf{X}$ is a given design region, and $M(\xi, \theta)$ is a normalized information matrix,

$$M(\xi, \theta) = \sum_{i=1}^{n} w_i \mu(x_i, \theta);$$

see, for instance, Atkinson and Donev (1992), Ch. 9.1; Fedorov (1972), Ch. 5; or Fedorov and Hackl (1997), Ch. 2.3. In the examples below, for the functional $\Psi$ we use the $D$-optimality criterion, $\Psi = -\log|M(\xi, \theta)|$. If $M$ depends on unknown parameters $\theta$, then a preliminary guess of $\theta$ is needed to construct locally optimal designs. If $\mu(x, \theta)$ does not depend on unknown parameters, then argument $\theta$ is dropped from both $\mu$ and $M$. For example, when matrices $\Lambda$ and $\Sigma$ are known and, therefore, for Model (1) matrix $S$ does not depend on unknown parameters, formula (2) reduces to $\mu(x) = F(x)S^{-1}(x)F^T(x)$. $D$-optimal designs in numerical examples were constructed in MATLAB using the first-order algorithm; see Fedorov and Hackl (1997), Ch. 2.

After discussing optimal designs for Model (1) in the next section, in Section 4 a special case of $\gamma_0 \equiv \gamma_1$ and $f(x) = (1, x)^T$ is considered, that is when a baseline measurement may be deemed identical to a placebo response. This reduced model leads to some unexpected designs, see examples in Section 4.

## 3 Optimal Designs for Model (1)

### 3.1 Equivalence of $D$-optimal Designs when only $\gamma$ are Unknown

In this section error variance $\Sigma$ and population variance $\Lambda$ are known (not parameters of the model) which leads to a known variance matrix $S$ in (4). Assume also that $S$ does not depend on dose $x$, that is $S(x, \theta) \equiv S$. This happens for a fixed effects model, $\Lambda = 0$, and constant error variance $\Sigma$, or even for a random effects model with $\lambda_{ij} = 0$ if $i$ or $j > 1$, that is if random variation is allowed for parameters $\gamma_0$ and $\gamma_1$ only, since in that case (4) implies

$$S = \begin{pmatrix} \Sigma_{11} + \lambda_0 & \Sigma_{12} + \lambda_{01} \\ \Sigma_{12} + \lambda_{01} & \Sigma_{22} + \lambda_1 \end{pmatrix}.$$

It turns out that when an intercept is included in the model for the 2nd measurement, then $D$-optimal designs for Model (1) do not depend on $S$ and coincide with those for a classic regression model with one measurement per individual and constant error variance,

$$y_i = f^T(x_i)\gamma_{(1)} + \varepsilon_i, \tag{5}$$

where $\varepsilon_i$ are i.i.d. $N(0, \sigma^2)$ random variables.

Let $M_0(\xi)$ be the information matrix, up to a multiplier $\sigma^{-2}$, of the design $\xi$ for Model (5), $M_0(\xi) = \sum_i w_i \mu_0(x_i)$, where $\mu_0(x) = f(x)f^T(x)$, and let $\psi(x, \xi)$ and $\psi_0(x, \xi)$ be sensitivity functions of the $D$-criterion for Models (1) and (5), respectively,

$$\psi(x,\xi) = \text{tr}\left[M^{-1}(\xi)\mu(x)\right], \quad \psi_0(x,\xi) = f^T(x)M_0^{-1}(\xi)f(x),$$

which are used in the proof of the generalized equivalence theorem; see Kiefer and Wolfowitz (1960), or Fedorov (1972), Ch. 5 for the multi-response case.

**Theorem 1.** *Let $f(x) = [1, g_2(x), \ldots, g_m(x)]^T$, and $S(x) \equiv S$ with an arbitrary non-singular $S$. Then D-optimal designs for Models (1) and (5) are the same. Moreover, for any nonsingular design $\xi$*

$$\psi(x,\xi) = \psi_0(x,\xi) + 1.$$

*Proof.* We write $g(x) = [g_2(x), \ldots, g_m(x)]^T$, so that

$$f^T(x) = \left[1, g^T(x)\right], \quad F^T(x) = \begin{bmatrix} 1 & 0 & \mathbf{0}_{1,m-1} \\ 0 & 1 & g^T(x) \end{bmatrix}.$$

If $W_\xi = \sum_i w_i g(x_i)$ and $H_0 = \sum_i w_i g(x_i) g^T(x_i) - W_\xi W_\xi^T$, then

$$M_0(\xi) = \begin{bmatrix} 1 & W_\xi^T \\ W_\xi & \sum_i w_i g(x_i) g^T(x_i) \end{bmatrix}, M_0^{-1}(\xi) = \begin{bmatrix} 1 + W_\xi^T H_0^{-1} W_\xi & -W_\xi^T H_0^{-1} \\ -H_0^{-1} W_\xi & H_0^{-1} \end{bmatrix}$$

(see the formula for partitioned matrix inverse in Harville (1997), Ch. 8.5e). Hence,

$$\psi_0(x,\xi) = 1 + W_\xi^T H_0^{-1}[W_\xi - 2g(x)] + g^T(x) H_0^{-1} g(x). \tag{6}$$

Let $V = S^{-1}$ and let $V_2 = (V_{12}, V_{22})^T$ be the second column of $V$. Note that for any nonsingular matrix $V$

$$V_2^T V^{-1} V_2 = V_{22}. \tag{7}$$

Then

$$\mu(x) = \begin{bmatrix} V & V_2 g^T(x) \\ g(x) V_2^T & V_{22} g(x) g^T(x), \end{bmatrix}, \; M(\xi) = \begin{bmatrix} V & V_2 W_\xi^T \\ W_\xi V_2^T & V_{22} \sum_i w_i g(x_i) g^T(x_i) \end{bmatrix}.$$

Inverting a partitioned matrix $M(\xi)$ together with (7) leads to

$$M^{-1}(\xi) = \begin{bmatrix} V^{-1} + V^{-1}V_2 W_\xi^T H^{-1} W_\xi V_2^T V^{-1} & -V^{-1}V_2 W_\xi^T H^{-1} \\ -H^{-1} W_\xi V_2^T V^{-1} & H^{-1} \end{bmatrix}, \quad (8)$$

where $H = V_{22}H_0$. Thus, the sensitivity function of the design $\xi$ for Model (1) is

$$\psi(x,\xi) = \text{tr}\left\{I_2 + V^{-1}V_2 W_\xi^T H^{-1}[W_\xi - 2g(x)]V_2^T + V_{22}H^{-1}g(x)g^T(x)\right\},$$

where $I_2$ is a $(2 \times 2)$ identity matrix. Therefore, the identity $\text{tr}(A_1A_2) = \text{tr}(A_2A_1)$ and (7) entail

$$\psi(x,\xi) = 2 + W_\xi^T H_0^{-1}[W_\xi - 2g(x)] + g^T(x)H_0^{-1}g(x),$$

which does not depend on $S$ and together with (6) proves the theorem. □

*Remark 1.* When the intercept is not included in the model, then $D$-optimal designs for model (1), in general, depend on $S$ if the latter is non-diagonal. The reason is that the information matrix may be presented as

$$\tilde{M}(\xi) = \begin{bmatrix} V_{11} & V_{12}W_\xi^T \\ V_{12}W_\xi & V_{22}\sum_i w_i g(x_i)g^T(x_i) \end{bmatrix},$$

and the analog of $H$ in (8) is $\tilde{H} = V_{22}\sum_i w_i g(x_i)g^T(x_i) - V_{11}^{-1}V_{12}^2 W_\xi W_\xi^T$, which introduces the dependence on $V_{12}$.

*Remark 2.* For linear optimality criteria $\Psi = \text{tr}\{AM^{-1}(\xi)\}$, where $A$ is a $(p \times p)$ non-negative definite matrix, the sensitivity function is

$$\psi_A(x,\xi) = \text{tr}\left[\mu(x)M^{-1}(\xi)AM^{-1}(\xi)\right].$$

Compared to $\psi(x,\xi)$, the function $\psi_A(x,\xi)$ involves additional multiplication by $M^{-1}(\xi)$, which introduces the dependence on $V_{12}$ as in Remark 1 and, in general, makes $A$-optimal designs dependent on $S$.

*Remark 3.* $D$-optimal designs for models (1) and (5) may be identical even when the variance matrix $S$ contains unknown parameters. For example, when $\Lambda = 0$ and elements $\Sigma_{11}, \Sigma_{22}, \Sigma_{12}$ of the matrix $\Sigma$ are unknown and independent of $x$, then $S \equiv \Sigma$ and formula (2) reduces to

$$\hat{M}(\xi,\theta) = \begin{bmatrix} M(\xi) & \mathbf{0}_{m+1,3} \\ \mathbf{0}_{3,m+1} & \mu_\Sigma \end{bmatrix}, \quad \mu_\Sigma = \begin{bmatrix} v_{11}^2/2 & v_{12}^2/2 & v_{11}v_{12} \\ v_{12}^2/2 & v_{22}^2/2 & v_{22}v_{12} \\ v_{11}v_{12} & v_{22}v_{12} & v_{12}^2 + v_{11}v_{22} \end{bmatrix},$$

where $v_{ij} = (\Sigma^{-1})_{ij}$ and $\mu_\Sigma$ is non-singular because

$$|\mu_\Sigma| = (v_{11}v_{22} - v_{12}^2)^3/4 > 0.$$

Therefore $|\hat{M}^{-1}(\xi)| = |M^{-1}(\xi)|\,|\mu_\Sigma|^{-1}$ and $\hat{\psi}(x,\xi) = \psi(x,\xi) + 3$.

### 3.2 Unknown $\gamma$ and Population Variance $\Lambda$

When the elements of the matrix $\Lambda$ are unknown, locally $D$-optimal designs for Model (1) become dependent on $S$ and, therefore, different from those for Model (5), even for diagonal $\Sigma = \sigma^2 I_2$ with known $\sigma^2$. We provide some numerical examples for the simple linear regression with $x \in \mathbf{X} = [0,1]$,

$$\eta_1(x,\gamma) = \gamma_1 + \gamma_2 x,\ \gamma = \begin{pmatrix} \gamma_0 \\ \gamma_1 \\ \gamma_2 \end{pmatrix}, \quad \Lambda = \begin{pmatrix} \lambda_0 & \lambda_{01} & \lambda_{02} \\ \lambda_{01} & \lambda_1 & \lambda_{12} \\ \lambda_{02} & \lambda_{12} & \lambda_2 \end{pmatrix},\ F(x) = \begin{pmatrix} 1 & 0 \\ 0 & 1 \\ 0 & x \end{pmatrix}. \tag{9}$$

The total number of unknown parameters in this subsection is $p = 9$ (three in response and six in $\Lambda$). Obviously, for model (5) with $\eta_1(x,\gamma) = \gamma_1 + \gamma_2 x$, a two-point design $\xi_1 = \{x_1 = 0, x_2 = 1, w_1 = w_2 = 1/2\}$ is $D$-optimal. This design remains optimal in the setting discussed in Remark 3, too. On the other hand, when the matrix $\Lambda$ is unknown, additional point(s) may appear in $D$-optimal designs. Note that (4) and (9) imply

$$F^T \Lambda F = \begin{pmatrix} \lambda_0 & \lambda_{01} + \lambda_{02}x \\ \lambda_{01} + \lambda_{02}x & \lambda_1 + 2\lambda_{12}x + \lambda_2 x^2 \end{pmatrix},\ \mu(x,\theta) = \begin{pmatrix} \sigma^{-2} F F^T & \mathbf{0}_{3,6} \\ \mathbf{0}_{6,3} & \sigma^{-4}\mu_{\lambda\lambda} \end{pmatrix} \tag{10}$$

where $F = F(x)$ and the $6 \times 6$-matrix $\mu_{\lambda\lambda}$ depends on $x$.

*Example 1.* If $\Sigma = I_2$, that is $\sigma^2 = 1$, and $\Lambda = 0.01\ I_3$, then the locally $D$-optimal design is $\xi_2 = \{x_1 = 0, w_1 = 0.4,\ x_2 = 0.5, w_1 = 0.2,\ x_3 = 1, w_3 = 0.4\}$; see its sensitivity function in Fig. 1, dash-dotted line. When elements of the matrix $\Lambda$ become larger, the central point $x_2$ moves away from the center of the design region. For example, when all diagonal elements are equal to 1 and non-diagonal - to 0.6, then $x_2 = 0.42$ while design weights remain essentially unchanged. The sensitivity function is given in Fig. 1, dashed line.

A similar situation holds when a non-diagonal matrix $\Sigma$ is introduced; the results do not change substantially from those discussed above.

*Example 2. Unknown $\Sigma$.* When the matrix $\Sigma$ does not depend on dose $x$, then (4) and (10) imply that rows/columns of the information matrix $\mu(x,\theta)$ which correspond to $\Sigma_{11}$ and $\lambda_0$, or $\Sigma_{22}$ and $\lambda_1$, are identical for all $x$. This leads to the singularity of $M(\xi,\theta)$ for any design $\xi$. Thus, parameters $\Sigma_{11}$ and $\lambda_0$, or $\Sigma_{22}$ and $\lambda_1$, are confounded and cannot be estimated separately. Therefore, a different parameterization is required in order to obtain meaningful results.

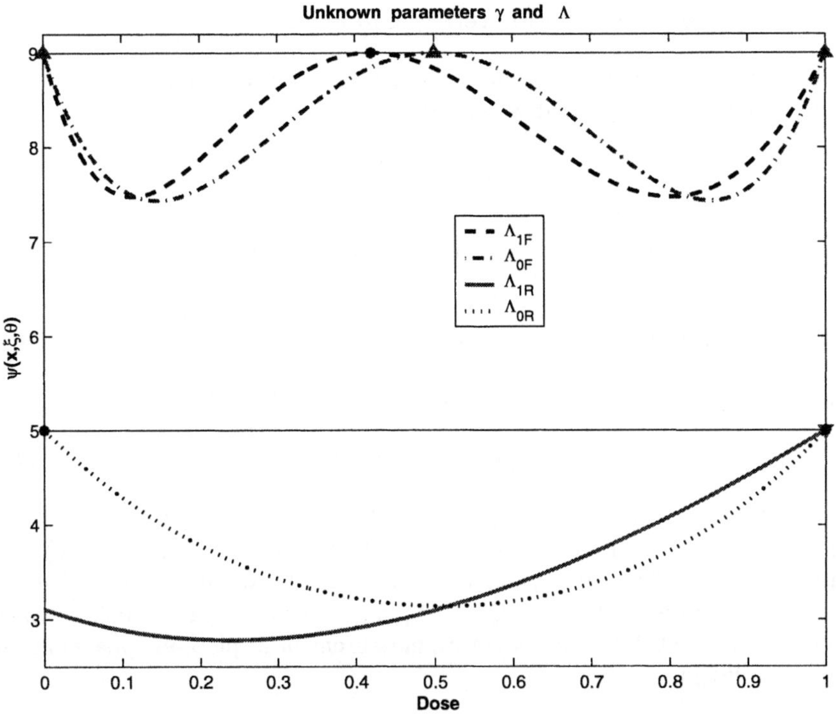


**Fig. 1.** Sensitivity functions of locally $D$-optimal designs, unknown $\gamma$ and $\Lambda$; $\Sigma = I_2$. Top two curves: model (9), $\Lambda = \Lambda_{1F}$ (dashed, $\lambda_{ii} = 1, \lambda_{ij} = 0.6$) and $\Lambda_{0F} = 0.01 I_3$ (dash-dotted). Bottom two curves: reduced model (11), $\Lambda = \Lambda_{1R}$ (solid, $\lambda_{11} = \lambda_{22} = 1, \lambda_{12} = 0.6$) and $\Lambda_{0R} = 0.01 I_2$ (dotted).

# 4 Case when Baseline and Placebo Responses Coincide

In this section we consider model (1), (9) with $\gamma_0 \equiv \gamma_1$, that is measurements at baseline and at $x = 0$ do not differ, and the first measurement is always taken at $x = 0$. Then an admissible design 'point' may be interpreted as a pair of two doses, of which the first dose, $x_0 = 0$, is forced in the design, so that the design region is $\mathbf{X_2} = \{X = (0, x),\ x \in [0, 1]\}$, and

$$\eta(X, \gamma) = \begin{bmatrix} \gamma_1 \\ \gamma_1 + \gamma_2 x \end{bmatrix}, \; F^T(X) = \begin{pmatrix} 1 & 0 \\ 1 & x \end{pmatrix}, \; \Lambda = \begin{pmatrix} \lambda_1 & \lambda_{12} \\ \lambda_{12} & \lambda_2 \end{pmatrix}. \tag{11}$$

We show that there is a noticeable difference between $D$-optimal designs for the reduced model (11), compared to the designs discussed in Section 3. In what follows, $\xi_1^* = \{X_1, w_1 = 1\}$ is a design supported on a single point $X_1 = (0, 1)$, which is an analog of a two-point $D$-optimal design $\xi_1$ for Models (1), (9). Similarly, let $X_0 = (0, 0)$, which corresponds to repeated measurements at placebo.

### 4.1 Only Response Parameters $\gamma$ Unknown

The number of unknown parameters is $p = 2$. Let $S$ be a non-singular matrix and without loss of generality we assume that

$$S = \begin{pmatrix} 1 & b \\ b & a^2 \end{pmatrix}.$$

Then it appears that $D$-optimal designs depend on $S$. Indeed,

$$\mu(X) = \frac{1}{a^2 - b^2} \begin{bmatrix} a^2 - 2b + 1 & x(1-b) \\ x(1-b) & x^2 \end{bmatrix}, \quad M^{-1}(\xi_1^*) = \begin{pmatrix} 1 & b-1 \\ b-1 & a^2 - 2b + 1 \end{pmatrix},$$

and $\psi(X, \xi_1^*) = 1 + x^2 + (b-1)^2(x-1)^2/(a^2 - b^2)$. When

$$(b-1)^2 \le a^2 - b^2, \tag{12}$$

then $\xi_1^*$ is $D$-optimal since $\psi(X_1, \xi_1^*) = 2$ and $\psi(X, \xi_1^*) \le 2 - 2x(1-x) \le 2$ for any $x \in [0,1]$. However, when (12) is violated, then $\psi(X_0, \xi_1^*) > 2$ and design $\xi_1^*$ loses its optimality. For example, when $b = 0$ and $a^2 = 1/2$, the $D$-optimal design is $\xi = \{X_1, w_1 = 0.75,\ X_0, w_0 = 0.25\}$, which seems counterintuitive since an extra measurement at placebo appears in the optimal design.

### 4.2 Unknown Parameters $\gamma$ and $\Lambda$

*Example 3.* Here the number of unknown parameters is $p = 5$ (two in response and three in $\Lambda$). Similar to Section 3.2, we take $\Sigma = I_2$. When $\Lambda = 0.01\ I_2$, the locally $D$-optimal design includes an extra measurement at placebo, $\xi_2^* = \{X_1 = (0,1), w_1 = 0.918, X_0 = (0,0), w_0 = 0.082\}$; its sensitivity function is given in Fig.1, dotted line. Once elements of $\Lambda$ become larger, point $X_0$ may be dropped from the optimal design. If $\lambda_1 = \lambda_2 = 1,\ \lambda_{12} = 0.6$, then design $\xi_1^*$ regains local $D$-optimality; see Fig. 1, solid line.

*Example 4. Unknown $\Sigma$.* Unlike Example 2, the parameter $\lambda_1$ is not confounded with any of $\Sigma_{ij}$ since $\partial S / \partial \lambda_1$ is now equal to a $(2 \times 2)$-matrix of ones. For example, in the case of diagonal $\Sigma = I_2$ and $\lambda_1 = \lambda_2 = 1, \lambda_{12} = 0.6$, the locally $D$-optimal design is supported on three points $\{0, 0.42, 1\}$ with weights 0.19, 0.27, 0.54, respectively.

## References

A.C. Atkinson and A. Donev. *Optimum experimental design.* Clarendon Press, Oxford, 1992.

A.C. Atkinson and V.V. Fedorov. The optimum design of experiments in the presence of uncontrolled variability and prior information. In *Optimal Design and Analysis of Experiments*, pages 327–344. North-Holland, Amsterdam, 1988.

V.V. Fedorov. *Optimum experimental design.* Academic Press, New York, 1972.

V.V. Fedorov, R. Gagnon, and S. Leonov. Design of experiments with unknown parameters in variance. *Appl. Stoch. Mod. Bus. Ind.*, 18(3):207–218, 2002.

V.V. Fedorov and P. Hackl. *Model-oriented design of experiments.* Springer Verlag, New York, 1997.

D.A. Harville. *Matrix algebra from a statistician's perspective.* Springer Verlag, New York, 1997.

J. Kiefer and J. Wolfowitz. The equivalence of two extremum problems. *Canad. J. Math.*, 12:363–366, 1960.

R. Muirhead. *Aspects of multivariate statistical theory.* Wiley, New York, 1982.

S. Piantadosi. *Clinical trials: a methodologic perspective.* Wiley, New York, 1997.

A.H. Roberts, D.G. Kewman, L. Mercier, and M. Hovell. The power of nonspecific effects in healing: implications for psychosocial and biological treatments. *Clin. Psychol. Rev*, 13:375–391, 1993.

# Small Size Designs in Nonlinear Models Computed by Stochastic Optimization

J.-P. Gauchi[1] and A. Pázman[2]

[1] Biometrics Unit, INRA-Jouy en Josas gauchi@banian.jouy.inra.fr
[2] Comenius University, Bratislava pazman@center.fmph.uniba.sk

**Summary.** Optimality criteria, that are functions of the mean square error matrix, are expressed as integrals of the density of the parameter estimator. The optimum design is obtained by an accelerated stochastic optimization method. The estimator is modified to reflect prior knowledge about the parameters, and to take into account the boundary of the parameter space. Results of Pázman and Pronzato (1992) are extended and improved by that. Computer results are presented on examples.

**Key words:** *A*- and *D*-optimality, distribution of estimators, Mean Square Error.

## 1 Introduction

We consider the nonlinear regression model

$$y_x = \eta(x,\theta) + \varepsilon_x \ , \quad \theta \in \Theta,$$

with $x \in \mathcal{X}$ = the design space, $\varepsilon_x$ normally distributed with mean zero and variance $\sigma^2(x)$. The vector of unknown parameters is $\theta \in \Theta \subset \mathbb{R}^p$. We compare different exact designs $\mathbb{X} = (x_1,\ldots,x_N)^T$ of a small fixed size $N \geq p$ and with uncorrelated observations $y_{x_1},\ldots,y_{x_N}$. For fixed $\mathbb{X}$ we use the vector notation

$$y = \eta_{\mathbb{X}}(\theta) + \varepsilon$$

with mean vector $\eta_{\mathbb{X}}(\theta) = (\eta(x_1,\theta),\ldots,\eta(x_N,\theta))^T$, and variance given by $\mathrm{Var}(y) = \Sigma_{\mathbb{X}} = \mathrm{diag}\{\sigma^2(x_1),\ldots,\sigma^2(x_N)\}$. Typically, one has to compare different designs according to the mean square error matrix

$$S(\mathbb{X},\theta) = E_\theta\left[\left(\hat{\theta}-\theta\right)\left(\hat{\theta}-\theta\right)^T\right],$$

where $\hat{\theta}$ is an estimator of $\theta$.

Optimality criteria are considered as real valued functions of $S$ of the form $\Phi[S(\mathbb{X},\theta)]$. We take $\Phi = \mathrm{tr}$, or $\Phi = \ln\det$ in case of a generalized $A$ - or $D$ - optimality criterion, etc. In the asymptotic case (with $N \to \infty$) the estimator

$\hat{\theta}$ is unbiased, and $S(\mathbb{X},\theta) = \text{Var}\left(\hat{\theta}\right)$, and we obtain standard criteria. Since the bias of $\hat{\theta}$ is related to the curvatures of the model, it has been proposed to find designs with restricted curvatures in Clyde and Chaloner (2002), which however may exclude many important designs a priori.

The stochastic optimization method has been used in Vila (1990) and Gauchi (1999) for designs minimizing the expected volume of a confidence region, and in Pázman and Pronzato (1992) where the generalized $A$−optimal designs have been computed with the use of a probability distribution of $\hat{\theta}$ for "flat" models (see Section 3). In the present paper the approach of the last paper is extended with important improvements: An accelerated stochastic optimization is involved, which with the present speed of computers makes the task realistic for a larger class of experiments. The "almost exact" density is extended to a larger class of "nonflat" nonlinear models, and a prior density is involved. The class of considered optimality criteria is enlarged and the "boundary effects" are reconsidered.

## 2 Optimality Criteria Expressed as an Integral

Any criterion function of the form $\Phi[S(\mathbb{X},\theta)] = \text{tr}\{US(\mathbb{X},\theta)\}$, where $U$ is a given p.s.d. matrix, can be expressed as one $p$−dimensional integral

$$\text{tr}\{U\ S(\mathbb{X},\theta)\} = \int_{\Theta} (z-\theta)^T U (z-\theta)\, dP_{\mathbb{X}}(z),$$

where the probability distribution $P_{\mathbb{X}}(\cdot)$ of $\hat{\theta}$ has typically two components. The first is given by a density over $\text{int}(\Theta)$, the second is a singular distribution over the boundary of $\Theta$. To smooth the second component one uses a penalized LSE

$$\tilde{\theta} = \arg\min_{\theta\in\Theta}\left\{[y-\eta_{\mathbb{X}}(\theta)]^T \Sigma^{-1}[y-\eta_{\mathbb{X}}(\theta)] + 2w^*(\theta)\right\} \tag{1}$$

The penalty may contain two additive terms, $w^*(\theta) = w(\theta) - \ln\pi(\theta)$, the first for smoothing the boundary of $\Theta$ and the second related to $\pi(\theta)$, a prior density reflecting experience obtained from some previous observations. The penalization $w^*(\theta)$ does not allow to have estimates on the boundary of $\Theta$. The probability distribution of $\tilde{\theta}$ is concentrated on $\text{int}(\Theta)$ , with a density $q_{\mathbb{X}}(z \mid \theta)$. Hence,

$$\text{tr}\{U\ S(\mathbb{X},\theta)\} = \int_{\text{int}(\Theta)} (z-\theta)^T U (z-\theta)\, q_{\mathbb{X}}(z\mid\theta)\, dz$$

Particular cases are the generalized $A$-optimality with $U = I$, the generalized $c$-optimality with $U = cc^T$. To generalize further, consider a mapping $H : \Theta \to \mathbb{R}^k$, $k \leq p$, and the criterion

$$\int_{\mathrm{int}(\Theta)} (H(z) - H(\theta))^T V (H(z) - H(\theta))\, q_{\mathbb{X}}(z \mid \theta)\, dz,$$

where $V$ is a p.s.d. $k \times k$ matrix. A particular case is when $k = 1, V = 1$: then the integral is the variance plus the squared bias of the penalized LSE $H\left(\tilde{\theta}\right)$ of a parameter function $H(\theta)$.

One can write also the $D$-optimality criterion in an integral form. Indeed, according to the definition of a determinant we have

$$\det\{S(\mathbb{X}, \theta)\} = \int_{\mathrm{int}(\Theta)} \cdots \int_{\mathrm{int}(\Theta)} \mathcal{L}\left(z^{(1)}, \ldots, z^{(p)}, \theta, \mathbb{X}\right) dz^{(1)} \ldots dz^{(p)},$$

where

$$\mathcal{L}\left(z^{(1)}, \ldots, z^{(p)}, \theta, \mathbb{X}\right) = \sum_{\rho \in \mathcal{S}_p} \left[\mathrm{sign}(\rho) \prod_{k=1}^{p} \left(z^{(k)} - \theta\right)_k \left(z^{(k)} - \theta\right)_{per(k)}\right] \times \prod_{k=1}^{p} q_{\mathbb{X}}\left(z^{(k)} \mid \theta\right)$$

and the sum is taken over $\mathcal{S}_p$, the set of all permutation $\rho$ of the set $\{1, \ldots, p\}$. However, the multiplicity of this integral increases very much with the increase of $p$.

## 3 Density of the Estimator

Here we consider a fixed design $\mathbb{X}$. We suppose that the model is regular under the design $\mathbb{X}$, that is that the matrix $J_{\mathbb{X}}(\theta) = \frac{\partial \eta_{\mathbb{X}}(\theta)}{\partial \theta^T}$ has full rank for every $\theta \in \mathrm{int}(\Theta)$. In common statistical models one can accept that the probability that there is more than one statistically significant local minimum in (1) is very small. Then an "almost exact" density of $\tilde{\theta}$ has the form (cf. Pázman (1984), Pázman (1993), Pázman (1993) for the case without penalty, Pázman and Pronzato (1992) for the case with penalty but in "flat" models, Gauchi and Pázman (2003) for the general case)

$$q_{\mathbb{X}}(z \mid \theta) = \frac{\mathcal{F}_{\mathbb{X}}(z, \theta)}{(2\pi)^{p/2} \det^{1/2}[M_{\mathbb{X}}(z)]} \times \exp\left\{-\frac{1}{2} \left\|P_{\mathbb{X}}^{z}\left[\eta_{\mathbb{X}}(z) + u_{\mathbb{X}}(z) - \eta_{\mathbb{X}}(\theta)\right]\right\|_{\Sigma_{\mathbb{X}}}^2\right\}.$$

where $M_{\mathbb{X}}(z) = [J_{\mathbb{X}}(z)]^T \Sigma_{\mathbb{X}}^{-1} J_{\mathbb{X}}(z)$ , $P_{\mathbb{X}}^{z} = J_{\mathbb{X}}(z) [M_{\mathbb{X}}(z)]^{-1} [J_{\mathbb{X}}(z)]^T \Sigma_{\mathbb{X}}^{-1}$, $u_{\mathbb{X}}(z) = J_{\mathbb{X}}(z) [M_{\mathbb{X}}(z)]^{-1} \frac{\partial w^*(z)}{\partial z}$. The term $\mathcal{F}_{\mathbb{X}}(z, \theta)$ depends on the structure of the model and on the level of approximation that we want to attain. We have

$$\begin{aligned}\mathcal{F}_{\mathbb{X}}(z,\theta) &= \det\left[T_{\mathbb{X}}(z,\theta)\right]\\ \left[T_{\mathbb{X}}(z,\theta)\right]_{ij} &= \left[M_{\mathbb{X}}(z)\right]_{ij} + \frac{\partial^2\eta_{\mathbb{X}}^T(z)}{\partial z_i \partial z_j}\Sigma_{\mathbb{X}}^{-1}[I-P_{\mathbb{X}}^z]\left[\eta_{\mathbb{X}}(z)-\eta_{\mathbb{X}}(\theta)\right]\\ &\quad -\frac{\partial^2\eta_{\mathbb{X}}^T(z)}{\partial z_i \partial z_j}\Sigma_{\mathbb{X}}^{-1}u_{\mathbb{X}}(z) + \frac{\partial^2 w^*(z)}{\partial z_i \partial z_j}\end{aligned}$$

if the Riemannian curvature tensor $R(\theta)$ with components

$$\begin{aligned}R_{hijk}(\theta) &= \tau_{hjik}(\theta) - \tau_{hkij}(\theta)\\ \tau_{hjik}(\theta) &= \frac{\partial^2\eta_{\mathbb{X}}^T(\theta)}{\partial\theta_h\partial\theta_j}\Sigma_{\mathbb{X}}^{-1}\left[I-P_{\mathbb{X}}^{\theta}\right]\frac{\partial^2\eta_{\mathbb{X}}(\theta)}{\partial\theta_i\partial\theta_k}\end{aligned}$$

vanishes identically yielding "flat" models. In particular this happens when $p=1$, or when the information matrix does not depend on $\theta$, or when the model is intrinsically linear (cf. Pázman (1993)).

When $R(\theta)$ does not vanish and the "almost exact" precision is required, we have for $\dim(\theta)=2$ (in an abbreviated notation) $\mathcal{F}=\det T + R_{1212}$. In general one can write

$$\mathcal{F}(z,\theta) = \sum_{\rho\in\mathcal{S}_p}\operatorname{sign}(\rho)\sum_{q\in\mathcal{P}(1,2)}\prod_{\{i\}\in q}T_{i,\pi(i)}(z,\theta)\prod_{\{k,\ell\}\in q}\tau_{k,\pi(k),\ell,\pi(\ell)}(z)$$

where, as above, the first sum is over all permutations $\rho$ of $\{1,\ldots,p\}$ and $\mathcal{P}(1,2)$ is the set of all partitions of the set $\{1,\ldots,p\}$ into singletons $\{i\}$ and pairs $\{k,\ell\}$. (*e.g.*, with $p=3$ the set $\mathcal{P}(1,2)$ consists of partitions $(\{1\},\{2\},\{3\})$, $(\{1\},\{2,3\})$, $(\{1,2\},\{3\})$, $(\{1,3\},\{2\})$). If we neglect partitions containing pairs, we obtain the "flat" case. If we neglect partitions containing more than one pair, we obtain an approximation which is linear in the components of $R(\theta)$, etc.

## 4 The Penalty for the Boundary of $\Theta$

The basic idea is that the penalty $w(\theta)$ must be equal to $+\infty$ on the boundary of $\Theta$, it must be twice continuously differentiable on $\operatorname{int}(\Theta)$, and it must be zero outside a narrow strip along the boundary. Since $\Theta$ is usually a $p$-dimensional interval $[a_1,b_1]\times\ldots\times[a_p,b_p]$, we may take $w(\theta)=\sum_{i=1}^p w_i(\theta_i)$ with $w_i(\theta_i)=0$ for $\theta_i\in[a_i+\Delta_i,\ b_i-\Delta_i]$, and $\Delta_i>0$ but small. In the examples we used for the upper boundary of $[a_i,b_i]$ the expression

$$w_i(\theta_i) = K\tan\frac{\pi}{2}x^5 \quad \text{with } x=\frac{\theta_i-b_i+\Delta_i}{\Delta_i} \tag{2}$$

which is similar to that used in Pázman and Pronzato (1992), and where $K$ is a constant fixed a priori. We proceed similarly for the lower boundary. However, if the curves $\{\eta_{\mathbb{X}}(\theta):\theta\in\Theta,\ (\theta_1,\ldots,\theta_{k-1},\theta_{k+1},\ldots,\theta_p)=const\}$ are very much curved on the expectation surface, then it may be better to relate the penalty not with the parameter space, but with the expectation surface itself. So alternatively we propose

$$x = \left\{ 1 - \frac{\left\| \eta\left(\theta_1, \ldots, \theta_i, \ldots, \theta_p\right) - \eta\left(\theta_1, \ldots, b_i, \ldots, \theta_p\right) \right\|_{\Sigma}^2}{\left\| \eta\left(\theta_1, \ldots, b_i - \Delta_i, \ldots, \theta_p\right) - \eta\left(\theta_1, \ldots, b_i, \ldots, \theta_p\right) \right\|_{\Sigma}^2} \right\} \tag{3}$$

for the upper boundary of $[a_i, b_i]$, etc. This is however more complicated to compute. A check of the right choice of $\Delta_i$ , $w_i(\theta_i)$ and of the approximate density is performed after that the optimum design is computed (cf. Gauchi and Pázman (2003) for details).

## 5 Accelerated Method of Stochastic Optimization

The aim is to solve the minimization problem

$$\mathbb{X}_{\text{opt}} = \arg\min_{\mathbb{X}} \int_{\Theta^p} \mathcal{L}\left(\bar{z}\,;\theta, \mathbb{X}\right) d\bar{z},$$

where the function $\mathcal{L}$ is given. Because of the complexity of this function we chose a method of stochastic optimization. Among the methods published, Vila (1990) selected, in the class of the multidimensional Kiefer-Wolfowitz processes (Blum (1954)), the accelerated process of Fabian (1967) for successfully solving his own optimization problem. A lot of numerical studies on this Fabian process have been made Gauchi (1999) which enable us to solve now much more complicated minimization problems such as the second example presented in this paper. Moreover new improvements are presented here. Without these improvements the computation in the examples is too difficult to manage (*e.g.*, an excessively long time). First of all, we remind the details of the original accelerated Fabian process.

graphOriginal accelerated Fabian process

At the $k$th iterative step the vector $\mathbb{X}^{(k+1)}$ is computed, its components being the elements of the set $\mathcal{X}$, as:

$$\mathbb{X}^{(k+1)} = \mathbb{X}^{(k)} - a^{(k)}\mathbb{W}^{(k)},$$

where $a^{(k)} = a_0/k^{\beta}$, $a_0 > 0$ and $0 < \beta \leq 1$. Experience showed us that we can take $\beta = 1$. The vector $\mathbb{W}^{(k)}$ is the stochastic quasigradient. Instead of derivatives used in a classical gradient, $\{\mathbb{W}^{(k)}\}_j$ contains differences of the function $\mathcal{L}$ of the form

$$\frac{\mathcal{L}\left(\bar{z}\,;\theta, \mathbb{X}^{(k)} + c_k\mathbb{E}^j\right) - \mathcal{L}\left(\bar{z}\,;\theta, \mathbb{X}^{(k)} - c_k\mathbb{E}^j\right)}{2c_k}$$

where $\mathbb{E}^j$ denotes the $j$th canonical vector, and $c_k$ are numbers defined by $c_k = c/k^{1/(4h+2)}$, $c > 0$, and $h$ is an integer explained hereafter. To express the changes of the function $\mathcal{L}$ we take the "weighted" mean of such differences

$$\sum_{i=1}^{h} \nu_i \frac{\mathcal{L}\left(\bar{z}\,;\theta,\mathbb{X}^{(k)}+u_i c_k \mathbb{E}^j\right)-\mathcal{L}\left(\bar{z}\,;\theta,\mathbb{X}^{(k)}-u_i c_k \mathbb{E}^j\right)}{2c_k} \tag{4}$$

with

$$u_i = \cos\left[\frac{(h-i)\,\pi}{2h-1}\right]$$
$$\nu_i = \frac{2}{2h-1}\left(-1\right)^{i-1}\frac{1}{u_i^2}.$$

The number $h$ is chosen so that the integral of the function $\mathcal{L}$ has bounded partial derivatives with respect to the components of $\mathbb{X}$ up to order $(2h+1)$, usually we take $h \leq 6$. Expression (4) is related to a given point $\bar{z}$. So, we apply a Monte-Carlo approach . We take simulations to obtain uniformly distributed random points $\bar{z}_{i,j,s} \in \Theta^p$, and to obtain the final form of the quasigradient

$$\{\mathbb{W}^{(k)}\}_j = \sum_{i=1}^{h}\frac{\nu_i}{N_i}\sum_{s=1}^{N_i}\frac{\mathcal{L}\left(\bar{z}_{\,i,j,s};\theta,\mathbb{X}^{(k)}+u_i c_k \mathbb{E}^j\right)-\mathcal{L}\left(\bar{z}_{\,i,j,s};\theta,\mathbb{X}^{(k)}-u_i c_k \mathbb{E}^j\right)}{2c_k}.$$

Here $N_i = \mu_i N_T$ with $N_T$ fixed a priori, and $\mu_i$ are given by :

$$\mu_i = \left[2h(h-1)+1/2\right]^{-1} u_i^{-2}(1-\delta_{hi}/2)$$

with $\delta_{hi}$ the Kronecker symbol.

The iteratively obtained sequence of designs $\mathbb{X}^{(1)}, \mathbb{X}^{(2)}, \ldots$ converges in the quadratic mean, and hence in probability to $\mathbb{X}_{\text{opt}}$ (see Fabian (1967)).

graphImprovements of the accelerated Fabian process

Three kinds of improvements have been made :

1. At the beginning of the process. Because of the crucial influence of the choice of $a_0$ on the process speed and on the behaviour of $\mathbb{W}^{(k)}$ we propose two alternative methods. The first method is an independent normalization of each stochastic gradient component by means of results of Polyak and Tsypkin (1973) and Pronzato (1986). In the second method instead of $a^{(k)}$ we take a diagonal matrix $A^{(k)}$ (for details, see Gauchi and Pázman (2003)).
2. In the sampling method. When the graph of the integrand $\mathcal{L}$ has narrow peaks, we proposed for another class of criteria Gauchi and Vila (1994) not to take uniform samples in $\Theta^p$ but make sampling with another measure. Our contribution has been to adapt this importance sampling method to the regulation of our optimization process (see Gauchi and Pázman (2003)).
3. For terminating the process. Some stopping rules proposed by Pflug (1988) have been adapted to this improved Fabian process and are given with details in Gauchi (1999).

## 6 Examples and Numerical Results

*Example 1.* The Michaelis-Menten model

As in Pázman and Pronzato (1992) we use the Michaelis-Menten model in exactly the same conditions: The response function of this model is $\eta(x,\theta) = \theta_1 x/(\theta_2 + x)$ and we have $x \in \mathcal{X} = [0\ ;\ 2]$, $\mathbb{X}^{(0)} = (0.5\ , 1.5)^T$, $\theta = (0.1, 1.7)^T$, $\Theta = [0\ ;\ 0.25] \times [0\ ;\ 10]$, $\sigma^2(x) = 25 \times 10^{-6}$, $\Delta_i = 0.1(b_i - a_i)$, $K = 0.01$.

For the generalized $A-$optimality criterion with the penalty chosen by the authors, we find by our stochastic optimization method (1000 iterations) the final optimal design $\mathbb{X}^*_{[T1]} = (0.706\ , 2.0)^T$, computed in about 4 minutes (on a Pentium IV-PC, with the IML module of SAS) instead of $(0.700, 2.0)^T$ computed in about 20 hours (in the year 1990). With the new penalties given by (2) and (3), and $\Delta_i = 0.05(b_i - a_i)$ we find $\mathbb{X}^*_{[T2]} = (0.628\ , 2.0)^T$ and $\mathbb{X}^*_{[T3]} = (0.636, 2.0)^T$, respectively. We put now some prior on the parameter $\theta_2$ represented by $N\left(1.7\ ,\ \sigma^2_{prior} = 25 \times 10^{-4}\right)$. The generalized $A-$optimality design with this prior and the penalty (2) is now $\mathbb{X}^*_{[T4]} = (0.96, 1.33)^T$.

For the generalized $D$-optimality criterion, with the penalty (2), we find the optimal design $\mathbb{X}^*_{[D1]} = (0.6\ , 2.0)^T$ and with the same prior on the parameter $\theta_2$ as above, we find the optimal design $\mathbb{X}^*_{[D2]} = (0.674\ , 1.578)^T$. From a practical point of view it is also relevant to examine the influence of the variance of the prior and the prior choice $\theta$ on the final optimal design. This is done in Gauchi and Pázman (2003).

*Example 2.* The Baranyi growth curve model

As a real example we consider here a well-known growth curve model from microbiology (see Baranyi and Roberts (1994, 1995)) which has been shown to have high performance in microbiological experiments. In this model $t > 0$ is the time of observation, the unknown parameters having a clear interpretation are denoted as $\theta = (x_0, x_{\max}, \mu_{\max}, lag)$, and

$$\eta(\theta; t) = \frac{1}{\ln(10)} \left[\log(x_0) + \mu_{\max} A - \log(B)\right],$$

where

$$\begin{aligned} A &= t + \frac{1}{\mu_{\max}} \log(A_0) \\ A_0 &= \exp(-\mu_{\max} t) + \exp(-\mu_{\max} lag) - \exp(-\mu_{\max} t - \mu_{\max} lag). \\ B &= 1 + \frac{[\exp(\mu_{\max} A)] - 1}{\frac{x_{\max}}{x_0}} \end{aligned}$$

We took the following conditions:
$t \in \mathcal{X} = [0, 50]$, $\mathbb{X}^{(0)} = (0, 7.2, 16.7, 50)^T$, $\theta = \left(100, 10^8, 1; 5\right)^T$, $\sigma^2(x) = 10^{-3}$

$\Theta = [80, 120] \times [8\,10^7, 1.2\,10^8] \times [0.8, 1.2] \times [4, 6]$, $\Delta_i = 0.05(b_i - a_i)$ and $K = 0.01$.

For the generalized $A$-optimality criterion with the penalty (2), we found, with 200 iterations, the final optimal design $\mathbb{X}^*_{[T5]} = (0.11, 1.80, 10.20, 41.86)^T$, computed in about 10 hours. For the generalized $D$-optimality criterion with the same penalty we obtain, with 50 iterations, the final optimal design $\mathbb{X}^*_{[D3]} = (3.35, 5.98, 24.85, 47.69)^T$, computed in about 40 hours.

## References

J. Baranyi and T.A. Roberts. A dynamic approach to predicting bacterial growth in food. *Int. J. Food. Microbiol.*, 23:277–294, 1994.

J. Baranyi and T.A. Roberts. Mathematics of predictive food microbiology. *Int. J. Food. Microbiol.*, 26:199–218, 1995.

J.R. Blum. Multidimensional stochastic approximation methods. *Ann. Stat.*, 25:737–744, 1954.

M. Clyde and K. Chaloner. Constrained designs strategies for improving normal approximations in nonlinear regression problems. *J. Stat. Plann. Inf.*, 104(1):175–196, 2002.

V. Fabian. Stochastic approximation of minima with improved asymptotic speed. *Ann. Stat.*, 38:191–200, 1967.

J.P. Gauchi. *Contribution à la l'étude et au calcul de critères de plans d'expériences optimaux pour des modèles de régression non linéaire.* PhD thesis, CNAM, Paris, 1999.

J.P. Gauchi and A. Pázman. Distribution of the least squares estimator, stochastic, optimization and optimum designs in nonlinear models. Technical Report 2003-7, INRA, Biometrics Unit, Jouy-en-Josas, 2003.

J.P. Gauchi and J.P. Vila. Réduction de variance par échantillonnage selon l'importance: application à l'optimisation stochastique d'un critère de plans d'expériences. In *Actes des XXVIèmes Journées de Statistique ASU, Neuchatel, 1994*, pages 317–320. ASU, 1994.

A. Pázman. Probability distribution of the multivariate nonlinear least squares estimates. *Kybernetika*, 20:209–230, 1984.

A. Pázman. Higher dimensional nonlinear regression - a statistical use of the Riemannian curvature tensor. *Statistics*, 25:17–25, 1993.

A. Pázman. *Nonlinear statistical models*, volume 254 of *Mathematics and its Applications*. Kluwer Academic Publishers Group, Dordrecht, 1993.

A. Pázman and L. Pronzato. Nonlinear experimental design based on the distribution of estimators. *J. Stat. Plann. Inf.*, 33:385–402, 1992.

G. Ch. Pflug. Stepsize rules, stopping times and their implementation in stochastic quasigradient algorithms. In Yu. Ermoliev and R.J.-B. Wets, editors, *Numerical techniques for stochastic optimization*, pages 353–372. Springer-Verlag, Heidelberg, 1988.

B.T. Polyak and Ya. Z. Tsypkin. Pseudogradient adaptation and training algorithms. *Aut. Remote Control*, 34(3):377–397, 1973.

L. Pronzato. *Synthèse d'expériences robustes pour modèles à paramètres incertains.* PhD thesis, Université de Paris-Sud, Orsay, 1986.

J.-P. Vila. Exact experimental designs via stochastic optimization for nonlinear regression models. In *COMPSTAT 1990 (Dubrovnik, 1990)*, pages 291–296, Heidelberg, 1990. Physica.

# Asymptotic Properties of Biased Coin Designs for Treatment Allocation

A. Giovagnoli

University of Bologna, via Belle Arti 41, 40126, Bologna, Italy
giovagno@stat.unibo.it

**Summary.** Biased coin designs ($BCD$) are a special type of randomized sequential experiments. A common feature of several $BCD$'s is their Markov property, which allows a direct study of the design asymptotics. Possible criteria for comparing different biased coin designs are discussed with reference to particular cases.

**Key words:** biased coin design, treatment allocation.

## 1 Introduction

Sequentially designed experiments occur in biomedical or pharmaceutical practice and are deemed to be more efficient than non-sequential ones for any given sample size. The experiment may be *response-adaptive*, if the choice of the next design point, that is the combination of experimental conditions under the experimenter's control, is driven by previous responses as well as previous allocations. Mathematically, a sequential experiment is a sequence $\{(X_n, Y_n), n \in \mathbb{N}\}$, where the $X$'s denote the design points and the $Y$'s the responses; a probabilistic model for the responses is assumed and the design consists in an assignment rule of $X_{n+1}$ given the $\sigma$-algebra generated by the "past", $\Im_n = \sigma\{X_1, Y_1, \ldots, X_n, Y_n\}$. This rule may be deterministic or, more likely, probabilistic, since in general a certain amount of randomization is desired to protect against accidental bias, selection bias, etc. A frequent assumption is conditional independence of the observations, namely, with obvious notation,

$$P(Y_{n+1} \mid X_{n+1}, \Im_n) = P(Y_{n+1} \mid X_{n+1}) \quad \text{for all } n \in \mathbb{N}. \tag{1}$$

A *Biased Coin Design* ($BCD$) is a type of randomized sequential experiment. The expression goes back to the authoritative paper Efron (1971), which was the first to propose a design for the comparison of two treatments aimed at a trade-off between balance and selection bias. Randomization is achieved given by the hypothetical tossing of a biased coin with probability $p$ $(1/2 \leq p \leq 1)$, namely

$$P(\delta_{n+1} = 1) = \begin{cases} p & \text{if } D_n < 0 \\ 1-p & \text{if } D_n > 0 \\ 1/2 & \text{if } D_n = 0. \end{cases} \tag{2}$$

Here $\delta_n = 1$ if the $n$th observation is allocated to treatment $T_1$, $\delta_n = -1$ otherwise, and $D_n = \sum_{i=1}^{n} \delta_i$ is the difference between the two allocations after $n$ assignments.

Efron's $BCD$ has been extended in many different ways, both with and without a response-adaptive allocation mechanism. However, an essential feature of all $BCD$'s is that at each step the randomization device (the "coin") is independent of the experimental outcomes, that is randomization is **not** response-driven: this may indeed be taken as the general definition of a $BCD$. Here I want to underline another common feature of several $BCD$'s, namely their Markov property, shared by the urn designs discussed in Baldi Antognini (2004), which allows the design asymptotics to be studied directly. Possible criteria for comparing large sample biased coin designs involve the optimality of the asymptotic experiment, the speed of convergence and the precision of the approximation of the empirical experiment to the asymptotic one, namely minimization of the asymptotic variances of asymptotically normal designs. This paper starts with a general discussion of Markov experiments and the results are then applied to $BCD$'s. Results are just stated and the proofs omitted or only sketched.

## 2 Markovian Experiments

A sequential experiment is understood to be *Markovian* if the allocation of the next design point depends only on the most recent outcome, that is

$$P(X_{n+1} \mid \Im_n) = P(X_{n1} \mid X_n, Y_n) \text{ for all } n \in \mathbb{N}. \tag{3}$$

**Lemma 1 (Fundamental Lemma).** *Under (1) and (3), given a starting point $x_0$, both $\{(X_n, Y_n)\}$ and $\{X_n\}$ are Markov chains.*

*Remark 1.* If the experiment is not response-adaptive there is almost nothing to prove: the proof for $\{(X_n, Y_n)\}$ is trivial and is only a little more elaborate for $\{X_n\}$. The importance of this lemma lies in its generality, since it holds for any pair of sequences $\{X_n\}$ and $\{Y_n\}$. In other words, $Y_n$ could be replaced by an observable quantity which *depends only on the history of the experiment,* and is not an actual physical "response", as in the urn design case. Instead of the first-order property (3) we could take into account second order Markovian designs, as in Bortot and Giovagnoli (2004), but the Fundamental Lemma does not extend to higher order dependence.

Definition (3) can be slightly generalized by replacing the sequence of design points $\{X_n\}$ by a sequence of design "statistics" $S_n = s(X_1, \ldots, X_n)$ such that $\sigma(X_1, \ldots, X_n) = \sigma(S_1, \ldots, S_n)$ for all $n \in \mathbb{N}$. Then condition (3) becomes

$$P(S_{n+1} \mid \Im_n) = P(S_{n+1} \mid S_n, Y_n) \text{ for all } n \in \mathbb{N} \tag{4}$$

and it can be shown that under (1) and (4) the results of the Fundamental Lemma still hold, in particular $\{S_n\}$ is a Markov chain.

If the transition probabilities (3) do not depend on $n$, not too severe conditions on the experiment usually ensure the existence of the stationary distribution $\pi$ of the chain $\{X_n\}$ (or $\{S_n\}$). On a discrete design space $\mathcal{D} = \{d_k, k \in \mathbb{Z}\}$, if $N_n(d_k)$ is the number of times the design point $d_k$ has been employed in $n$ steps, then for all $k$

$$\bar{N}_n(d_k) = N_n(d_k)/n \to \pi(d_k) \text{ as } n \to \infty \text{ a.s.} \tag{5}$$

by the Law of Large Numbers for Markov chains. Thus, the asymptotic behaviour of the experiment is described by $\pi$, which can be referred to as the *asymptotic experiment.* By (5), for large samples the asymptotic experiment is approximated by the design point frequencies $\bar{N}_n(x)$ for all $x$. Besides, from the CLT for Markov chains (see Roberts (1997)) there follows

**Proposition 1.**

$$\sqrt{n}\{\bar{N}_n(d_k) - \pi(d_k)\} \to N(0, v_k), \quad \textit{for all } k \textit{ as } n \to \infty.$$

Hence, the approximation is the more efficient the smaller the asymptotic variance of $\bar{N}_n(x)$ is, for all $x \in \mathcal{D}$.

A class of Markov chains with wide applications to Design Theory are the non-homogeneous random walks (homogeneous in time, non-homogeneous with respect to the state space). The transition probabilities are usually denoted by

$$\begin{aligned} p_k &= P(X_{n+1} = d_{k+1} \mid X_n = d_k) > 0 \\ q_k &= P(X_{n+1} = d_{k-1} \mid X_n = d_k) > 0 \\ r_k &= P(X_{n+1} = \ d_k \mid X_n = d_k) = 1 - p_k - q_k \geq 0, \end{aligned}$$

with some readjustments if there are barriers.

**Proposition 2.** *The stationary distribution of a random walk* $\pi = \{\pi_k = \pi(d_k)\}$ *on* $\mathcal{D}$ *is given by the equilibrium equations*

$$\pi_k = \pi_{k-1}\lambda_k \textit{ for all } k, \textit{ with } \pi_0 = \left[1 + \sum_{i=1}^{\infty}\prod_{k=1}^{i}\lambda_k + \sum_{i=1}^{\infty}\prod_{k=1}^{i} 1/\lambda_{-k+1}\right]^{-1} \tag{6}$$

*where* $\lambda_k = p_{k-1}/q_k$ *for all* $k$, *if and only if*

$$\sum_{i=1}^{\infty}\prod_{k=1}^{i}\lambda_k + \sum_{i=1}^{\infty}\prod_{k=1}^{i} 1/\lambda_{-k+1} < \infty,$$

*which is satisfied when there is only a finite number of states.*

We may try to modify the sequential algorithm in order to improve $\pi$. For instance, the following result holds true.

**Proposition 3.** *If $\{\lambda_k\}$ is non-increasing, in particular if $\{p_k\}$ is non-increasing and $\{q_k\}$ is non-decreasing, then $\pi$ is unimodal, so there is a design point $d_{m_0}$ or a value of the design statistic $S$, which tends to occur most frequently.*

This value may be controlled by the experimenter.

**Proposition 4 (Giovagnoli and Pintacuda (1998b)).** *Given two random walks $\{X_n\}_{n\in\mathbb{N}}$ and $\left\{\tilde{X}_n\right\}_{n\in\mathbb{N}}$ with $\{\lambda_k\}$ and $\{\tilde{\lambda}_k\}$ non-increasing such that $\lambda_k \geq \tilde{\lambda}_k > 1$ for all $k < m_0$ and $\lambda_k \leq \tilde{\lambda}_k \leq 1$ for all $k > m_0$, then the stationary distributions have the same mode $d_{m_0}$ and $\left\{\tilde{X}_n\right\}_{n\in\mathbb{N}}$ is asymptotically more peaked around $d_{m_0}$ than $\{X_n\}_{n\in\mathbb{N}}$.*

In the MCMC literature the performance of Markov chains is compared through the speed of convergence in distribution. However, there are differences between the MCMC and the experimental design viewpoint. For designs, convergence in distribution of $X_n$ (or $S_n$) to $\pi$ may or may not take place and in general is not relevant, whereas the convergence that matters is (5), which is easier to achieve. In the literature theoretical results on the speed of convergence (5) are scarce, and the use of simulations is widespread. Stochastically monotone chains (chain such that $X_0 \leq_{st} X_1$ implies $\{X_n\} \leq_{st} \{X_{n+1}\}$ for all $n \in \mathbb{N}$, see Shaked and Shanthikumar (1994)) can be compared through the speed of convergence of the monotone sequence of their expected values. In some cases we can make use of the following condition.

**Proposition 5.** *A random walk $\{X_n\}$ is stochastically monotone if and only if $p_k \leq 1 - q_{k+1}$ for all $k \in \mathbb{N}$.*

In order to compare asymptotic variances other concepts of stochastic orderings of Markov chains are also useful, as is shown in the last section of the paper.

## 3 Different Types of Biased Coin Designs

### 3.1 BCD's for Comparing Two Treatments

A possible extension of (2) is the Adjustable Biased Coin Design ($ABCD$) of Baldi Antognini and Giovagnoli (2004), which consists in making $P(\delta_{n+1} = 1)$ a decreasing function of $D_n$, so that the tendency towards balance is stronger the more we move away from it. Let $F(\cdot)$ be a non-increasing function $F : \mathbb{R} \to [0, 1]$, such that $F(-x) = 1 - F(x)$ and let

$$P(\delta_{n+1} = 1) = F(D_n) \tag{7}$$

so that $P(\delta_{n+1} = 1) = 1/2$ in case of balance. Different choices of $F(\cdot)$ give rise to different properties of the algorithm. Apart from Efron's design, special cases of the $ABCD$ are the *Biased Coin Design With Imbalance Tolerance* $BCDWIT(p, b)$ of Chen (1999)

$$P(\delta_{n+1} = 1) = \begin{cases} 1 & \text{if } D_n = -b \\ p & \text{if } -b < D_n < 0 \\ 1/2 & \text{if } D_n = 0 \\ q & \text{if } 0 < D_n < b \\ 0 & \text{if } D_n = b \end{cases}, \qquad b \in \mathbb{N}.$$

and the *Big Stick Design* $BSD(b)$ introduced in Soares and Wu (1983), which corresponds to $BCDWIT(1/2, b)$. On the other hand definition of an Adaptive Biased Coin in Wei (1978) is

$$P(\delta_{n+1} = 1) = F\left(\frac{D_n}{n}\right), \tag{8}$$

instead of (7). Two special cases have been discussed in Atkinson (1982), viz.

$$P(\delta_{n+1} = 1) = N_2/n \quad \text{and} \quad P(\delta_{n+1} = 1) = \frac{N_2^2}{N_1^2 + N_2^2}$$

(where $N_1$ and $N_2$ are the number of times treatments $T_1$ and $T_2$ are assigned), and in Smith (1984), viz.

$$F(x) = \frac{(1-x)^t}{(1-x)^t + (1+x)^t}.$$

## 3.2 Biased Coin Design for Targeting a Quantile

The expression *biased coin designs* has also been used by Durham and Flournoy. Let $\mathcal{D} = \{d_k,\ k = 0, 1, \ldots, M\}$ denote a finite set of possible treatments in increasing order. Let $X_n$ be the treatment applied at the $n$th step and $Y_n$ be the corresponding observation taking values 0 and 1 for negative and positive responses respectively. Assume the probability of positive response to be an increasing function $Q(x)$ of the treatment level. The biased coin designs $BCD-I$ and $BCD-II$ of Durham and Flournoy (1995) aimed at targeting a quantile $x_{(p)}$ with $0 \leq p \leq 1/2$ (now more commonly known now as *Up-and-Down* ($U\&D$) designs) are defined as follows.

Choose a starting point $x_0 \in \mathcal{D}$. If $X_n = d_k$ $(n \geq 1)$, then:

**BCD-I** Toss a biased coin with probability of "heads" $b = \frac{p}{1+p}$. If heads is observed, then $X_{n+1} = d_{k+1}$. If tails is observed, then
Conditionally on $Y_n = 0$, $X_{n+1} = d_k$
Conditionally on $Y_n = 1$, $X_{n+1} = d_{k-1}$.

**BCD-II** Conditionally on $Y_n = 0$, toss a biased coin with probability of "heads" $b = \frac{p}{1-p}$. If heads occurs, $X_{n+1} = d_{k+1}$, if tails occurs then $X_{n+1} = d_k$. Conditionally on $Y_n = 1$, let $X_{n+1} = d_{k-1}$.

The main area of application of these algorithms is for escalating the dose of a potentially toxic drug in Phase I clinical trials. Durham and Flournoy's and other Up-and-Down designs have been extended in Giovagnoli and Pintacuda (1998b), where a general U&D Rule has been defined as follows:

Choose a starting point $x_0 \in \mathcal{D} = \{d_k, k \in I \subseteq \mathbb{Z}\}$. If $n \geq 1$ and $X_n = d_k$, then

| Conditionally on $Y_n = 0$: | Conditionally on $Y_n = 1$: |
|---|---|
| $X_{n+1} = d_{k+1}$ with pr. $\alpha_k > 0$ | $X_{n+1} = d_{k-1}$ with pr. $\beta_k > 0$ |
| $X_{n+1} = d_{k-1}$ with pr. $\beta'_k \geq 0$ | $X_{n+1} = d_{k+1}$ with pr. $\alpha'_k \geq 0$ |
| $X_{n+1} = d_k$ with pr. $1 - \alpha_k - \beta'_k \geq 0$ | $X_{n+1} = d_k$ with pr. $1 - \beta_k - \alpha'_k \geq 0$ |

where $\alpha_k \geq \alpha'_k$, $\beta_k \geq \beta'_k$, $\beta_k + \alpha'_k \leq 1$, $\alpha_k + \beta'_k \leq 1$. The $\alpha_k, \alpha'_k, \beta_k, \beta'_k$ can be chosen so as to modify the shape of the asymptotic design, using Proposition 4; condition $\frac{\alpha_k - \beta'_k}{\alpha_k + \beta_k - \alpha'_k - \beta'_k} = p$ $(0 < p < 1)$ targets the process to the quantile $x_{(p)}$ of interest.

## 4 Asymptotic Properties of the Adjustable Biased Coin Designs

For the *ABCD*, $\{D_n\}$ is the design statistic of interest and the sequence $\{D_n,\ n = 0, 1, \ldots\}$ is a random walk over the integers $\mathbb{Z}$, with transition probabilities $p_j = F(j)$, $q_j = F(-j)$ for all $j \in \mathbb{Z}$ and $\lambda_j = F(j-1)/F(-j)$. Applying results of Section 2, we see that the asymptotic distribution $\pi_D(\cdot)$ of $\{D_n\}$ exists, given by (6), is symmetric, that is $\pi_D(j) = \pi_D(-j)$. From a balance point of view, the asymptotic behaviour of the *ABCD* is identical to the completely randomized design, namely

$$D_n/n \to 0 \text{ almost surely as } n \to \infty.$$

However, we are also interested in randomness. An inverse measure of randomness is *predictability*, that is the expected proportion $G_n$ of correct guesses at stage $n$ using an optimal guessing strategy. For the *ABCD* the probability of guessing correctly at stage $n+1$ equals $F(-|D_n|)$, the probability for the under-represented treatment to be selected, with no preference in case of a tie. Thus $G_n = \frac{1}{n}\sum_{h=0}^{n-1} F(-|D_h|)$ and by (5) we have

$$\lim_{n\to\infty} G_n = \sum_{h=0}^{\infty} F(-h)\pi_{|D|}(h) \quad \text{a.s.,}$$

where $\pi_{|D|}$ is the stationary distribution of $\{|D_n|\}$:

$$\pi_{|D|}(0) = \pi_D(0), \quad \pi_{|D|}(h) = 2\pi_D(h) \quad \text{for all } h > 0.$$

Furthermore the sequences $\{|D_{2m}|\}_{m\in\mathbb{N}}$ and $\{|D_{2m+1}|\}_{m\in\mathbb{N}}$ are stochastically monotone random walks, by Proposition 5. This allows investigation of the speed of convergence of $\mathrm{E}(|D_n|)$, a measure of imbalance of the $ABCD$, as the function $F$ varies and provides a theoretical justification of why even for small values of $n$ ($n \leq 10$), $P(|D_n|)$ is very near its limit, as found in Baldi Antognini and Giovagnoli (2004). Observe that Wei's (8) biased coin method gives rise to a Markov chain whose transition probabilities are nonstationary, so the above results cannot be applied.

## 5 Asymptotic Properties of Up-and-Down Designs

Given a General $U\&D$, by the Fundamental Lemma, $\{X_n\}$ is a random walk over $\mathcal{D}$ with transition probabilities

$$p_k = \alpha_k(1 - F(d_k)) + \alpha'_k F(d_k), \quad q_k = \beta_k F(d_k) + \beta'_k(1 - F(d_k)).$$

We can make the algorithm a more efficient approximation of the asymptotic design, as defined in Mira (2001):

**Proposition 6.** *When $\alpha'_k = \beta'_k = 0$ for all $k$, choosing $\alpha_k$ and $\beta_k$ as large as possible, namely $\alpha_k = \min\{p/(1-p), 1\}$ and $\beta_k = \alpha_k(1-p)/p$, as suggested by Durham and Flournoy's $BCD-II$, minimizes the asymptotic variance of $\bar{N}_n(d_k)$ as an estimator of $\pi(d_k)$.*

This result is proved for $\mathcal{D} = \{d_k,\ k = 0, 1, \ldots, M\}$ in Bortot and Giovagnoli (2004). It can be extended to the case of an unlimited set of design points as follows: under the very mild assumption that $\inf_k F(d_k) = 0$ and $\sup_k F(d_k) = 1$ the asymptotic design exists (see Giovagnoli and Pintacuda (1998a)); such a design is the same for every $\alpha_k$ and $\beta_k$ because of $\lambda_k = \frac{\alpha_k(1-F(d_k))}{\beta_k F(d_k)}$ in (6). Then we can observe that increasing $\alpha_k$ improves the transition probabilities of the Markov chain $\{X_n\}$ w.r.t. Peskun's ordering and apply Tierney's extension (see Tierney (1998)).

*Acknowledgement.* This research was supported partly by the 2003 Research Project: "Asymptotic Properties of Stochastic Processes: methodological tools and physical, medical and engineering applications" of the Department of Statistical Sciences, University of Bologna, and partly by the 2003 Italian PRIN "Statistical methods for stochastic ordering with social, medical and environmental applications".

## References

A.C. Atkinson. Optimum biased coin designs for sequential clinical trials with prognostic factors. *Biometrika*, 69:61–67, 1982.

A. Baldi Antognini. Extensions of Ehrenfest's urn designs for comparing two treatments. In A. Di Bucchianico, H. Läuter, and H.P. Wynn, editors, *mODa 7—Advances in Model-Oriented Design and Analysis (Heeze, June 14-18, 2004)*, Contributions to Statistics, pages 23–31, Heidelberg, 2004. Physica Verlag.

A. Baldi Antognini and A. Giovagnoli. A new "biased coin design" for the sequential allocation of two treatments. *J. R. Statist. Soc. A*, 53:1–14, 2004.

P. Bortot and A. Giovagnoli. Up-and-down experiments of first and second order. *J. Statist. Planning Infcrcncc*, Invited revision, 2004.

Y. Chen. Biased coin design with imbalance tolerance. *Commun. Statist. - Stochastic Models*, 15:953–975, 1999.

S.D. Durham and N. Flournoy. Up-and-down designs I: stationary treatment distributions. In N. Flournoy and W.F. Rosenberger, editors, *Adaptive Designs*, IMS Lecture Notes, pages 139–157. Hayward CA, 1995.

B. Efron. Forcing sequential experiments to be balanced. *Biometrika*, 58: 403–417, 1971.

A. Giovagnoli and N. Pintacuda. Markovian experiments that approximate optimal designs for quantal response curves. *Metron*, 56:77–96, 1998.

A. Giovagnoli and N. Pintacuda. Properties of the frequency distribution induced by a general up-and-down method for estimating quantiles. *J. Statist. Planning Inference*, 74:51–63, 1998.

A. Mira. Ordering and improving the performance of Monte Carlo Markov chains. *Statistical Science*, 16:340–350, 2001.

G.O. Roberts. Markov chain concepts related to sampling algorithms. In W.R. Gilk, S. Richardson, and D.J. Spiegelhalter, editors, *Markov Chain Monte Carlo in Practice*, pages 45–57, London, 1997. Chapman and Hall.

M. Shaked and J.G. Shanthikumar. *Stochastic Orders and their Applications.* Academic Press, 1994.

R. Smith. Sequential treatment allocation using biased coin designs. *Journal of the Royal Statistical Society B*, 46:519–543, 1984.

J.F. Soares and C.F.J. Wu. Some restricted randomization rules in sequential designs. *Commun. Statist. - Theor. Meth*, 12:2017–2034, 1983.

L. Tierney. A note on Metropolis-Hastings kernels for general state spaces. *Ann. Appl. Probab.*, 8:1–9, 1998.

L.J. Wei. The adaptive biased coin design for sequential experiments. *Annals of Statistics*, 6:92–100, 1978.

# Lower Bounds on Efficiency Ratios Based on $\Phi_p$-Optimal Designs

R. Harman

Department of Probability and Statistics
Faculty of Mathematics, Physics and Informatics
Comenius University, Bratislava, Slovakia
harman@fmph.uniba.sk

**Summary.** Suppose that we intend to perform linear regression experiments with uncorrelated errors according to a given asymptotic design $\xi$. The problem which we address is the question of performance-stability of $\xi$ under change of optimality criterion. More precisely, we describe a method of how to calculate lower bounds on the minimal possible efficiency of $\xi$ with respect to any orthogonally invariant information function. The bounds constructed depend only on the eigenvalues of the information matrix of a known regular $\Phi_p$-optimal design. We also point out some theoretical consequences of the bounds and illustrate the use of the results on the model of spring balance weighing.

**Key words:** $\Phi_p$-optimal design, efficiency, orthogonally invariant criteria, spring balance weighing.

## 1 Introduction

A common practice in design of experiments is to select a fixed criterion $\Phi$ measuring the "largeness" of information matrices and giving us an unambiguous method of choosing a best performance design which we then call $\Phi$-optimal. Often, though, realistic considerations force us to admit that there are more criteria which reflect useful statistical aspects of designs. In this situation, a natural question is whether the performance of the $\Phi$-optimal design is at least acceptable with respect to other reasonable criteria.

One of the first systematic attempts to solve a similar problem appeared in Kiefer (1975) and Galil and Kiefer (1977), where the notion of (in)efficiency ratio was used to reflect the efficiency of a given $\Phi_r$-optimal design relative to different criteria of $\Phi_p$-optimality. The obtained theoretical bounds on the $E$-efficiency of $\Phi_r$-optimal designs were very general but pessimistic in comparison to the efficiencies numerically computed for a specific model.

The main aim of this paper is to describe a method of how to calculate lower bounds on the minimal efficiency ratio of any given design $\xi$ under all orthogonally invariant criteria. These bounds depend only on the eigenvalues of the information matrix associated with a known $\Phi_p$-optimal design and are simple to calculate numerically. Moreover, the new bounds on the $E$-efficiency imply the ones in Galil and Kiefer (1977).

We consider the usual linear regression model on a compact experimental domain $X \subseteq \mathbb{R}^d$. For each design point $x \in X$ we can observe a random variable $Y = f^T(x)\beta + \varepsilon$, where $f : X \to \mathbb{R}^m$ is a vector of linearly independent continuous regression functions, $\beta \in \mathbb{R}^m$ is an unknown vector of parameters, and $\varepsilon$ is an unobservable random error. For different observations, the errors are uncorrelated, with zero mean and the same variance which is assumed to be 1 without loss of generality.

For a model of this form, an asymptotic design is a probability measure $\xi$ finitely supported on $X$ (see *e.g.*, (Pázman, 1986, Section II.2 ) or (Pukelsheim, 1993, Section 1.24)). The symbol $\Xi$ denotes the set of all designs on $X$. The performance of a design $\xi \in \Xi$ is based on the information matrix associated with $\xi$, which is a positive semidefinite matrix

$$\mathbf{M}(\xi) = \sum_{x \in X; \xi(x) > 0} \xi(x) f(x) f^T(x).$$

On the set $\mathcal{S}_+^m$ of all positively semidefinite matrices of type $m \times m$ we define an optimality criterion which measures the largeness of information matrices, that is the quality of corresponding designs. An information function is a criterion $\Phi : \mathcal{S}_+^m \to [0, \infty)$ which is isotonic with respect to the Loewner ordering, concave, upper semicontinuous, positively homogeneous and nonconstant (Pukelsheim, 1993, Section 5.8).

In the sequel, we will work with the class $OIC$ of all orthogonally invariant criteria, by which we understand information functions $\Phi$ with a property of orthogonal invariance: $\Phi(\mathbf{U}\mathbf{M}\mathbf{U}^T) = \Phi(\mathbf{M})$ for all $\mathbf{M} \in \mathcal{S}_+^m$ and orthogonal matrices $\mathbf{U}$, that is $\Phi(\mathbf{M})$ depends only on the eigenvalues of $\mathbf{M}$.

The class $OIC$ contains all the concave versions of the Kiefer's criteria of $\Phi_p$-optimality for $p \in [-\infty, 1]$ parameterized as in (Pukelsheim, 1993, Chapter 6). In particular, $T$-, $D$-, and $E$ -optimality criteria are defined $\Phi_1(\mathbf{M}) = m^{-1}\,\mathrm{tr}(\mathbf{M})$, $\Phi_0(\mathbf{M}) = (\det(\mathbf{M}))^{1/m}$, and $\Phi_{-\infty}(\mathbf{M}) = \lambda_1(\mathbf{M})$, that is the minimal eigenvalue of $\mathbf{M}$. The criterion of $A$-optimality is defined $\Phi_{-1}(\mathbf{M}) = m(\mathrm{tr}(\mathbf{M}^{-1}))^{-1}$ for a regular $\mathbf{M}$ and $\Phi_{-1}(\mathbf{M}) = 0$ for a singular $\mathbf{M}$. Notice that $OIC$ also contains all the characteristic polynomial criteria written in their positively homogeneous and concave form (see *e.g.*, Rodríguez-Díaz and López-Fidalgo (2003)). Moreover, any convex combination, or a minimum of a finite set of criteria from $OIC$ is again an orthogonally invariant criterion.

A design $\xi^*$ is $\Phi$-optimal with respect to a given criterion $\Phi \in OIC$ if and only if $\Phi(\mathbf{M}(\xi^*)) = \max\{\Phi(\mathbf{M}(\xi)) : \xi \in \Xi\}$. The matrix $\mathbf{M}(\xi^*)$ is then called $\Phi$-optimal and $\Phi(\mathbf{M}(\xi^*))$ is called the $\Phi$-optimal value. The $\Phi$-efficiency of an arbitrary design $\xi$ is defined as (Pukelsheim (1993), p. 132)

$$\mathrm{eff}(\xi|\Phi) = \Phi(\mathbf{M}(\xi))/\Phi(\mathbf{M}(\xi^*)).$$

By the minimal efficiency ratio of $\xi$ we understand the value $\min_{\Phi \in OIC} \mathrm{eff}(\xi|\Phi)$.

## 2 Bounds on the $E_k$-optimal Values Based on the Eigenvalues of $\Phi_p$-optimal Information Matrices

Following Harman (2004) we define the criterion $\Phi_{E_k}$ of $E_k$ -optimality by

$$\Phi_{E_k}(\mathbf{M}) = \sum_{i=1}^{k} \lambda_i(\mathbf{M})$$

for $k = 1, \ldots, m$ and all $\mathbf{M} \in \mathcal{S}_+^m$, where $\lambda_1(\mathbf{M}) \leq \lambda_2(\mathbf{M}) \leq \ldots \leq \lambda_m(\mathbf{M})$ are the eigenvalues of $\mathbf{M}$. Notice that $\Phi_{E_1}$ is the criterion of $E$-optimality, $\Phi_{E_m}$ is a constant multiple of the criterion of $T$-optimality, and $\Phi_{E_k} \in OIC$ for any $k$. The criteria of $E_k$-optimality play an important role for minimal efficiency ratios, because it can be proved that (see Harman (2004)):

**Theorem 1.** *For any $\xi \in \Xi$ we have*

$$\min_{\Phi \in OIC} \text{eff}(\xi|\Phi) = \min_{k=1,\ldots,m} \text{eff}(\xi|\Phi_{E_k}).$$

Clearly, the $E_k$-efficiencies $\text{eff}(\xi|\Phi_{E_k})$ can be directly calculated once we know the $E_k$-optimal values

$$v_k = \max_{\xi \in \Xi} \Phi_{E_k}(\mathbf{M}(\xi)).$$

Although the $E_k$-optimal values are usually difficult to find, for an informative lower bound on the minimal efficiency ratio, upper bounds on $v_k$'s sometimes suffice. The main result of this article is that such bounds can be calculated from the eigenvalues of any regular $\Phi_p$-optimal information matrix.

**Theorem 2.** *Let $p \in (-\infty, 1]$, let $\xi_p^*$ be a $\Phi_p$-optimal design and let $0 < \lambda_1 \leq \lambda_2 \leq \ldots \leq \lambda_m$ be the eigenvalues of $\mathbf{M}(\xi_p^*)$. Then for any $k \in \{1, \ldots, m\}$:*

$$v_k \leq \max_{r=1,\ldots,k} \frac{r \sum_{i=1}^{m} \lambda_i^p}{\sum_{i=k+1-r}^{m} \lambda_i^{p-1}}.$$

We remark that a similar bound as the one in Theorem 2 can also be based on designs which are optimal with respect to other differentiable orthogonally invariant criteria, than the Kiefer's criteria of $\Phi_p$-optimality.

## 3 Bounds on the $E$-efficiency and the Minimal Efficiency Ratio of $\Phi_p$-optimal Designs

The formula given in Theorem 2 substantially simplifies for $k = 1$, which entails the following bounds on the $E$-efficiency of $\Phi_p$-optimal designs.

**Theorem 3.** *Let $p \in (-\infty, 1]$, let $\xi_p^*$ be a $\Phi_p$-optimal design and let $0 < \lambda_1 \leq \lambda_2 \leq \ldots \leq \lambda_m$ be the eigenvalues of $\mathbf{M}(\xi_p^*)$. Then*

$$\text{eff}\left(\xi_p^* | \Phi_{E_1}\right) \geq \frac{\sum\limits_{i=1}^{m} (\lambda_i/\lambda_1)^{p-1}}{\sum\limits_{i=1}^{m} (\lambda_i/\lambda_1)^{p}}.$$

For a $D$-optimal design $\xi_0^*$, the bound given in the previous theorem has a particularly simple form: $\text{eff}\left(\xi_0^* | \Phi_{E_1}\right) \geq \frac{1}{m} \sum_{i=1}^{m} \lambda_1/\lambda_i$. Moreover, for fixed $p$ the bound from Theorem 3 can not drop under a certain positive level depending only on the multiplicity of the minimal eigenvalue of $\mathbf{M}(\xi_p^*)$. This leads to the following generalization of Theorem 5.1. from Galil and Kiefer (1977).

**Theorem 4.** *Let $m \geq 1$, $p \in (-\infty, 0]$ and let $\xi_p^*$ be a $\Phi_p$-optimal design. If $h$ is the multiplicity of the minimal eigenvalue of $\mathbf{M}(\xi_p^*)$ then*

$$\text{eff}\left(\xi_p^* | \Phi_{E_1}\right) \geq \min_{x \in [0,1]} \frac{1 + (m/h - 1)x^{1-p}}{1 + (m/h - 1)x^{-p}}$$

*In particular, for a $D$-optimal design $\xi_0^*$ and for an $A$-optimal design $\xi_{-1}^*$:*

$$\text{eff}\left(\xi_0^* | \Phi_{E_1}\right) \geq \frac{h}{m}, \text{eff}\left(\xi_{-1}^* | \Phi_{E_1}\right) \geq \frac{2}{1 + \sqrt{m/h}}.$$

Although for $h = 1$ Theorem 4 gives the same results as Theorem 5.1 from Galil and Kiefer (1977), the difference in favour of Theorem 4 can be considerable in the case of a high multiplicity $h$ of the minimal eigenvalue.

By Theorem 2, every known $\Phi_p$-optimal design yields upper bounds on the $E_k$-optimal values and, consequently, lower bounds on the $E_k$-efficiencies of any given design $\xi$. By Theorem 1, we can thus calculate a lower bound on the efficiency of $\xi$ which is simultaneously valid with respect to all orthogonally invariant criteria. Moreover, Theorem 1 in combination with Theorem 2 lead to several general theoretical conclusions regarding the minimal efficiency ratio of $\Phi_p$-optimal designs themselves. For example, by Theorem 5.1 from Galil and Kiefer (1977) we know that the $E$-efficiency of a $D$-optimal design is at least $1/m$. We can strengthen this result as follows.

**Theorem 5.** *The efficiency of a $D$-optimal design $\xi_0^*$ is at least $1/m$ with respect to any orthogonally invariant criterion.*

We remark that the previous property of $D$-optimality is unique in the class of criteria $\Phi_p$, because for $p \neq 0$ and a $\Phi_p$-optimal design $\xi_p^*$ not only the minimal efficiency ratio, but even the $D$-efficiency of $\xi_p^*$ can be arbitrarily low in some models (cf. Galil and Kiefer (1977)).

The following theorems give sufficient conditions on "universal" optimality of $\Phi_p$-optimal designs under the class $OIC$.

**Theorem 6.** *If a design $\xi_p^*$ is $\Phi_p$-optimal for some $p \in (-\infty, 1]$ and if $\mathbf{M}(\xi_p^*)$ is proportional to the identity matrix, then $\xi_p^*$ is optimal with respect to all orthogonally invariant criteria.*

**Theorem 7.** *If a design $\xi^*$ is $\Phi_p$-optimal for all $p \in (-\infty, 1)$ and if $\mathbf{M}(\xi^*)$ has at most two distinct eigenvalues, then $\xi^*$ is optimal with respect to all orthogonally invariant criteria.*

Notice that Theorem 7 directly implies that the equispaced support design $\xi^*$ in the standard trigonometric regression is universally optimal under all criteria from $OIC$, because it is known that $\xi^*$ is $\Phi_p$-optimal for all $p \in (-\infty, 1)$ and $\mathbf{M}(\xi^*)$ has only two distinct eigenvalues (see *e.g.*, (Pukelsheim, 1993, Section 9.16), cf. Rodríguez-Díaz and López-Fidalgo (2003) and the proof based on subgradients of $\Phi_{E_k}$'s in Harman (2004).)

# 4 Example: the Minimal Efficiency Ratio of the $D$-optimal Design for the Model of Spring Balance Weighing

Consider the $m$-way multivariate linear model of the first degree over the unit cube

$$Y = x_1\beta_1 + \ldots + x_m\beta_m + \varepsilon,$$

where $x = (x_1, \ldots, x_m)^T \in X = \{0,1\}^m$. This model is called the model of spring balance weighing if $\beta_1, \ldots, \beta_m$ represent unknown weights, and the components of $x$ mean the presence or the absence of items in a weighing on a balance measuring weight by the tension of a spring.

The model in consideration has been extensively analyzed from the perspective of optimal design, and all $\Phi_p$-optimal designs are known (Cheng (1987), (Pukelsheim, 1993, Section 14.10), see also Dette and Studden (1993)). To illustrate the use of the bounds given by Theorems 2 and 4 we will only employ the $D$-optimal design.

First, if $m$ is even, then the $D$-optimal design $\xi_0^*$ puts the same weight to all the points from $X$ with either $m/2$ or $m/2+1$ components equal to 1. One can directly calculate that the information matrix $\mathbf{M}(\xi_0^*)$ has the smallest eigenvalue $(m+2)/(4m+4)$ of multiplicity $m-1$ and the largest eigenvalue $(m+2)/4$ of multiplicity 1. Secondly, if $m$ is odd, then the $D$-optimal design $\xi_0^*$ is uniform on all the unit cube vertices with $(m+1)/2$ components equal to 1. In this case, the information matrix $\mathbf{M}(\xi_0^*)$ has the smallest eigenvalue $(m+1)/4m$ of multiplicity $m-1$ (if $m > 1$) and the largest eigenvalue $(m+1)^2/4m$ of multiplicity 1.

Notice that Theorem 4 immediately entails that the $E$-efficiency of $\xi_0^*$ is at least $(m-1)/m$, that is very high for large dimensions $m$. More importantly, using Theorem 2 we are able to find the *exact* minimal efficiency ratio of the

$D$-optimal design. Obviously, the vector $1_m \in X$ consisting of all ones has the largest norm, which means that the design which puts mass 1 to $1_m$ is $E_m$-optimal (that is formally $T$-optimal, cf. with Pukelsheim (1993), Section 9.15.) and the $E_m$-optimal value is $m$. This gives $\mathrm{eff}(\xi_0^*|\Phi_{E_m}) = (1/2)\ (m+2)/(m+1)$ for an even $m$ and $\mathrm{eff}(\xi_0^*|\Phi_{E_m}) = (1/2)\ (m+1)/m$ for an odd $m$.

For $k < m$, lower bounds on the $E_k$-efficiencies of $\xi_0^*$ can be calculated from the eigenvalues $\lambda_1 \leq \ldots \leq \lambda_m$ of $\mathbf{M}(\xi_0^*)$ making use of Theorem 2. For any $m > 1$ (odd or even) we thus obtain:

$$\mathrm{eff}(\xi_0^*|\Phi_{E_k}) = \frac{1}{v_k}\sum_{i=1}^{k}\lambda_i \geq \min_{r=1,\ldots,k}\frac{\sum_{i=1}^{k}\lambda_i\sum_{i=k+1-r}^{m}\frac{1}{\lambda_i}}{r\cdot m} \geq \frac{m}{m+1},$$

where checking the last inequality is a matter of elementary algebra. Consequently $\mathrm{eff}(\xi_0^*|\Phi_{E_m}) \leq \mathrm{eff}(\xi_0^*|\Phi_{E_k})$ for all $1 \leq k \leq m$, which means

$$\min_{\Phi\in OIC}\mathrm{eff}(\xi_0^*|\Phi) = \begin{cases}\frac{1}{2}\frac{m+2}{m+1} & \text{if } m \text{ even}\\ \frac{1}{2}\frac{m+1}{m} & \text{if } m \text{ odd}\end{cases}$$

by Theorem 1. Hence, the efficiency of the $D$-optimal design is more than 50% with respect to any orthogonally invariant criterion and for any dimension of the model. Moreover, from among all the orthogonally invariant criteria, the $D$-optimal design performs worst with respect to the criterion of $T$-optimality.

Using the model of spring balance weighing we can also make some additional comments on Theorems 6 and 7. For $m = 2$, the information matrix of the $E$-optimal design $\xi_{-\infty}^*$ is proportional to the identity matrix, but $\xi_{-\infty}^*$ is *not* optimal with respect to all criteria from $OIC$ (*e.g.*, $\xi_{-\infty}^*$ is not $D$-optimal, see (Pukelsheim, 1993, Section 8.6)). This means that Theorem 6 cannot be extended to include the case of $E$-optimality ($p = -\infty$). It is also known that for odd $m > 1$ the design $\xi_0^*$ is $\Phi_p$-optimal for all $p \in [-\infty, 0]$, the information matrix $\mathbf{M}(\xi_0^*)$ has only two distinct eigenvalues, yet $\xi_0^*$ is not optimal with respect to all orthogonally invariant criteria (*e.g.*, $\xi_0^*$ is not $T$-optimal, Pukelsheim (1993), p. 377). In combination with Theorem 7 this fact means that $\Phi_p$-optimality for all the most studied $p \in [-\infty, 0]$ is generally much weaker than $\Phi_p$-optimality for the complete class including also the less popular positive values of $p$.

## 5 Appendix - Proofs

In this section, we will give proofs of Theorems 2 and 7. Theorem 3 is a direct consequence of Theorem 2 and the proof of Theorem 4 can be based on Theorem 3 and standard techniques of multidimensional calculus. Theorems 5 and 6 are simple corollaries of Theorems 1 and 2.

We now define the following cones: $\mathbb{R}_+^m = \{x \in \mathbb{R}^m : x_1 \geq 0, \ldots, x_m \geq 0\}$, $\mathbb{R}_{++}^m = \{x \in \mathbb{R}^m : x_1 > 0, \ldots, x_m > 0\}$, and $\mathbb{R}_{\leq}^m = \{x \in \mathbb{R}^m : x_1 \leq \ldots \leq x_m\}$.

Next, let $0_k = (0, \ldots, 0)^T \in \mathbb{R}^k$ and $1_k = (1, \ldots, 1)^T \in \mathbb{R}^k$. If $\lambda_1(\mathbf{M}) \leq \lambda_2(\mathbf{M}) \leq \ldots \leq \lambda_m(\mathbf{M})$ are the eigenvalues of $\mathbf{M} \in \mathcal{S}_+^m$, then we will use the $m$-dimensional vectors

$$\lambda_\uparrow(\mathbf{M}) = (\lambda_1(\mathbf{M}), \ldots, \lambda_m(\mathbf{M}))^T \text{ and } \lambda_\downarrow(\mathbf{M}) = (\lambda_m(\mathbf{M}), \ldots, \lambda_1(\mathbf{M}))^T.$$

For $\mu \in \mathbb{R}_{++}^m$ we define $z_0(\mu) = 0_m$, $z_m(\mu) = \left(\sum_{i=1}^m \mu_i\right)^{-1} 1_m$, and $z_s(\mu) = \left(\sum_{i=m-s+1}^m \mu_i\right)^{-1} \left(0_{m-s}^T, 1_s^T\right)^T$ for $s = 1, \ldots, m-1$.

**Lemma 1.** *Let $\mu \in \mathbb{R}_{++}^m$. Then*

$$\left\{\lambda \in \mathbb{R}_+^m \cap \mathbb{R}_\leq^m : \lambda^T \mu \leq 1\right\} = \operatorname{conv}\left\{z_0(\mu), \ldots, z_m(\mu)\right\}.$$

*Proof.* Let $\mu \in \mathbb{R}_{++}^m$, $\lambda \in \mathbb{R}_+^m \cap \mathbb{R}_\leq^m$, and $\lambda^T \mu \leq 1$. Set $\alpha_0 = 1 - \lambda^T \mu$, $\alpha_m = \left(\sum_{i=1}^m \mu_i\right) \lambda_1$ and $\alpha_s = \left(\sum_{i=m-s+1}^m \mu_i\right) \left(\lambda_{m-s+1} - \lambda_{m-s}\right)$ for all $s = 1, \ldots, m-1$. It is simple to check that $\alpha_s \geq 0$ for all $s = 0, 1, \ldots, m$, $\sum_{s=0}^m \alpha_s = 1$, and $\sum_{s=0}^m \alpha_s z_s(\mu) = \lambda$. This proves that the set on the left-hand side is included in the set on the right-hand side. The opposite inclusion is obvious. □

**Proof of Theorem 2.** Let $p \in (-\infty, 1]$, $\xi_p^*$ be a $\Phi_p$-optimal design, $\mathbf{M}_p = \mathbf{M}\left(\xi_p^*\right)$, and $\lambda = \lambda_\uparrow(\mathbf{M}_p) \in \mathbb{R}_{++}^m$. The function $\Phi_p$ is differentiable in $\mathbf{M}_p$ with gradient $\nabla\Phi_p(\mathbf{M}_p) = \frac{1}{m}\Phi_p^{1-p}(\mathbf{M}_p)\mathbf{M}_p^{p-1}$. Let $\mathbf{N}_k$ be any $\Phi_{E_k}$-optimal information matrix. Using a lemma of Fan (see *e.g.*, Borwein and Lewis (2000) Lemma 1.2.1. and Exercise 13 p. 12) and the subgradient theorem ((Pukelsheim, 1993, Section 7.4)) we obtain

$$\lambda_\uparrow^T(\mathbf{N}_k)\,\lambda_\downarrow(\nabla\Phi_p(\mathbf{M}_p)) \leq \operatorname{tr}\left(\mathbf{N}_k \nabla\Phi_p(\mathbf{M}_p)\right) \leq \operatorname{tr}\left(\mathbf{M}_p \nabla\Phi_p(\mathbf{M}_p)\right) = \Phi_p(\mathbf{M}_p).$$

Therefore $\lambda_\uparrow^T(\mathbf{N}_k)\,\mu \leq 1$, where

$$\mu = \Phi_p^{-1}(\mathbf{M}_p)\lambda_\downarrow(\nabla\Phi_p(\mathbf{M}_p)) = \left(\sum\nolimits_{i=1}^m \lambda_i^p\right)^{-1} \left(\lambda_1^{p-1}, \ldots, \lambda_m^{p-1}\right)^T \in \mathbb{R}_{++}^m.$$

By Lemma 1 we obtain

$$\lambda_\uparrow(\mathbf{N}_k) \in \operatorname{conv}\left\{z_0(\mu), \ldots, z_m(\mu)\right\}.$$

For $\gamma \in \mathbb{R}_+^m$ let $\varphi_k(\gamma)$ be the sum of the $k$ smallest components of $\gamma$. Notice that the function $\varphi_k$ is linear and nonnegative on the convex and compact set $\operatorname{conv}\left\{z_0(\mu), \ldots, z_m(\mu)\right\} \subset \mathbb{R}_+^m \cap \mathbb{R}_\leq^m$, which together with $z_0(\mu) = 0_m$ entails that the maximum of $\varphi_k$ is attained on $z_1(\mu)$, $z_2(\mu)$, …or $z_m(\mu)$. Therefore

$$\begin{aligned} v_k = \Phi_{E_k}(\mathbf{N}_k) &= \varphi_k(\lambda_\uparrow(\mathbf{N}_k)) \leq \max_{s=1,\ldots,m} \varphi_k(z_s(\mu)) \\ &= \sum\nolimits_{i=1}^m \lambda_i^p \max_{s=1,\ldots,m} \varphi_k\left(z_s\left(\lambda_1^{p-1}, \ldots, \lambda_m^{p-1}\right)\right) \\ &= \sum\nolimits_{i=1}^m \lambda_i^p \max_{r=1,\ldots,k} \left(r\left(\sum\nolimits_{i=k+1-r}^m \lambda_i^{p-1}\right)^{-1}\right). \end{aligned}$$

□

**Proof of Theorem 7.** Let $\xi^*$ be $\Phi_p$-optimal for all $p \in (-\infty, 1)$ and let $\mathbf{M}(\xi^*)$ have at most two distinct eigenvalues. If all the eigenvalues of $\mathbf{M}(\xi^*)$ are same then Theorem 7 follows from Theorem 6. Therefore we can assume that $\lambda_\uparrow(\mathbf{M}(\xi^*)) = (\lambda, \ldots, \lambda, \gamma, \ldots, \gamma)^T$, $\lambda < \gamma$ and the multiplicity of $\lambda$ is $s \leq m-1$. Denote by $u_k(p)$ the upper bound on $v_k$ given by Theorem 2 taken in $p \in (-\infty, 1)$. It is possible to verify that:

1. For $k \in \{1, \ldots, s\}$: $\lim_{p \to -\infty} u_k(p) = k\lambda$
2. For $k \in \{s+1, \ldots, m-1\}$: $u_k(1 - \log_{\gamma/\lambda} \frac{m-s}{k-s}) = s\lambda + (k-s)\gamma$
3. $\lim_{p \to 1} u_m(p) = s\lambda + (m-s)\gamma$.

By the assumption of Theorem 7 we thus have $v_k \leq \inf_{p \in (-\infty,1)} u_k(p) = \Phi_{E_k}(\mathbf{M}(\xi^*))$ which means that $\xi^*$ must bc $E_k$-optimal for all $k = 1, \ldots, m$. Using Theorem 1 this implies optimality of $\xi^*$ with respect to all orthogonally invariant criteria. □

*Acknowledgement.* The author would like to thank anonymous referees for helpful comments. This work was supported by VEGA grant No. 1/0264/03 of the Slovak Agency.

## References

J.M. Borwein and A.S. Lewis. *Convex analysis and nonlinear optimization, Theory and examples.* Springer-Verlag, New York, 2000.

C.S. Cheng. An application of the Kiefer-Wolfowitz equivalence theorem to a problem in Hadamard transform optics. *Ann. Stat.*, 15:1593–1603, 1987.

H. Dette and W.J. Studden. Geometry of $E$-optimality. *Ann. Stat.*, 21: 416–433, 1993.

Z. Galil and J. Kiefer. Comparison of simplex designs for quadratic mixture models. *Technometrics*, 19(4):445–453, 1977.

R. Harman. Minimal efficiency of designs under the class of orthogonally invariant information criteria. *Metrika*, 2004.

J. Kiefer. Optimal design: Variation in structure and performance under change of criterion. *Biometrika*, 62(2):277–288, 1975.

A. Pázman. *Foundations of optimum experimental design.* Reidel, Dordrecht, 1986.

F. Pukelsheim. *Optimal design of experiments.* John Wiley & Sons, New York, 1993.

J.M. Rodríguez-Díaz and J. López-Fidalgo. A bidimensional class of optimality criteria involving $\Phi_p$ and characteristic criteria. *Statistics*, 37:325–334, 2003.

# On a Functional Approach to Locally Optimal Designs

V.B. Melas

St.-Petersburg State University, University av., 28, St.Petersburg, 198904, Russia, v.melas@pobox.spbu.ru

**Summary.** This paper concerns locally optimal experimental designs for nonlinear regression models. It is based on the functional approach introduced in Melas (1978). In this approach locally optimal design points and weights are studied as implicitly given functions of the nonlinear parameters included in the model. Representing these functions in a Taylor series enables analytical solution of the optimal design problem for many nonlinear models. A wide class of such models is here introduced. It includes, in particular, the three-parameter logistic model. For this model we construct the analytical solution and use it for studying the efficiency of locally optimal designs. As a criterion of optimality the well known $D$-criterion is considered.

**Key words:** nonlinear models, locally optimal experimental designs, $D$-criterion, functional approach, three-parameter logistic distribution, microbiological applications.

## 1 Introduction

Modern optimal design theory, established to a great extent by J. Kiefer, see Kiefer (1985), relates mainly to linear (in parameters) regression models. This paper is devoted to the analytical study of locally optimal designs for nonlinear (in parameters) regression models. Up to now such designs have been investigated mainly for cases where these designs can be found in an explicit form or as a result of numerical procedures, see Han and Chaloner (2003) and references in that paper. A functional approach has been introduced by Melas (1978) for hyperexponential models. It is based on the study of optimal design points as implicitly given functions of values of parameters. Recently this approach was applied to rational models Melas (2001). Convenient recurrent formulas for calculating Taylor coefficients of implicit functions were introduced in Dette, Melas, and Pepelyshev (2004a). The present paper develops the functional approach for a wider class of nonlinear models.

The main results of the present paper consist of the following. First, we establish that for a wide class of nonlinear models support points of locally $D$-optimal designs are real analytic functions of the nonlinear parameters. Secondly, we demonstrate for an example that the recurrent formulas mentioned above allow us to calculate the designs with a high precision. As example we take three parameters logistic model. Thirdly, we use the Taylor expansions to estimate the minimal efficiency of locally optimal designs if the nonlinear parameters vary in a given set. Also we show that the minimal efficiency can be substantially improved by an optimal choice of the initial values inside the set. In this way the locally optimal designs become very close to maximin efficient designs Müller (1995), Dette, Melas, and Wong (2004b). A formal outline of the problem and a basic equation for the support points are given in Section 2. The main analytical results are described in Section 3. Section 4 contains the example.

## 2 Outline of the Problem

Let the experimental results $y_1, \ldots, y_N \in \mathbb{R}^1$ be described by the model

$$y_j = \eta(x_j, \Theta) + \varepsilon_j, \; j = 1, \ldots, N, \tag{1}$$

where $\eta(x, \Theta)$ is a function of a known form, $\Theta = (\theta_1, \ldots, \theta_m)^T$ is the vector of unknown parameters, $x_j \in \mathcal{X}$ for a given set $\mathcal{X}$, $\{\varepsilon_j\}$ are random errors such that

$$E\varepsilon_j = 0, \; E\varepsilon_i\varepsilon_j = \delta_{ij}\sigma^2 h^2(x_i), \; i, j, = 1, \ldots, N, \tag{2}$$

$\delta_{ij}$ is the Kronecker symbol, $\sigma^2$ is unknown constant, and $h(x)$ is a known function, $h(x) > 0$, $x \in \mathcal{X}$.

A discrete probability measure given by the table

$$\xi = \begin{pmatrix} x_1, & \ldots, & x_n \\ \omega_1, & \ldots, & \omega_n \end{pmatrix},$$

where $x_i \in \mathcal{X}$, $x_i \neq x_j \; (i \neq j)$, $\omega_i > 0$, $\sum_{i=1}^{n}, \omega_i = 1$ will be called, as usual, an approximate experimental design.

Let approximately $\omega_i \; N$ experiments be performed at the point $x_i$ , $i = 1, \ldots, n$ according to a design $\xi$, and now consider the least squares estimate $\Theta_{(N)}$ of the parameter vector $\Theta$. Let us denote by $\Theta_{tr}$ the true value of the parameter vector $\Theta$ appearing in our model (1)-(2). It is proved in Jennrich (1969) that under some regularity conditions that $\Theta_{(N)} \to \Theta_{tr}$ almost surely and the vector $(\Theta_{(N)} - \Theta_{tr})/\sqrt{N}$ has an asymptotically normal distribution with zero expectation and the variance matrix

$$\mathcal{D} = \sigma^2 M^{-1}(\xi, \Theta),$$

where $M(\xi, \Theta)$ is the information matrix,

$$M(\xi, \Theta) = \left( \sum_{\ell=1}^{n} f_i(x_\ell, \Theta) f_j(x_\ell, \Theta) \omega_{\ell h}^{-2}(x_\ell) \right)_{ij=1}^{m},$$

$f_i(x_\ell, \Theta) = \frac{\partial \eta(x_\ell, \Theta)}{\partial \theta_i}$, $\Theta = \Theta_{tr}$.

A design $\xi^*$ that maximizes $\det M(\xi, \Theta)$ under a fixed value $\Theta = \Theta^{(0)}$ in the class of all approximate designs will be called a locally optimal design. Usually a locally optimal design depends only on a part of parameters (see Section 4). Without loss of generality assume that these parameters are $\theta_{m-r+1}, \ldots, \theta_m$ and denote $\Theta_1 = (\theta_1, \ldots, \theta_{m-r})^T$, $\Theta_2 = (\theta_{m-r+1}, \ldots, \theta_m)^T$. Our next goal is to study the dependence of $\xi^*$ on $\Theta_2^0$. Assuming that $\Theta_1$ is fixed, we will consider the matrix $M(\xi, \Theta_2) = M(\xi, \Theta)$.

In many practical problems $\mathcal{X} = [a, b]$ and we will restrict our attention to this case. Without loss of generality assume that

$$a \leq x_1 < x_2 < \ldots < x_n \leq b.$$

A triple $(n_1, n_2, n_3)$, where $n_1 (n_3)$ is the number of support points at the left (right) bound, $n_1, n_3 = 0$ or $1$, $n_2 = n - n_1 - n_3$, will be called the type of design.

Let $\Theta_2 \in \Omega$, where $\Omega$ is a given open one-connected set of possible values of the parameters.

**Assumption 1.** The functions

$$\frac{f_1(x, \Theta)}{h(x)}, \ldots, \frac{f_m(x, \Theta)}{h(x)} \tag{3}$$

generate an extended Chebyshev system of the first order on $[a, b]$ for any $\Theta_2 \in \Omega$ (see (Karlin and Studden, 1966, Chapter 1), for the definition).

**Assumption 2.** The functions (3) are real analytic functions with fixed $\Theta_1$ and $(x, \Theta_2) \in (a, b) \times \Omega$.

Consider designs that are locally $D$-optimal in the class of designs with minimal support (that is with $n = m$). In this paper these designs will be called saturated locally $D$-optimal. Such designs often prove to be locally $D$-optimal in the class of all approximate designs.

**Assumption 3.** Assume that with $\Theta_2^0 \in \Omega$ all saturated locally $D$-optimal designs are of one and the same type $(n_1, n_2, n_3)$, $n_1 + n_2 + n_3 = m$.

**Assumption 4.** For any $\Theta_2^0 \in \Omega$ there exists a unique saturated locally $D$-optimal design.

Note that Assumptions 1-4 hold for many models used in microbiology (see Dette, Melas, Pepelyshev, and Strigul (2003) and an example in Section 4).

We will consider the case $n_1 = 0$ and $n_3 = 1$; all other cases can be considered in a similar way.

## 3 Basic Analytical Results

It is easy to check (see Fedorov (1972)) that all weight coefficients in a saturated locally $D$-optimal design are the same. Thus we can restrict our attention by designs

$$\xi_\tau = \begin{pmatrix} x_1 & \dots & x_{m-1} & b \\ 1/m & \dots & 1/m & 1/m \end{pmatrix},$$

where $\tau = (x_1, \dots, x_{m-1})^T$, $a < x_1 < \dots < x_{m-1} < b$.

Let $\Psi$ be a real-analytic function such that

$$\Theta_2 \to z = \Psi(\Theta_2) \in \mathbb{R}^r$$

is one-to-one and therefore the inverse function $\Psi^{-1}(z)$ is well-defined on the set

$$Z = \{z \mid \exists \Theta_2 \in \Omega \colon z = \Psi(\Theta_2)\}.$$

Define the function

$$\varphi(\tau, z) = [\det M(\xi_\tau, \Theta(z)]^{1/m},$$

where $\Theta^T(z) = (\Theta_1^T, (\Psi^{-1}(z))^T)$ if $r < m$ and $\Theta(z) = \Psi^{-1}(z)$ if $r = m$.

If $\xi = \xi_\tau$ is a saturated locally $D$-optimal design, then

$$\frac{\partial \varphi(\tau, z)}{\partial \tau_i} = 0, \; i = 1, \dots, m-1.$$

This system of equations implicitly determines the support points of a locally saturated $D$-optimal design as functions of the vector $z = \Psi(\Theta_2)$. Denote by $J(\tau, z)$ the Jacobi matrix

$$J(\tau, z) = \left( \frac{\partial^2}{\partial \tau_i \partial \tau_j} \varphi(\tau, z) \right)_{i,j=1}^{m-1}$$

and set

$$g(\tau, z) = \frac{\partial}{\partial \tau} \varphi(\tau, z).$$

Our main result is the following theorem.

**Theorem 1.** *Let Assumptions 1–4 be satisfied for model (1)–(2), $n_1 = 0$, $n_3 = 1$ and the function $\Psi$ be described above. Let $z_{(0)} \in Z$ and $\tau_{(0)}$ be such that $\xi_{\tau_0}$ is a saturated locally D-optimal design for $\Theta^0 = \Theta(z_{(0)})$. Then $\det J(\tau_{(0)}, z_{(0)}) \neq 0$ and there exists a real-analytic vector function $\tau^*(z)$ on $Z$ and such that*

1. *$\tau^*(z_{(0)}) = \tau_{(0)}$,*
2. *the design $\xi_{\tau^*(z)}$ is a saturated locally D-optimal design for $\Theta^0 = \Theta(z)$ with any $z \in Z$.*

*Proof.* Due to the Implicit Function Theorem (see *e.g.*, Gunning and Rossi (1965)) it suffices to prove that the Jacobi matrix is invertible for any $z \in Z$ and $\tau$ such that $\xi_\tau$ is a saturated locally $D$-optimal design. Details can be found in the technical report Melas (2004). □

Let $Z \subset \mathbb{R}$, $z_{(0)} \in Z$. If we are able to calculate (analytically or numerically) a saturated locally $D$-optimal design for $\Theta^0 = \Theta(z_0)$ then due to Theorem 1 we have

$$\tau^*(z) = \sum_{s=0}^{\infty} \tau_{(s)}(z - z_{(0)})^s$$

in some neighbourhood of the point $z_{(0)}$, where the $\tau_{(s)}$ are the Taylor coefficients of the function $\tau(z) = \tau^*(z)$.

These coefficients can be calculated by the recurrent formulas of Dette, Melas, and Pepelyshev (2004a):

$$\begin{aligned} \tau_{(s)} &= -\frac{1}{s!} J^{-1} \left( g(\tau_{(s-1)}(z), z) \right)_{(s)}, \; J = J(\tau_{(0)}, z_{(0)}), \\ \tau_{(s-1)}(z) &= \sum_{t=0}^{s-1} \tau_{(t)}(z - z_{(0)})^t, \; s = 1, 2, \ldots \end{aligned} \qquad (4)$$

These formulas can be extended to the case $Z \subset \mathbb{R}^r$, $r > 1$. They can be easily computed in the Maple package.

## 4 The Three-Parameter Logistic Distribution

Consider the function

$$\eta(t, \alpha, \beta, \gamma) = \frac{\alpha e^{\gamma t + \beta}}{1 + e^{\gamma t + \beta}}.$$

It is called the three-parameter logistic distribution. By the substitutions $x = e^t$, $\theta_1 = \alpha$, $\theta_2 = \gamma$, $\theta_3 = e^{-\beta}$, this function reduces to

$$\eta(x, \Theta) = \frac{\theta_1 x^{\theta_2}}{\theta_3 + x^{\theta_2}}, \qquad (5)$$

which is called the Hill equation in microbiological studies (see, *e.g.*, Bezeau and Endrenyi (1986)).

We will construct locally $D$-optimal designs for model (5) using the functional approach described above. Assume that $x \in [a, b]$, $a \geq 0$, $\theta_1 \neq 0$, $\theta_3 > 0$, $h(x) \equiv 1$. A direct calculation yields

$$\det M(\xi, \Theta) = \theta_1^4 \theta_3^2 \det \overline{M}(\eta, \theta_3),$$

where

$$\xi = \begin{pmatrix} x_1 & x_2 & x_3 \\ 1/3 & 1/3 & 1/3 \end{pmatrix}, \quad \eta_\xi = \begin{pmatrix} t_1 & t_2 & t_3 \\ 1/3 & 1/3 & 1/3 \end{pmatrix},$$
$$t_i = x_i^{\theta_2}, \; i = 1, 2, 3,$$
$$\overline{M}(\eta_\xi, \theta_3) = \sum_{i=1}^{3} f(t_i, \theta_3) f^T(t_i, \theta_3)/3,$$
$$f(t, \theta) = \left( \frac{t}{\theta + t}, \frac{t}{(\theta + t)^2}, \frac{t \ln t}{(\theta + t)^2} \right)^T.$$

Set

$$z = 1/\theta_3, \; r = 1, \; \Omega = (0, \infty).$$

Assumption 2 follows here from the properties of elementary functions, Assumptions 1 and 3 follows from the results of Dette, Melas, and Wong (2004b). It was also proved there that a locally $D$-optimal design is of type $(0, 2, 1)$ and is unique. This means that Assumption 4 holds for the model under consideration. Theorem 1 thus yields that the support points of locally $D$-optimal designs are real-analytic functions of $z$ with $z \in (0, 1)$.

Let us consider the case $[a, b] = [0, 1]$, $\theta_2 = 1$. For arbitrary $0 \leq a < b, \theta_2$ optimal designs can be calculated by a scale transformation. Using Formula (4), we constructed expansions of the vector function $\tau^*(z) = (x_1^*(z), x_2^*(z))^T$ in a neighbourhood of the point $z_{(0)} = 1$ in terms of powers of $(z - 1)$ and $1/(z - 1)$. The corresponding coefficients are presented in Tables 1 and 2, respectively. Note that $\tau^*(1)$ was found numerically.

Let $\xi_{<n>}(z)$ be the design constructed by using the first $n$ coefficients. The following inequality determines closeness of the design to the locally optimal one:

$$\begin{aligned} &\max_{x \in [0,1]} |d(x, \xi_{<n>}(z)) - 3| \leq 10^{-5}, \\ &d(x, \xi) = f^T(x) \, M^{-1}(\xi, z) \, f(x) \end{aligned} , \tag{6}$$

where

$$f(x) = \frac{\partial \eta(x, \Theta)}{\partial \Theta}, \; M(\xi, z) := M(\xi, \Theta(z)), \Theta = \Theta(z) = (1, 1, 1/z)^T.$$

It proves that for the first expansion with twenty coefficients the inequality (6) holds with $0 < z \leq 2.7$. For the second expansion with the same number of the coefficients, the inequality holds for $0.6 \leq z \leq 13.8$.

The behaviour of the design points for $0 \leq z \leq 10$ is presented in Fig. 1. We used the first expansion for $z \leq 1$ and the second for $1 \leq z \leq 10$ to construct the figure.

Note also that the efficiency of a given design for $\Theta^0 = \Theta(z)$ can be measured by the quantity

$$I(\xi, z) = \left( \frac{\det M(\xi, z)}{\det M(\xi_{\tau^*(z)}, z)} \right)^{1/3}.$$

**Table 1.** Coefficients of the Taylor expansions for $x_1^*$ and $x_2^*$ in a neighbourhood of $z = 1$ in terms of powers of $(z - 1)$.

| | 0 | 1 | 2 | 3 | 4 | 5 | 6 |
|---|---|---|---|---|---|---|---|
| $x_1^*$ | 0.09723 | -0.03401 | 0.01308 | -0.00530 | 0.00222 | -0.00095 | 0.00041 |
| $x_2^*$ | 0.47233 | -0.10533 | 0.02743 | -0.00791 | 0.00245 | -0.00080 | 0.00027 |

**Table 2.** Coefficients of the Taylor expansions for $x_1^*$ and $x_2^*$ in a neighbourhood of $z = 1$ in powers of $1/(z - 1)$.

| | 0 | 1 | 2 | 3 | 4 | 5 | 6 |
|---|---|---|---|---|---|---|---|
| $x_1^*$ | 0.09723 | 0.03401 | -0.02093 | 0.01314 | -0.00844 | 0.00555 | -0.00375 |
| $x_2^*$ | 0.47233 | 0.10533 | -0.07790 | 0.05838 | -0.04431 | 0.03404 | -0.02647 |

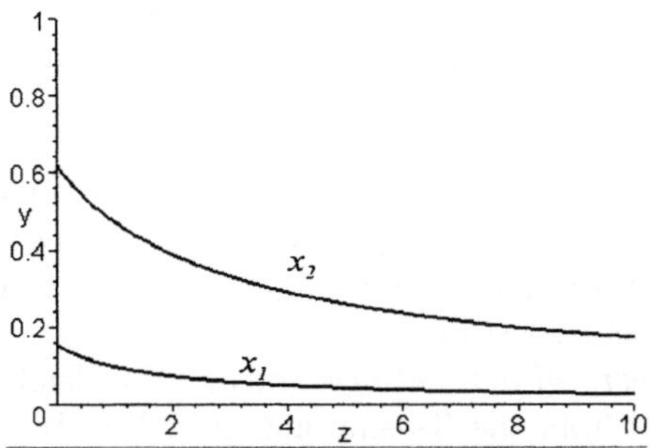


**Fig. 1.** The dependence of the support points $x_1^*$ and $x_2^*$ on $z$.

Numerical calculations show that the design $\xi(z^*) := \xi_{\tau^*(z^*)}$ with $z^* = 0.5$ has maximum of the minimal efficiency at the interval $[0.1, 1]$ among locally $D$-optimal designs at points $z = 0.1, \ldots, 0.9, 1$. Its minimal efficiency is equal to 0.981. Note that a maximin efficient $D$-optimal design, that is a design maximizing the minimum by $z \in [0.1, 1]$ of the efficiency among all (approximate) designs, was constructed numerically in Dette, Melas, and Wong (2004b). This design is very close to $\xi(0.5)$ and has minimal efficiency 0.982.

A similar calculation was performed for the interval $[1, 10]$ for $z$. It showed that the design $\xi(4)$, the best design among $\xi(1), \xi(2), \ldots, \xi(10)$, has minimal efficiency 0.8407. The maximin efficient design calculated in Dette, Melas, and Wong (2004b) has four support points with unequal weights and its minimal efficiency equals 0.885. But, for example, design $\xi(1)$, the locally optimal design for $z = 1$, has minimal efficiency 0.5430 on $[1, 10]$. This design is rather bad! It requires almost twice as much observations as $\xi(4)$ to achieve the same accuracy of the parameters if the true value of $z$ is equal to 10. Thus we see that the approach allows very exact calculation of locally $D$-optimal designs and secure a study of their efficiency. We conclude also that locally $D$-optimal designs could be very efficient if the initial values are chosen optimally inside given intervals of possible values.

*Acknowledgement.* The author thanks Patrick Laycock and an anonymous referee for useful comments on an earlier version of the paper.

## References

M. Bezeau and L. Endrenyi. Design of experiments for the precise estimation of dose-response parameters: the Hill equation. *J. Theoret. Biol.*, 123(4): 415–430, 1986.

H. Dette, V.B. Melas, and A. Pepelyshev. Optimal designs for estimating individual coefficients - a functional approach. *J. Statist. Plann. Inference*, 118:201–209, 2004.

H. Dette, V.B. Melas, A. Pepelyshev, and N. Strigul. Efficient design of experiments in the Monod model. *J. R. Stat. Soc. Ser. B Stat. Methodol.*, 65(3):725–742, 2003.

H. Dette, V.B. Melas, and W.K. Wong. Optimal design for goodness-of-fit of the Michaelis–Menten enzyme kinetic function. Preprint Ruhr-Universität Bochum. 2004. http://www.ruhr-uni-nochum.de/mathematik3/preprint.htm.

V.V. Fedorov. *Theory of optimal experiments.* Academic Press, New York, 1972. Translated from the Russian and edited by W. J. Studden and E. M. Klimko, Probability and Mathematical Statistics, No. 12.

R.C. Gunning and H. Rossi. *Analytic Functions of Several Complex Variables.* Prentice-Hall Inc., Englewood Cliffs, N.J., 1965.

C. Han and K. Chaloner. *D*- and *c*-optimal designs for exponential regression models used in viral dynamics and other applications. *J. Statist. Plann. Inference*, 115(2):585–601, 2003.

R.I. Jennrich. Asymptotic properties of non-linear least squares estimators. *Ann. Math. Statist.*, 40:633–643, 1969.

S. Karlin and W.J. Studden. *Tchebycheff systems: With applications in analysis and statistics.* Pure and Applied Mathematics, Vol. XV. Interscience Publishers John Wiley & Sons, New York-London-Sydney, 1966.

J.C. Kiefer. *Collected Papers.* Springer Verlag, New York, 1985.

V.B. Melas. Optimal designs for exponential regression. *Math. Operationsforsch. Statist. Ser. Statist.*, 9(1):45–59, 1978.

V.B. Melas. Analytical properties of locally $D$-optimal designs for rational models. In A.C. Atkinson, P. Hackl, and W.J. Müller, editors, *mODa 6—advances in model-oriented design and analysis (Puchberg/Schneeberg, 2001)*, Contrib. Statist., pages 201–209. Physica-Verlag, Heidelberg, 2001.

V.B. Melas. On the functional approach to optimal designs for nonlinear models. Preprint Ruhr-Universität Bochum. 2004. http://www.ruhr-uni-nochum.de/mathematik3/preprint.htm.

C.H. Müller. Maximin efficient designs for estimating nonlinear aspects in linear models. *J. Statist. Plann. Inference*, 44(1):117–132, 1995.

# Optimal Design Criteria Based on Tolerance Regions

C.H. Müller[1] and C.P. Kitsos[2]

[1] Carl von Ossietzky University Oldenburg, Institute for Mathematics, Postfach 2503, D-26111 Oldenburg, Germany, mueller@math.uni-oldenburg.de
[2] Department of Mathematics, Technological Educational Institute of Athens, Ag. Spyridonos, GR-12210 Egaleo, Greece, xkitsos@teiath.gr

**Summary.** New design criteria are derived by minimizing the tolerance regions in linear models. Since the regions usually are not completely ordered a complete ordering is provided by regarding the volume, the longest axis or the sum of the axes of the region as design criteria leading to $TD$-, $TE$- and $TA$-optimal designs. It is shown that the $TE$- and the $TA$-optimal designs coincide with the classical $E$- and $A$-optimal designs. However, the $TD$-optimal designs and $D$-optimal designs can be different which is shown by an example.

**Key words:** $\beta$-expectation tolerance region, prediction region, Bayesian tolerance region, linear model, optimal designs.

## 1 Introduction

For the vector $Y = (Y_1, \ldots, Y_N)^\top$ of observations we assume a general linear model of the form

$$Y = X\gamma + \sigma\,\varepsilon \qquad \text{with } \varepsilon \sim \mathcal{N}(0_N, I_{N\times N}),$$

where $\varepsilon = (\varepsilon_1, \ldots, \varepsilon_N)$ is the vector of errors, $\gamma \in \mathbb{R}^r$ the unknown parameter vector, $\sigma \in \mathbb{R}^+$ an unknown scale parameter, $X = (x_1, \ldots, x_N)^\top \in \mathbb{R}^{N\times r}$ the known design matrix, briefly also called design, and $\mathcal{N}(0_N, I_{N\times N})$ is the normal distribution with mean vector $0_N$ of zeros and covariance matrix $I_{N\times N}$, the identity matrix. Given a realization $y$ of $Y$, the aim is to construct a region $R(X,y) \subset \mathbb{R}^M$ in which the vector $V$ of future observations $V_1, \ldots, V_M$ will fall with a high probability. For that we assume that the vector $V = (V_1, \ldots, V_M)^\top$ of future observations will follow the same probability law as $Y$, namely

$$V = W\gamma + \sigma\,\varepsilon_0 \qquad \text{with } \varepsilon_0 \sim \mathcal{N}(0_M, I_{M\times M}),$$

where $W$ is the design matrix of the future observations. Hence, the distribution of $V$ is given by the same parameter $\theta = (\gamma, \sigma) \in \Omega := \mathbb{R}^r \times \mathbb{R}^+$ as the distribution of $Y$. Given $\theta$, $Y$ and $V$ shall be independent.

It is impossible to construct a region $R(X,y)$ so that $V$ lies in $R(X,y)$ with high probability for all $\theta$ and all realizations $y$ of $Y$, that is $P_{V|\theta}(R(X,y))$ is large for all $\theta$ and $y$. One compromise are $\beta$-expectation tolerance regions or mean coverage tolerance regions where $R(X,y)$ satisfies

$$\int P_{V|\theta}(R(X,y))\, f_{Y|\theta}(y)\, dy = \beta \tag{1}$$

for all $\theta \in \mathbb{R}^r \times \mathbb{R}^+$ (see *e.g.*, Aitchison and Dunsmore (1975), Guttman (1981), Jílek (1981), Jílek and Ackermann (1989)). But this means that $R(X,y)$ is also a prediction region (see Paulson (1943)) since

$$\int P_{V|\theta}(R(X,y))\, f_{Y|\theta}(y)\, dy = P_{(Y,V)|\theta}(\{(y,v);\ v \in R(X,y)\}).$$

By standard arguments used in linear models, it is easy to see that $R(X,y)$ is given by

$$\begin{aligned} R(X,y) = \Big\{ v \in \mathbb{R}^M;\ & \frac{1}{\hat{\sigma}(X,y)^2} \\ & \cdot (v - W\hat{\gamma}(X,y))^\top S(X)^{-1} (v - W\hat{\gamma}(X,y)) \le \frac{M}{N-r} F_{M,N-r,\beta} \Big\}, \end{aligned} \tag{2}$$

where

$$\begin{aligned} S(X) &:= I_{M\times M} + W(X^\top X)^{-1} W^\top, \\ \hat{\gamma}(X,y) &:= (X^\top X)^{-1} X^\top y, \\ \hat{\sigma}(X,y)^2 &:= (y - X\hat{\gamma}(X,y))^\top (y - X\hat{\gamma}(X,y)), \end{aligned}$$

and $F_{M,N-r,\beta}$ is the $\beta$ quantile of the $F$-distribution with $M$ and $N-r$ degrees of freedom. But $R(X,y)$ given by (2) is also a $\beta$-expectation tolerance region with respect to $\theta$ in a Bayesian sense as studied in Haq and Rinco (1976), that is it satisfies

$$\begin{aligned} \beta = P_{V|Y=y}(R(X,y)) &= \int_{R(X,y)} f_{V|Y=y}(v)\, dv \\ &= \int_\Omega \int_{R(X,y)} f_{V|\Theta=\theta}(v)\, dv\, f_{\Theta|Y=y}(\theta)\, d\theta = \int_\Omega P_{V|\Theta=\theta}(R(X,y))\, P_{\Theta|Y=y}(d\theta). \end{aligned} \tag{3}$$

for all $y \in \mathbb{R}^N$. This has the advantage that the mean coverage property holds for all $y$. Property (3) can be seen by the structural approach of Fraser (1968). A modern derivation bases on invariant measures and is given in Section 2.

In Section 3 we investigate designs $X$ that minimize the $\beta$-expectation tolerance regions of the form (2). In particular, we study design criteria for minimizing $R(X,y)$ and compare the new design criteria with the classical criteria based on minimizing confidence regions for $W\gamma$.

## 2 $\beta$-expectation Bayesian Tolerance Regions

At first note the following representation of $S(X)^{-1}$.

**Lemma 1.** $S(X)^{-1} = I_{M\times M} - W\,(X^\top X + W^\top W)^{-1} W^\top$.

*Proof.* Set $L := X^\top X$. Then we have

$$(L + W^\top W)^{-1} = L^{-1} - L^{-1}W^\top (I_{M\times M} + W\,L^{-1}W^\top)^{-1} W\,L^{-1}.$$

This implies

$$\begin{aligned}
&I_{M\times M} - W(L + W^\top W)^{-1}W^\top \\
&= I_{M\times M} - W\,L^{-1}W^\top \\
&\quad + W\,L^{-1}W^\top (I_{M\times M} + W\,L^{-1}W^\top)^{-1} W\,L^{-1}W^\top \\
&= I_{M\times M} - W\,L^{-1}W^\top (I_{M\times M} + W\,L^{-1}W^\top)^{-1} (I_{M\times M} + W\,L^{-1}W^\top) \\
&\quad + W\,L^{-1}W^\top (I_{M\times M} + W\,L^{-1}W^\top)^{-1} W\,L^{-1}W^\top \\
&= I_{M\times M} - W\,L^{-1}W^\top (I_{M\times M} + W\,L^{-1}W^\top)^{-1} \\
&= (I_{M\times M} + W\,L^{-1}W^\top)(I_{M\times M} + W\,L^{-1}W^\top)^{-1} \\
&\quad - W\,L^{-1}W^\top (I_{M\times M} + W\,L^{-1}W^\top)^{-1} \\
&= (I_{M\times M} + W\,L^{-1}W^\top)^{-1} = S(X)^{-1}.
\end{aligned}$$
□

The structural inference, as described by Fraser (1968), is based on the underlying group structures of the parameter space $\Omega = \mathbb{R}^r \times \mathbb{R}^+$ and the set of transformations of the observations. If we define the operator $\circ$ on $\Omega$ as $\theta \circ \theta_* = (\gamma, \sigma) \circ (\gamma_*, \sigma_*) := (\gamma + \sigma\,\gamma_*, \sigma\,\sigma_*)$ then $(\Omega, \circ)$ is a non-abelian group. Let

$$A_\theta = A_{(\gamma,\sigma)} := \begin{pmatrix} I_{r\times r} & \gamma \\ 0_{1\times r} & \sigma \end{pmatrix}$$

and $g_\theta : \mathbb{R}^{N\times r} \times \mathbb{R}^N \to \mathbb{R}^{N\times r} \times \mathbb{R}^N$ be defined by

$$g_\theta(X, y) := (X, y)\,A_\theta = (X, X\gamma + \sigma y).$$

Then $\mathcal{G} := \{g_\theta;\ \theta \in \Omega\}$ with the composition $\circ$ of functions is also a group. Since $X\gamma + \sigma Y \sim \mathcal{N}(X(\gamma + \sigma\,\gamma_*), \sigma^2\,\sigma_*^2\, I_{N\times N})$ if $Y \sim \mathcal{N}(X\gamma_*, \sigma_*^2\, I_{N\times N})$ the linear model is invariant with respect to $(\mathcal{G}, \circ)$ and $(\Omega, \circ)$. Then, taking the right invariant measure (right Haar measure) on $\Omega$ as prior distribution the posterior distribution has a simple form (see Fraser (1968), p. 127). Note, that the right Haar measure $\rho$ on $\Omega$ is given by $\frac{\rho(d(\gamma,\sigma))}{d(\gamma,\sigma)} = \frac{1}{\sigma} 1_{(0,\infty)}(\sigma)$.

**Lemma 2.** *Let the right Haar measure $\rho$ on $\Omega$ be the prior distribution of $\Theta$. Then the posterior distribution of $\Theta$ given $Y = y$ has a Lebesgue density of the form*

$$f_{\Theta|Y=y}(\theta) \propto \exp\Big\{ -\frac{1}{2\sigma^2}\left[(\hat{\gamma}(X,y)-\gamma)^\top X^\top X(\hat{\gamma}(X,y)-\gamma) - \hat{\sigma}(X,y)^2\right]\Big\}$$
$$\cdot\ \hat{\sigma}(X,y)^{N-r}\left(\frac{1}{\sigma}\right)^{N+1} 1_{(0,\infty)}(\sigma), \tag{4}$$

*where* $\propto$ *means equality with the exception of constants.*

*Proof.* Set

$$T(X,y) := (\hat{\gamma}(X,y), \hat{\sigma}(X,y)) \in \mathbb{R}^r \times \mathbb{R}^+,$$
$$U(X,y) := \hat{\sigma}(X,y)^{-1}(Y - X\hat{\gamma}(X,y)) \in \mathcal{U} := \mathbb{R}^N,$$
$$W(X,y) := (T(X,y), U(X,y)),$$

where the mapping $W$ maps the observation space $\mathbb{R}^{N\times r}\times\mathbb{R}^N$ into $\Omega\times\mathcal{U}$. The mapping $T : \mathbb{R}^{N\times r}\times\mathbb{R}^N \to \Omega$ is equivariant, that is $T(g_\theta(X,y)) = \theta\circ T(X,y)$, and the mapping $U : \mathbb{R}^{N\times r}\times\mathbb{R}^N \to \mathcal{U}$ is invariant, that is $U(g_\theta(X,y)) = U(X,y)$. The transformations $T$ and $U$ together with $\varepsilon \sim \mathcal{N}(0_N, I_{N\times N})$ provide the structural linear model (see Fraser (1968)).

Moreover, $T(X,Y)$ and $U(X,Y)$ are independent according to the Basu Theorem, since $T$ is a complete sufficient statistic, and also $\hat{\gamma}(X,Y)$ and $\hat{\sigma}(X,Y)$ are independent (see *e.g.*, Schervish (1995)). If $\Theta = \theta = (\gamma,\sigma)$, then $\hat{\gamma}(X,Y)$ has a $\mathcal{N}(\gamma, \sigma^2 X^\top X)$ distribution and $\frac{1}{\sigma^2}\hat{\sigma}(X,Y)^2$ has a $\chi^2$ distribution with $N-r$ degrees of freedom. Hence $P_{W(X,Y)|\Theta=\theta}$ has a Lebesgue density of the form $f_{W(X,Y)|\Theta=\theta}(\hat{\gamma},\hat{\sigma},\hat{u}) = f_{\hat{\gamma}(X,Y)|\Theta=\theta}(\hat{\gamma})\cdot f_{\hat{\sigma}(X,Y)|\Theta=\theta}(\hat{\sigma})\cdot f_{U(X,Y)}(\hat{u})$, where

$$f_{\hat{\gamma}(X,Y)|\Theta=\theta}(\hat{\gamma}) \propto \left(\frac{1}{\sigma}\right)^r \exp\left\{-\frac{1}{2\sigma^2}(\hat{\gamma}-\gamma)^\top X^\top X(\hat{\gamma}-\gamma)\right\}$$

and

$$f_{\hat{\sigma}(X,Y)|\Theta=\theta}(\hat{\sigma}) \propto \left(\frac{\hat{\sigma}}{\sigma}\right)^{N-r-1} \exp\left\{-\frac{1}{2}\left(\frac{\hat{\sigma}}{\sigma}\right)^2\right\}\frac{1}{\sigma}\,1_{(0,\infty)}(\hat{\sigma}).$$

Since the left Haar measure (or left invariant measure) $\lambda$ on $\Omega = \mathbb{R}^r\times\mathbb{R}^+$ has a Lebesgue-density of the form $\frac{\lambda(d(\hat{\gamma},\hat{\sigma}))}{d(\hat{\gamma},\hat{\sigma})} = \frac{1}{\hat{\sigma}^{r+1}}1_{(0,\infty)}(\hat{\sigma})$ we can express the Lebesgue density $f_{\hat{\gamma}(X,Y)|\Theta=\theta}\cdot f_{\hat{\sigma}(X,Y)|\Theta=\theta}$ also as a density $\tilde{f}_{\hat{\gamma}(X,Y)|\Theta=\theta}\cdot \tilde{f}_{\hat{\sigma}(X,Y)|\Theta=\theta}$ with respect to the left Haar measure $\lambda$. Let $\nu$ be the Lebesgue measure on $\mathbb{R}^N$; then $\tilde{f}_{W(X,Y)|\Theta=\theta}(\hat{\gamma},\hat{\sigma},\hat{u}) = \tilde{f}_{\hat{\gamma}(X,Y)|\Theta=\theta}(\hat{\gamma})\cdot\tilde{f}_{\hat{\sigma}(X,Y)|\Theta=\theta}(\hat{\sigma})\cdot f_{U(X,Y)}(\hat{u})$ is the $\lambda\otimes\nu$ density of $W(X,Y)$ given $\Theta = \theta$. The right Haar measure which is related to $\lambda$ is the prior distribution $\rho$. Hence, the conditions of Assumption 6.58 in Schervish (1995), p. 368–369, are satisfied so that Lemma 6.65 on p. 371 in Schervish (1995) holds which provides the posterior distribution of $\Theta$ given $W(X,Y) = (\hat{\gamma}, \hat{\sigma}, \hat{u})$. In particular, using the relation between $\lambda$ and $\rho$, we have

$$\int \tilde{f}_{T(X,Y)|\Theta=\theta}(\hat{\gamma}, \hat{\sigma}) \rho(d\theta) = \hat{\sigma}^r$$

so that $\tilde{f}_{W(X,Y)}(\hat{\gamma}, \hat{\sigma}, \hat{u}) = \hat{\sigma}^r \, f_{U(X,Y)}(\hat{u})$ is the marginal density. Then Bayes Theorem yields that the posterior distribution has a density with respect to $\rho$ of the form

$$\begin{aligned} \tilde{f}_{\Theta|W(X,Y)=(\hat{\gamma},\hat{\sigma},\hat{u})}(\theta) &= \frac{\tilde{f}_{W(X,Y)|\Theta=\theta}(\hat{\gamma}, \hat{\sigma}, \hat{u})}{\hat{\sigma}^r \, f_{U(X,Y)}(\hat{u})} \\ &= \frac{1}{\hat{\sigma}^r} \tilde{f}_{\hat{\gamma}(X,Y)|\Theta=\theta}(\hat{\gamma}) \, \tilde{f}_{\hat{\sigma}(X,Y)|\Theta=\theta}(\hat{\sigma}). \end{aligned}$$

Using the relations of the Haar measures $\rho$ and $\lambda$ to the Lebesgue measure on $\mathbb{R}^r \times \mathbb{R}^+$ we see that the posterior distribution has a Lebesgue density of the form

$$\begin{aligned} & f_{\Theta|W(X,Y)=(\hat{\gamma},\hat{\sigma},\hat{u})}(\theta) = \frac{1}{\hat{\sigma}^r} f_{\hat{\gamma}(X,Y)|\Theta=\theta}(\hat{\gamma}) \, f_{\hat{\sigma}(X,Y)|\Theta=\theta}(\hat{\sigma}) \, \hat{\sigma}^{r+1} \, \frac{1}{\sigma} \\ & \propto \exp\left\{ -\frac{1}{2\sigma^2} \left[ (\hat{\gamma} - \gamma)^\top X^\top X (\hat{\gamma} - \gamma) - \hat{\sigma}^2 \right] \right\} \hat{\sigma}^{N-r} \left( \frac{1}{\sigma} \right)^{N+1} 1_{(0,\infty)}(\sigma). \end{aligned}$$

Note that $f_{\Theta|W(X,Y)=(\hat{\gamma},\hat{\sigma},\hat{u})} = f_{\Theta|Y=y}$ if $\hat{\gamma} = \hat{\gamma}(X,y)$, $\hat{\sigma} = \hat{\sigma}(X,y)$, $\hat{u} = U(X,y)$, since $W$ is an injective mapping. Hence the assertion is proved. □

From the posteriori density, we can deduce the $\beta$-expectation tolerance region as proposed by Haq and Rinco (1976).

**Theorem 1.** *The region $R(X,y) \subset \mathbb{R}^M$ given by (2) is a $\beta$-expectation tolerance region in the sense of (3), that is it satisfies $P_{V|Y=y}(R(X,y)) = \beta$ for all $y \in \mathbb{R}^N$.*

*Proof.* Since $V$ given $\Theta = \theta$ has a $\mathcal{N}(W\gamma, \sigma^2 I_{M\times M})$ distribution the Lebesgue density of $V$ given $Y = y$ is with (4)

$$\begin{aligned} f_{V|Y=y}(v) &= \int_\Omega f_{V|\Theta=\theta}(v) \; f_{\Theta|Y=y}(\theta) \, d\theta \\ \propto & \int_\Omega \exp\left\{ -\frac{1}{2\sigma^2} \left[ (\hat{\gamma} - \gamma)^\top X^\top X (\hat{\gamma} - \gamma) + (v - W\gamma)^\top (v - W\gamma) \right] \right\} \\ & \cdot \exp\left\{ -\frac{1}{2\sigma^2} \hat{\sigma}^2 \right\} \hat{\sigma}^{N-r} \left( \frac{1}{\sigma} \right)^{N+M+1} d\theta. \end{aligned}$$

Set

$$h(X,y,v) := (X^\top X + W^\top W)^{-1} (X^\top X \, \hat{\gamma} + W^\top v).$$

Then we have with Lemma 1

$$\begin{aligned}
&(\hat{\gamma}-\gamma)^\top X^\top X(\hat{\gamma}-\gamma)+(v-W\gamma)^\top(v-W\gamma)\\
&=(\gamma-h(X,y,v))^\top(X^\top X+W^\top W)(\gamma-h(X,y,v))\\
&\quad+(v-W\hat{\gamma})^\top S(X)^{-1}(v-W\hat{\gamma})
\end{aligned}$$

which implies by integration of the density of the normal and $\chi^2$ distribution

$$\begin{aligned}
f_{V|Y=y}(v) \propto &\int_{\mathbb{R}+}\left[\int_{\mathbb{R}^r}\left(\frac{1}{\sigma}\right)^r \exp\left\{-\frac{1}{2\sigma^2}\left[(\gamma-h(X,y,v))^\top(X^\top X\right.\right.\right.\\
&\left.\left.\left.+\ W^\top W)(\gamma-h(X,y,v))\right]\right\}d\gamma\right]\cdot\\
&\exp\left\{-\frac{1}{2\sigma^2}\left[(v-W\hat{\gamma})^\top S(X)^{-1}(v-W\hat{\gamma})+\hat{\sigma}^2\right]\right\}\hat{\sigma}^{N-r}\left(\frac{1}{\sigma}\right)^{N-r+M+1}d\sigma\\
&\propto\left(1+\hat{\sigma}^{-2}\,(v-W\hat{\gamma})^\top S(X)^{-1}(v-W\hat{\gamma})\right)^{-(N-r+M)/2}\hat{\sigma}^{-M}.
\end{aligned}$$

This means that $\frac{\sqrt{N-r}}{\hat{\sigma}}\,(V-W\hat{\gamma})^\top S(X)^{-1/2}$ given $Y=y$ has a multivariate $T$-distribution with $N-r$ degrees of freedom. Thus $\frac{N-r}{M}\frac{1}{\hat{\sigma}^2}\,(V-W\hat{\gamma})^\top S(X)^{-1}(V-W\hat{\gamma})$ given $Y=y$ has an $F$-distribution with $M$ and $N-r$ degrees of freedom (see Tiao and Guttman (1965)). Hence, the assertion follows. □

## 3 Optimal Designs

Optimal designs for $\beta$-expectation tolerance regions will be those designs that provide that $R(X,y)$ given by (2) is as small as possible. The best situation is that a design $X_*$ provides a region $R(X_*,y)$ with $R(X_*,y)\subset R(X,y)$ for all other designs $X$ and all $y\in\mathbb{R}^N$. This means that $S(X_*)\le S(X)$ for all $X$ in the positive-semidefinite sense (that is, $A\le B$ if and only if $c^\top Ac\le c^\top Bc$ for all $c\in\mathbb{R}^M$). Such designs are called uniform optimal for $\beta$-expectation tolerance regions, shortly $TU$-optimal. But as in classical design theory such designs do not exist except in degenerate cases since $S(X)$ and thus $R(X,y)$ are not completely ordered. To get a complete ordering we can regard the volume of the region $R(X,y)$ which is a function of $\det(S(X))$, the determinant of $S(X)$. Since $R(X,y)$ is an ellipsoid we can also consider the sum of the axes of the ellipsoid expressed by $\mathrm{tr}(S(X))$, the trace of $S(X)$, or the longest axis expressed by $\lambda_{\max}(S(X))$, the maximum eigenvalue of $S(X)$.

**Definition 1.** *Let $\Delta$ be a set of competing designs. A design $X_*\in\Delta$ is called*

1. *$TU$-optimal if $S(X_*)\le S(X)$ for all $X\in\Delta$,*
2. *$TD$-optimal if $\det(S(X_*))\le\det(S(X))$ for all $X\in\Delta$,*
3. *$TA$-optimal if $\mathrm{tr}(S(X_*))\le\mathrm{tr}(S(X))$ for all $X\in\Delta$,*

*4. TE-optimal if* $\lambda_{\max}(S(X_*)) \leq \lambda_{\max}(S(X))$ *for all* $X \in \Delta$.

We will compare the new design criteria with the classical design criteria which are motivated by minimizing the confidence region for $W\gamma$. The confidence region for $W\gamma$ is given by

$$C(X,y) = \left\{ v \in \mathbb{R}^M; \; \frac{1}{\hat{\sigma}(X,y)^2} \right.$$
$$\left. \cdot \left(v - W\hat{\gamma}(X,y)\right)^\top \left(W(X^\top X)^{-1}W^\top\right)^{-1} \left(v - W\hat{\gamma}(X,y)\right) \leq \frac{M}{N-r} F_{M,N-r,\beta} \right\}$$

(see *e.g.*, Christensen (1987)). Hence minimizing the set $C(X,y)$, the volume, the sum of the axes, or the longest axis of $C(X,y)$ leads to the classical design criteria of $U$-, $D$-, $A$-, and $E$-optimality based on $W(X^\top X)^{-1}W^\top$, $\det(W(X^\top X)^{-1}W^\top)$, $\mathrm{tr}(W(X^\top X)^{-1}W^\top)$, and $\lambda_{\max}(W(X^\top X)^{-1}W^\top)$ (see *e.g.*, Pázman (1986) or Pukelsheim (1993)).

It is clear from the form of $S(X)$ that the tolerance regions are always larger than the confidence regions which is due to the variability of the future observations. Nevertheless almost all new design criteria are equivalent to the classical criteria.

**Theorem 2.** *Let* $\Delta$ *be a set of competing designs and* $X_* \in \Delta$. *Then we have:*

1. $X_*$ *is TU-optimal if and only if* $X_*$ *is U-optimal.*
2. $X_*$ *is TA-optimal if and only if* $X_*$ *is A-optimal.*
3. $X_*$ *is TE-optimal if and only if* $X_*$ *is E-optimal.*

*Proof.* Assertion 1. is obvious. We denote the eigenvalues of $W(X^\top X)^{-1}W^\top$ by $\lambda_1, \ldots, \lambda_M$. Then $1+\lambda_1, \ldots, 1+\lambda_M$ are the eigenvalues of $S(X)$. This implies assertion 3. Since the trace of a matrix is the sum of its eigenvalues also assertion 2. follows. □

Theorem 2 shows that the $TU$-, $TA$-, and $TE$-optimal designs can be constructed via the methods which were developed for the classical criteria (see *e.g.*, Pázman (1986) or Pukelsheim (1993)). The only exception is the $TD$-criterium.

Although the determinant of a matrix is the product of the eigenvalues of the matrix the proof of Theorem 2 cannot be adapted for the $D$-criterium. Namely, if $\lambda_1, \ldots, \lambda_M$ are the eigenvalues of $W(X^\top X)^{-1}W^\top$ then $\det(W(X^\top X)^{-1}W^\top) = \prod_{m=1}^M \lambda_m$ and $\det(S(X)) = \prod_{m=1}^M (1+\lambda_m)$. Hence, a minimum of $\prod_{m=1}^M \lambda_m$ is not equivalent to a minimum of $\det(S(X)) = \prod_{m=1}^M (1+\lambda_m)$. The following example shows that there are really situations where the $D$- and $TD$-optimal designs are different.

*Example 1.* Consider a simple linear regression model where any observation $y_n$ has the form $y_n = (1, t_n)\gamma + \sigma\varepsilon_n$ with $t_n \in \mathbb{R}$ and $\gamma \in \mathbb{R}^2$. Every future observation $v_m$ follows the same linear regression model, that is

$v_m = (1, u_m)\gamma + \sigma\varepsilon_{0m}$ with $u_m \in \mathbb{R}$. Assume that we have $\alpha N$ observations at $t_n = 1$ and $(1-\alpha) N$ observations at $t_n = 0$ so that the design matrix depends on $\alpha$ and shall be denoted by $X_\alpha$. The proportion $\alpha$ shall be chosen so that the tolerance region for two future observations $V_1$ and $V_2$, one at $u_1 = 2$ and the other at $u_2 = 3$, is as small as possible. Then we have

$$W\,(X_\alpha^\top X_\alpha)^{-1} W^\top = \begin{pmatrix} 1 & 2 \\ 1 & 3 \end{pmatrix} \begin{pmatrix} N & \alpha N \\ \alpha N & \alpha N \end{pmatrix}^{-1} \begin{pmatrix} 1 & 2 \\ 1 & 3 \end{pmatrix}^\top$$

so that

$$\det(W\,(X_\alpha^\top X_\alpha)^{-1} W^\top) = \frac{1}{N^2} \frac{1}{\alpha(1-\alpha)} \tag{5}$$

and

$$\begin{aligned} \det(S(X_\alpha)) &= \det(I_{2\times 2} + W\,(X_\alpha^\top X_\alpha)^{-1} W^\top) \\ &= \frac{1}{N^2} \frac{1}{\alpha - \alpha^2} \left( N^2(\alpha - \alpha^2) + N(13 - 8\alpha) + 1 \right). \end{aligned} \tag{6}$$

Quantity (5) is minimized by $\alpha = 0.5$ while the minimum of quantity (6) depends on $N$ and is minimized, for example, by $\alpha = 0.616$ for $N = 10$ and by $\alpha = 0.617$ for $N = 50$. Hence, the $TD$-optimal designs are different from the classical $D$-optimal designs if $\Delta$ is the set of all designs with observations at 0 and 1, that is $t_n \in \{0, 1\}$. The $TD$-optimal design puts more observations at 1 than the $D$-optimal design. The same holds if the set of competing designs is $\Delta^*$, the set of all designs with $t_n \in [0, 1]$.

## References


J. Aitchison and I. Dunsmore. *Statistical Prediction Analysis.* University Press, Cambridge, 1975.

R. Christensen. *Plane Answers to Complex Questions - The Theory of Linear Models.* Springer, New York, 1987.

D.A.S. Fraser. *The Structure of Inference.* Wiley, New York, 1968.

I. Guttman. *Statistical Tolerance Regions. Classical and Bayesian.* Griffin, London, 1981.

M.S Haq and S. Rinco. $\beta$-expectation tolerance regions for a generalized multivariate model with normal error variables. *J. Mult. Anal.*, 6:414–421, 1976.

M. Jílek. A bibliography of statistical tolerance regions. *Math. Operationsforschung und Statistik - Series Statistics*, 12:441–456, 1981.

M. Jílek and H. Ackermann. A bibliography of statistical tolerance regions II. *Statistics*, 20:165–172, 1989.

E. Paulson. A note on control limits. *Ann. Math. Statist.*, 14:90–93, 1943.

A. Pázman. *Foundations of Optimum Experimental Design.* Reidel, Dordrecht, 1986.

F. Pukelsheim. *Optimal Design of Experiments.* Wiley, New York, 1993.

M.J. Schervish. *Theory of Statistics.* Springer, New York, 1995.

G.G. Tiao and I. Guttman. The inverted Dirichlet distribution with applications. *J. Amer. Statist. Assoc.*, 60:793–805, 1965.

# Simultaneous Choice of Design and Estimator in Nonlinear Regression with Parameterized Variance

A. Pázman[1] and L. Pronzato[2]

[1] Comenius University, Department of Probability and Statistics, Mlyns-ká Dolina, 84215 Bratislava, Slovakia pazman@center.fmph.uniba.sk
[2] Laboratoire I3S, Université de Nice Sophia-Antipolis/CNRS, Bat. Euclide, Les Algorithmes, BP 121, 2000 route des Lucioles, 06903 Sophia Antipolis Cedex, France pronzato@i3s.unice.fr

**Summary.** In some nonlinear regression problems with parameterized variance both the design and the method of estimation have to be chosen. We compare asymptotically two methods of estimation: the penalized weighted LS (PWLS) estimator, which corresponds to maximum likelihood estimation (MLE) under the assumption of normal errors, and the two-stage LS (TSLS) estimator. We show that when the kurtosis $\kappa$ of the distribution of the errors is zero, the asymptotic covariance matrix of the estimator is smaller for PWLS than for TSLS, which may not be the case when $\kappa$ is not zero. We then suggest to construct two optimum designs, one for PWLS under the assumption $\kappa = 0$, the other for TSLS (with arbitrary $\kappa$), and compare their properties for different values of $\kappa$. All developments are made under the assumption of a randomized design, which allows rigourous proofs for the asymptotic properties of the estimators while avoiding the technical difficulties encountered in classical references such as Jennrich (1969).

**Key words:** Nonlinear regression, least squares estimation, penalized least squares, two-stage least squares, iterative squares, robustness, kurtosis.

## 1 Introduction

We consider a nonlinear regression problem, with observations

$$Y_k = \eta(x_k, \bar{\theta}) + \varepsilon_k, \ \mathrm{E}(\varepsilon_k) = 0, \quad k = 1, \ldots, N, \tag{1}$$

where $\bar{\theta}$ denotes the unknown true value of the model parameters. The observation errors $\varepsilon_k = \varepsilon(x_k)$ are assumed to be independently distributed. It frequently happens that the full parameterized probability distribution of the errors $\varepsilon_k$ is not available, whereas their variance is a known function of the design variable $x$ and of (some of) the parameters $\theta$ of the mean response, that is,

$$\mathrm{E}\left(\varepsilon_k^2\right) = \lambda(x_k, \bar{\theta}), \quad k = 1, \ldots, N. \tag{2}$$

The parameter estimation problem in this case is called method of fitting expectations in Jennrich and Ralston (1979), see also Malyutov (1988). The (ordinary) LS estimator, which ignores the information contained in the variance function, is strongly consistent and asymptotically normally distributed under standard assumptions. However, using the information on $\theta$ provided by the variance may yield a more precise estimation, hence the importance of choosing a suitable estimation method. We consider two approaches in Section 3, first a penalized weighted least-squares (PWLS) estimator, which corresponds to maximum likelihood estimation under the assumption of normal errors, and secondly a two-stage least-squares (TSLS) estimator, where ordinary LS is used at the first stage, and the estimator is substituted into the variance function, to be used for weighted LS estimation at the second stage. The asymptotic properties of the estimators are obtained under the assumption of a *randomized design*, introduced in Section 2, which allows relatively simple although rigourous proofs. Section 4 presents a design strategy based on the asymptotic properties derived in Section 3. We compute an optimum design for PWLS estimation under the assumption of a zero kurtosis for the errors $\varepsilon_k$, together with an optimum design for TSLS estimation. Then we compare their performance (and that of the associated estimation methods) when the kurtosis varies. We show that in some particular situations the conclusion is design-free. One estimation method is always preferable, independently of the design, depending on the value of the kurtosis and the magnitude of the errors.

## 2 Randomized Designs and Uniform Strong Law of Large Numbers

**Definition 1.** *We call a* randomized design *with measure $\xi$ on the design space $\mathcal{X}$, $\int_{\mathcal{X}} \xi(dx) = 1$, a sequence $\{x_i\}$ of design points independently sampled from the measure $\xi$ on $\mathcal{X}$.*

The following assumptions will be used throughout the paper.

**H1** $\Theta$ is a compact subset of $\mathbb{R}^d$, $\bar{\theta} \in \Theta$.

**H2** $\eta(x,\theta)$ and $\lambda(x,\theta)$ are continuous functions of $\theta \in \Theta$ for any $x \in \mathcal{X}$, with $\eta(x,\theta)$ and $\lambda^{-1}(x,\theta)$ bounded on $\mathcal{X} \times \Theta$, $\lambda(x,\bar{\theta})$ bounded on $\mathcal{X}$.

**H3** $\eta(x,\theta)$ and $\lambda(x,\theta)$ are twice continuously differentiable with respect to $\theta \in \mathrm{int}(\Theta)$ for any $x \in \mathcal{X}$, their derivatives are bounded on $\mathcal{X} \times \mathrm{int}(\Theta)$.

Our proofs are based on uniform convergence with respect to $\theta$ of the criterion function $J_N(\theta)$ defining the estimator $\hat{\theta}^N = \arg\min_\theta J_N(\theta)$. We shall thus need a uniform Strong Law of Large Numbers (SLLN). Note that the proper definition of the estimator as a random variable is ensured by Lemma 2 in Jennrich (1969) (see also Bierens (1994), p. 16). In the following $\hat{\theta}^N$ will refer to the measurable choice from $\arg\min_{\theta\in\Theta} J_N(\theta)$. The asymptotic results of the next sections are based on the following lemma, which is a simplified version of Theorem 2.7.1 in Bierens (1994).

**Lemma 1 (Uniform SLLN).** *Let $\{z_i\}$ be a sequence of i.i.d. random vectors, and $a(z,\theta)$ be a Borel measurable real function of $(z,\theta) \in \mathbb{R}^r \times \Theta$, continuous in $\theta$ for any $z$, with $\Theta$ a compact subset of $\mathbb{R}^p$. Suppose that*

$$\mathrm{E}\left(\sup_{\theta\in\Theta} |a(z,\theta)|\right) < \infty, \tag{3}$$

*then $\mathrm{E}(a(z,\theta))$ is continuous in $\theta \in \Theta$ and $\frac{1}{N}\sum_{i=1}^{N} a(z_i,\theta) \overset{\theta}{\rightrightarrows} \mathrm{E}(a(z,\theta))$ a.s. when $N \to \infty$, where $\overset{\theta}{\rightrightarrows}$ means uniform convergence with respect to $\theta$.*

Once the almost sure uniform convergence of the criterion function $J_N(\cdot)$ is obtained, the almost sure convergence of the estimator will follow from the next lemma. The proof is a straightforward application of the continuity and uniform convergence properties.

**Lemma 2 (Consistency from uniform convergence of the estimation criterion).** *Assume that the sequence of functions $\{J_N(\theta)\}$ converges uniformly on $\Theta$ to the function $J(\theta)$, with $J_N(\theta)$ continuous with respect to $\theta \in \Theta$ for any $N$, $\Theta$ a compact set of $\mathbb{R}^p$, and $J(\theta)$ such that*

$$J(\bar{\theta}) = \min_{\theta\in\Theta} J(\theta) \text{ and } J(\theta) > J(\bar{\theta}) \text{ for all } \theta \neq \bar{\theta} \in \Theta\,.$$

*Then $\lim_{N\to\infty} \hat{\theta}^N = \bar{\theta}$, where $\hat{\theta}^N \in \arg\min_{\theta\in\Theta} J_N(\theta)$. When the functions $J_N(\cdot)$ are random, and the uniform convergence to $J(\cdot)$ is almost sure, then convergence of $\hat{\theta}^N$ to $\bar{\theta}$ is also almost sure.*

## 3 Penalized Weighted LS and Two-Stage LS Estimation

Since the optimum weights $w(x) = \sigma^{-2}(x,\bar{\theta}) = \lambda^{-1}(x,\bar{\theta})$ cannot be used for weighted LS estimation ($\bar{\theta}$ is unknown), it is tempting to use the weights $\lambda^{-1}(x,\theta)$, that is, to choose $\hat{\theta}^N$ that minimises the criterion

$$J_N(\theta) = \frac{1}{N}\sum_{k=1}^{N} \frac{[y(x_k) - \eta(x_k,\theta)]^2}{\lambda(x_k,\theta)}\,. \tag{4}$$

However, this approach in not recommended since $\hat{\theta}^N$ is generally not consistent. Indeed, using Lemma 1 with $z_k = (x_k,\varepsilon_k)$, $a(z_k,\theta) = [\eta(x_k,\bar{\theta}) + \varepsilon_k - \eta(x_k,\theta)]\lambda^{-1}(x_k,\theta)$ and Lemma 2 we easily obtain the following.

**Theorem 1.** *Let $\{x_i\}$ be a randomized design with measure $\xi$ on $\mathcal{X} \subset \mathbb{R}^d$. Assume that H1 and H2 are satisfied. Then the estimator $\hat{\theta}_{LS}^N$ that minimises (4) in the model (1,2) converges a.s. to the set $\bar{\Theta}$ of values of $\theta$ that minimise*

$$J(\theta) = \int_{\mathcal{X}} \lambda(x,\bar{\theta})\lambda(x,\theta)^{-1}\,\xi(dx) + \int_{\mathcal{X}} \lambda(x,\theta)^{-1}[\eta(x,\theta) - \eta(x,\bar{\theta})]^2\,\xi(dx)\,.$$

Notice that, in general, $\bar{\theta} \notin \bar{\Theta}$.

### 3.1 Penalized Weighted LS Estimation

Consider now the following modification of the criterion (4),

$$J_N(\theta) = \frac{1}{N}\sum_{k=1}^{N} \frac{[y(x_k) - \eta(x_k, \theta)]^2}{\lambda(x_k, \theta)} + \frac{1}{N}\sum_{k=1}^{N} \log \lambda(x_k, \theta) \,. \tag{5}$$

The estimator $\hat{\theta}^N_{PWLS}$ that minimises (5) can be considered as a *penalized* weighted LS (PWLS) estimator, where the term $(1/N)\sum_{k=1}^N \log \lambda(x_k, \theta)$ penalizes large variances. It corresponds to maximum likelihood estimation under the assumption that the errors $\varepsilon_k$ are normally distributed. It can be obtained numerically by direct minimization of (5) using a nonlinear optimisation method, or by solving an infinite sequence of weighted LS problems as suggested in Downing, Fedorov, and Leonov (2001).

The next theorem shows that this estimator is strongly consistent and asymptotically normally distributed without the assumption of normal errors. The proof follows directly from Lemmas 1 and 2 and the Central Limit Theorem (CLT).

**Theorem 2.** *Let $\{x_i\}$ be a randomized design with measure $\xi$ on $\mathscr{X} \subset \mathbb{R}^d$. Assume that H1 and H2 are satisfied and that for any $\theta, \theta'$ in $\Theta$,*

$$\left.\begin{array}{r} \int_{\mathscr{X}} \lambda^{-1}(x,\theta)[\eta(x,\theta) - \eta(x,\theta')]^2\, \xi(dx) = 0 \\ \int_{\mathscr{X}} |\lambda^{-1}(x,\theta)\lambda(x,\theta') - 1|\, \xi(dx) = 0 \end{array}\right\} \Leftrightarrow \theta = \theta' \,. \tag{6}$$

*Then the estimator $\hat{\theta}^N_{PWLS}$ that minimises (5) in the model (1,2) converges a.s. to $\bar{\theta}$. If, moreover, the errors $\varepsilon_k$ have finite fourth-order moment $\mathrm{E}\left(\varepsilon_i^4\right)$ and $\mathrm{E}\left(\varepsilon_i^3\right) = 0$ for all k, $\bar{\theta} \in \mathrm{int}(\Theta)$, H3 is satisfied and the matrix*

$$\begin{aligned} \mathbf{M}_1(\xi, \bar{\theta}) = & \int_{\mathscr{X}} \lambda^{-1}(x, \bar{\theta}) \left.\frac{\partial \eta(x,\theta)}{\partial \theta}\right|_{\bar{\theta}} \left.\frac{\partial \eta(x,\theta)}{\partial \theta^\top}\right|_{\bar{\theta}} \xi(dx) + \\ & \frac{1}{2}\int_{\mathscr{X}} \lambda^{-2}(x, \bar{\theta}) \left.\frac{\partial \lambda(x,\theta)}{\partial \theta}\right|_{\bar{\theta}} \left.\frac{\partial \lambda(x,\theta)}{\partial \theta^\top}\right|_{\bar{\theta}} \xi(dx) \end{aligned} \tag{7}$$

*is nonsingular, then $\hat{\theta}^N_{PWLS}$ satisfies*

$$\sqrt{N}(\hat{\theta}^N_{PWLS} - \bar{\theta}) \xrightarrow{\mathrm{d}} z \sim \mathscr{N}(0, \mathbf{M}_1^{-1}(\xi,\bar{\theta})\mathbf{M}_2(\xi,\bar{\theta})\mathbf{M}_1^{-1}(\xi,\bar{\theta}))$$

*as $N \to \infty$, with*

$$\mathbf{M}_2(\xi, \bar{\theta}) = \mathbf{M}_1(\xi, \bar{\theta}) + \frac{1}{4}\int_{\mathscr{X}} \lambda^{-2}(x, \bar{\theta}) \left.\frac{\partial \lambda(x,\theta)}{\partial \theta}\right|_{\bar{\theta}} \left.\frac{\partial \lambda(x,\theta)}{\partial \theta^\top}\right|_{\bar{\theta}} \kappa(x)\, \xi(dx), \tag{8}$$

*where $\kappa(x) = \mathrm{E}\left(\varepsilon^4(x)\right)\lambda^{-2}(x,\bar{\theta}) - 3$ is the kurtosis of the distribution of $\varepsilon(x)$.*

One may notice that when the errors $\varepsilon_k$ are normally distributed, $\kappa(x) = 0$, $\mathbf{M}_2(\xi, \bar{\theta}) = \mathbf{M}_1(\xi, \bar{\theta})$ and

$$\sqrt{N}(\hat{\theta}^N_{PWLS} - \bar{\theta}) \xrightarrow{\mathrm{d}} z \sim \mathscr{N}(0, \mathbf{M}_1^{-1}(\xi,\bar{\theta})) \,, \; N \to \infty \,. \tag{9}$$

### 3.2 Two-stage LS Estimation

By two stage LS, we mean using first some estimator $\hat{\theta}_1^N$, and then plugging the estimate into the weight function $\lambda(x, \theta)$. The second-stage estimator $\hat{\theta}_{TSLS}^N$ is then obtained by minimizing

$$J_N(\theta, \hat{\theta}_1^N) = \frac{1}{N} \sum_{k=1}^{N} \frac{[y(x_k) - \eta(x_k, \theta)]^2}{\lambda(x_k, \hat{\theta}_1^N)} \tag{10}$$

with respect to $\theta \in \Theta$. Again, using Lemmas 1 and 2 and the CLT we can show that $\hat{\theta}_{TSLS}^N$ is consistent when $\hat{\theta}_1^N$ converges (it does not need to be consistent, that is, convergence to $\bar{\theta}$ is not required), and, when $\hat{\theta}_1^N$ is $\sqrt{N}$-consistent, that is, when $\sqrt{N}(\hat{\theta}_1^N - \bar{\theta})$ is bounded in probability ($\forall \varepsilon > 0\ \exists A$ and $N_0$ such that $\forall N > N_0$, $P\{\sqrt{N}(\hat{\theta}_1^N - \bar{\theta}) > A\} < \varepsilon$), $\hat{\theta}_{TSLS}^N$ is asymptotically normally distributed.

**Theorem 3.** *Let $\{x_i\}$ be a randomized design with measure $\xi$ on $\mathcal{X} \subset \mathbb{R}^d$. Assume that H1 and H2 are satisfied, that $\hat{\theta}_1^N$ converges to some $\hat{\theta} \in \Theta$ and that for any $\theta, \theta'$ in $\Theta$,*

$$\int_{\mathcal{X}} \lambda^{-1}(x, \hat{\theta})[\eta(x, \theta) - \eta(x, \theta')]^2 \, \xi(dx) = 0 \Leftrightarrow \theta = \theta'. \tag{11}$$

*Then the estimator $\hat{\theta}_{TSLS}^N$ that minimises (10) in the model (1,2) converges a.s. to $\bar{\theta}$. If, moreover, H3 is satisfied, the matrix*

$$\mathbf{M}(\xi, \bar{\theta}) = \int_{\mathcal{X}} \lambda^{-1}(x, \bar{\theta}) \left.\frac{\partial \eta(x, \theta)}{\partial \theta}\right|_{\bar{\theta}} \left.\frac{\partial \eta(x, \theta)}{\partial \theta^\top}\right|_{\bar{\theta}} \xi(dx) \tag{12}$$

*is nonsingular and the first-stage estimator $\hat{\theta}_1^N$ substituted into (10) is $\sqrt{N}$-consistent, with $\bar{\theta} \in \mathrm{int}(\Theta)$, then $\hat{\theta}_{TSLS}^N$ satisfies*

$$\sqrt{N}\left(\hat{\theta}_{TSLS}^N - \bar{\theta}\right) \xrightarrow{\mathrm{d}} z \sim \mathcal{N}(0, \mathbf{M}^{-1}(\xi, \bar{\theta})), \ N \to \infty .$$

Note that a natural candidate for the first-stage estimator $\hat{\theta}_1^N$ is the ordinary LS estimator, which is $\sqrt{N}$-consistent under the assumptions of Theorem 3. Also note that $\mathbf{M}^{-1}(\xi, \bar{\theta})$ is the asymptotic covariance matrix of the weighted LS estimator in the case where *the variance function (2) is known.*

Increasing the number of stages leads to iteratively re-weighted LS estimation, which relies on sequence of estimators constructed as follows:

$$\hat{\theta}_k^N = \arg\min_{\theta \in \Theta} J_N(\theta, \hat{\theta}_{k-1}^N) , \ k = 1, 2 \ldots \tag{13}$$

where $J_N(\theta, \theta')$ is defined by (10) and where $\hat{\theta}_1^N$ can be taken equal to the LS estimator. Using Theorem 3, a simple induction shows that, for any *fixed* $k$,

$\hat{\theta}_k^N$ is strongly consistent and asymptotically normally distributed, $\sqrt{N}(\hat{\theta}_k^N - \bar{\theta}) \xrightarrow{d} z \sim \mathcal{N}(0, \mathbf{M}^{-1}(\xi, \bar{\theta}))$. We simply mention the following property, which relies on a classical result in fixed point theory, see (Stoer and Bulirsch, 1993, p. 267), and states that the recursion (13) converges a.s. for $N$ large enough.

**Theorem 4.** *If the conditions of Theorem 3 are satisfied, the iteratively re-weighted LS estimator defined by (13) in the model (1,2) converges a.s. for $N$ large enough:* $P\left(\forall N > N_0,\ \lim_{k\to\infty} \hat{\theta}_k^N = \hat{\theta}_\infty^N\right) \to 1$ *when* $N_0 \to \infty$.

When the errors $\varepsilon_k$ are normally distributed, the asymptotic covariance matrix of the two-stage LS estimator, see (12), is larger than that of the penalized WLS estimator, see (7). This advantage of $\hat{\theta}_{PWLS}^N$ over $\hat{\theta}_{TSLS}^N$ may disappear when the distribution of the errors has a positive kurtosis. In general, the conclusion depends on the design $\xi$, which raises the issue of choosing simultaneously the method of estimation and the design. This is discussed in the next section.

## 4 Choosing the Design and the Estimator

In the rest of the paper we assume that the kurtosis $\kappa$ is constant (does not depend on $x$). It corresponds to the rather common situation where the distributions of the errors at different $x$ are similar and only differ by a scaling factor.

We start by a simple example where the choice between $\hat{\theta}_{PWLS}^N$ and $\hat{\theta}_{TSLS}^N$ does not depend on the design: depending on the kurtosis and the magnitude of the errors, one estimation method is uniformly better than the other.

*Example 1.* Suppose that in the model (1) the variance of the errors satisfies $\mathrm{Var}\,(\varepsilon_k) = \lambda(x_k, \theta) = \alpha\,(\eta(x_k, \theta) + \beta)^2$, for any $k$. In particular, when $\beta = 0$ it corresponds to the situation where the relative precision of the observations is constant (and equal to $\alpha > 0$). Direct calculations then gives for $\hat{\theta}_{PWLS}^N$ that

$$\mathbf{M}_1^{-1}(\xi, \bar{\theta})\mathbf{M}_2(\xi, \bar{\theta})\mathbf{M}_1^{-1}(\xi, \bar{\theta}) = \rho(\kappa, \alpha)\mathbf{M}^{-1}(\xi, \bar{\theta}),$$

where $\rho(\kappa, \alpha) = (1 + 2\alpha + \kappa\alpha)/(1 + 2\alpha)^2$, $\mathbf{M}_1(\xi, \bar{\theta})$ and $\mathbf{M}_2(\xi, \bar{\theta})$ are given by (7) and (8), and $\mathbf{M}^{-1}(\xi, \theta)$ is the asymptotic covariance matrix for $\hat{\theta}_{TSLS}^N$, see (12). Whatever the design $\xi$, TSLS should thus be preferred to PWLS when $\rho(\kappa, \alpha) > 1$, that is, when $\kappa > 2$ and $\alpha < \alpha^* = (\kappa - 2)/4$, and vice-versa otherwise. Generally speaking, it means that for observations with constant relative precision the two-stage procedure is always preferable when $\kappa > 2$, provided that the errors are small enough. For instance, when the errors have the exponential distribution $\varphi(\varepsilon) = (1/2)\exp(-|\varepsilon|)$, $\kappa = 3$ and the limiting value for $\alpha$ is $\alpha^* = 1/4$. Figure 1 gives the evolution of $\rho(\kappa, \alpha)$ as a function of $\alpha$ for $\kappa = 3$.

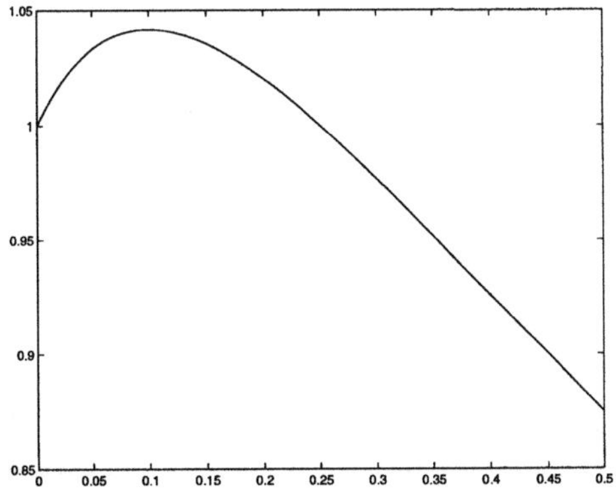


**Fig. 1.** Evolution of $\rho(\kappa, \alpha)$ as a function of $\alpha$ in Example 1: $\kappa = 3$ and $\hat{\theta}^N_{TSLS}$ should be preferred to $\hat{\theta}^N_{PWLS}$ for $\alpha < 1/4$.

In more general situations the estimator and the design must be chosen simultaneously. The following approach may be used: (i) determine the optimum designs $\xi^*_{PWLS}$ for the PWLS estimator *under the assumption of zero kurtosis* and $\xi^*_{TSLS}$ for the TSLS estimator; (ii) compare the values of the design criteria for both estimators at different values of the kurtosis $\kappa$. Note that the asymptotic covariance matrix of $\hat{\theta}_{PWLS}$ is linear in $\kappa$. Therefore, for any design criterion $\varphi(\cdot)$ such that $\varphi(\mathbf{M})$ is monotonic in $\mathbf{M}$, a value $\kappa^*$ exists such that $(\xi^*_{TSLS}, \hat{\theta}_{TSLS})$ should be preferred to $(\xi^*_{PWLS}, \hat{\theta}_{PWLS})$ for $\kappa > \kappa^*$.

The optimum design for TSLS estimation corresponds to a standard design problem with homoscedastic errors where the derivative $\partial\eta(x,\theta)/\partial\theta$ is replaced by $(1/\sqrt{\lambda(x,\theta)})\partial\eta(x,\theta)/\partial\theta$, see (12). In the case of PWLS with $\kappa = 0$, the information matrix to be used is $\mathbf{M}_1(\xi, \theta)$, see (9). Optimum designs for such situations can be constructed with the same algorithms as for the standard case, see Downing, Fedorov, and Leonov (2001). In particular, when $\lambda(x,\theta) = g[\eta(x,\theta)]$, where $g(\cdot)$ is a differentiable function from $\mathbb{R}$ into $\mathbb{R}^+$ with derivative $g'(\cdot)$,

$$\mathbf{M}_1(\xi,\theta) = \int_{\mathcal{X}} G(x,\theta)[\partial\eta(x,\theta)/\partial\theta][\partial\eta(x,\theta)/\partial\theta^\top]\xi(dx)$$

with

$$G(x,\theta) = \lambda^{-1}(x,\theta) + \{g'[\eta(x,\theta)]\}^2\lambda^{-2}(x,\theta)/2,$$

and standard approaches for optimum design with homoscedastic errors can be used.

*Example 2.* We take $\eta(x,\theta) = \theta_0 + \theta_1 x + \theta_2 x^2$, $\lambda(x,\theta) = 5/\eta(x,\theta)^2$, $x \in \mathcal{X} = [-1,1]$. Although this is a linear regression model, heteroscedasticity implies that numerical values must be specified for the parameters $\theta$ to design an optimum experiment. We take $\theta^\top = (\theta_0, \theta_1, \theta_2) = (2, 1, 1/2)$. The

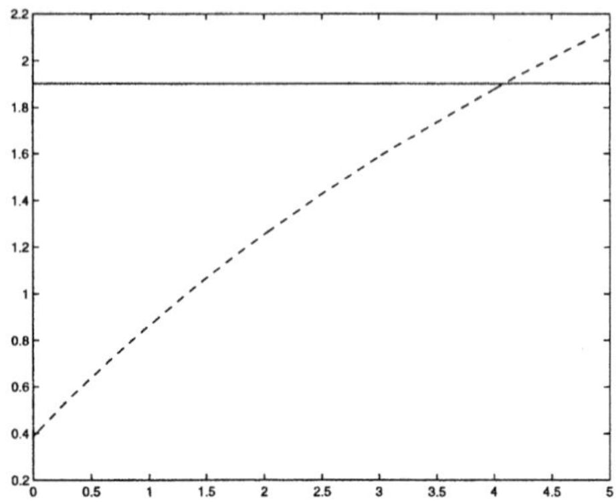


**Fig. 2.** Evolution of $\log\det[\mathbf{M}_1^{-1}(\xi^*_{PWLS},\theta)\mathbf{M}_2(\xi^*_{PWLS},\theta)\mathbf{M}_1^{-1}(\xi^*_{PWLS},\theta)]$ as a function of $\kappa$ (dashed line) and $\log\det\mathbf{M}^{-1}(\xi^*_{TSLS},\theta)$ (full line) in Example 2.

design criterion is $D$-optimality. The $D$-optimum design $\xi^*_{TSLS}$ for the TSLS estimator has three support points $-1, 0.256, 1$ that receive equal weight $1/3$. The $D$-optimum design $\xi^*_{PWLS}$ for the PWLS estimator with normal errors is supported at $-1, 0.078, 1$, with weight $1/3$ at each point. Figure 2 shows that TSLS estimation with $\xi^*_{TSLS}$ should be preferred to PWLS estimation with $\xi^*_{PWLS}$ when $\kappa > \kappa^* \simeq 4.09$.

## References

H.J. Bierens. *Topics in Advanced Econometrics.* Cambridge University Press, Cambridge, 1994.

D. Downing, V.V. Fedorov, and S. Leonov. Extracting information from the variance function: optimal design. In A.C. Atkinson, P. Hackl, and W.G. Müller, editors, *mODa6 – Advances in Model–Oriented Design and Analysis*, pages 45–52, Heidelberg, 2001. Physica-Verlag.

R.I. Jennrich. Asymptotic properties of nonlinear least squares estimation. *Annals of Math. Stat.*, 40:633–643, 1969.

R.I. Jennrich and M.L. Ralston. Fitting nonlinear models to data. *Annals Rev. Biophys. Bioeng.*, 8:195–238, 1979.

M.B. Malyutov. Design and analysis in generalized regression model $F$. In V.V. Fedorov and H. Laüter, editors, *Model-Oriented Data Analysis, (Eisenach, 1987)*, volume 297 of *Lecture Notes in Econom. and Math. Systems*, pages 72–76. Springer, Berlin, 1988.

J. Stoer and R. Bulirsch. *Introduction to Numerical Analysis.* Springer, Heidelberg, 2nd edition, 1993.

# Minimum Entropy Estimation in Semi-Parametric Models: a Candidate for Adaptive Estimation?

L. Pronzato, É. Thierry, and É. Wolsztynski

Laboratoire I3S, Université de Nice Sophia-Antipolis/CNRS, Bat. Euclide, Les Algorithmes, BP 121, 2000 route des Lucioles, 06903 Sophia Antipolis Cedex, France {pronzato,et,wolsztyn}@i3s.unice.fr

**Summary.** In regression problems with errors having an unknown density $f$, least squares or robust $M$-estimation is the usual alternative to maximum likelihood, with the loss of asymptotic efficiency as a consequence. The search for efficiency in the absence of knowledge of $f$ (adaptive estimation) has motivated a large amount of work, see in particular Stein (1956), Stone (1975), Bickel (1982) and the review paper Manski (1984). The present paper continues the work initiated in Pronzato and Thierry (2001a,b). The estimator is obtained by minimizing (an estimate of) the entropy of the symmetrized residuals. Connections and differences with previous work are indicated. The focus is mainly on the location model but we show how the results can be extended to nonlinear regression problems, in particular when the design consists of replications of a fixed design. Numerical results illustrate that asymptotic efficiency is not necessarily in conflict with robustness.

**Key words:** Adaptive estimation, efficiency, entropy, parameter estimation, semi-parametric models.

## 1 Introduction

Consider a regression problem, with observations

$$Y_i = \eta(\bar{\theta}, X_i) + \varepsilon_i\,, \quad i = 1, \ldots, n, \tag{1}$$

where $\bar{\theta}$ is the unknown value of the model parameters $\theta \in \Theta \subset \mathbb{R}^p$, $(\varepsilon_i)$ forms a sequence of i.i.d. random variables with p.d.f. $f$ and $\eta(\theta, x)$ is a known function of $\theta$ and $x$, the design variable. Most of the paper is devoted to the simplest case, the location model, where $\eta(\theta, x) = \theta$, but extension to nonlinear regression is considered in Section 4.

If $f$ is known, Maximum Likelihood estimation can be used and, under standard assumptions, is *asymptotically efficient*: $\sqrt{n}(\hat{\theta}^n_{ML} - \bar{\theta}) \xrightarrow{\text{d}} \mathcal{N}(\mathbf{0}, \mathbf{M}_F^{-1})$, with $\mathbf{M}_F$ the Fisher information matrix. Suppose now that the density $f$ is only known to be symmetric about zero (further regularity assumptions about $f$ will be used in Section 3). The model can then be termed *semi-parametric*, with $\theta$ and $f$ respectively its parametric and non-parametric parts, and $f$ can be considered as an infinite-dimensional nuisance parameter for the estimation of $\theta$. This corresponds to the approach taken in the seminal paper Stein (1956). The absence of knowledge of $f$ usually induces a loss of efficiency. An estimator that remains asymptotically efficient in these conditions is called *adaptive* (see Bickel (1982) for a precise definition and Begun et al. (1983) for a necessary condition for adaptive estimation). Beran (1974) and Stone (1975) proved that adaptive estimation in the location model was possible, using respectively adaptive rank estimates, and an approximation of the score function based on a kernel estimate of $f$ from residuals obtained with a preliminary $\sqrt{n}$-consistent estimator. It is this second approach which has been further developed in Bickel (1982), see also Manski (1984). Here we shall follow the approach taken in Pronzato and Thierry (2001a,b) (PT-2001a,b in what follows) and minimize the entropy of a kernel estimate of $f$ based on the $2n$ symmetrized residuals $e_i(\theta), -e_i(\theta)$, $i = 1, \ldots, n$, with $e_i(\theta) = Y_i - \eta(\theta, x_i)$ the $i$th residual in the regression model (1). The motivation is given in Section 2, where the connections and differences with the Stone-Bickel approach are detailed. The simplest case of a location problem is considered in Section 3 where adaptivity of a particular minimum-entropy estimator is proved. Extension to nonlinear regression is considered in Section 4, with an example illustrating the robustness properties of the estimator.

## 2 Minimizing Entropy

The ML estimator $\hat{\theta}^n_{ML}$ (for $f$ known) minimizes

$$\bar{H}_n(\theta) = -\frac{1}{n}\sum_{i=1}^{n} \log f[e_i(\theta)] \tag{2}$$

with respect to $\theta$. The initial motivation in (PT-2001a,b) comes from the simple observations that (i) $\bar{H}_n(\bar{\theta}) = -(1/n)\sum_{i=1}^{n} \log f(\varepsilon_i)$ is an empirical version of $H(f) = -\int \log[f(x)]f(x)dx$, (ii) the entropy of a distribution is a measure of its dispersion, (iii) the entropy of the true distribution of the symmetrized residuals is minimum at $\theta = \bar{\theta}$ (symmetrized residuals are used because the entropy is shift-invariant). A possible construction when $f$ is unknown is as follows: construct a kernel estimator $\hat{f}^\theta_n$ from the symmetrized residuals (so that $\hat{f}^\theta_n$ is symmetric); compute its entropy, which forms the estimation criterion $\hat{H}_n(\theta)$ to be minimized. In (PT-2001a,b), $\hat{H}_n(\theta)$ is given by

$$\hat{H}_n(\theta) = -\int_{-A_n}^{A_n} \log[\hat{f}^\theta_n(x)]\hat{f}^\theta_n(x)dx \tag{3}$$

with $(A_n)$ a suitably increasing sequence of positive numbers (to be chosen in accordance with the decrease of the bandwidth $h_n$ of the kernel estimate $\hat{f}^\theta_n$). Other estimators of the entropy of $\hat{f}^\theta_n$ will be used in what follows.

In the location model, this construction can be justified following an approach similar to Beran (1978). The distribution of the observations $Y_i$ has the density $g(y) = f(y - \bar{\theta})$. Define $\hat{\beta} = (\hat{\theta}, \hat{f})$ in the semi-parametric model, where $\hat{\theta}$ is a postulated value for $\bar{\theta}$ and $\hat{f}$ a postulated symmetric p.d.f., and let $g_{\hat{\beta}}(y)$ be the associated density for the observations, $g_{\hat{\beta}}(y) = \hat{f}(y - \hat{\theta})$. Assume that a kernel estimator $\hat{g}_n$ of $g$ has been formed, based on the observations $Y_1, \ldots, Y_n$. The estimator in Beran (1978) is based on the Hellinger distance between $\hat{g}_n$ and $g_{\hat{\beta}}$. Here we use the Kullback-Leibler divergence $L(\hat{g}_n, g_{\hat{\beta}}) = \int \log[\hat{g}_n(y)/g_{\hat{\beta}}(y)]\hat{g}_n(y)dy$. Straightforward calculation shows that the symmetric $\hat{f}$ and parameter $\hat{\theta}$ that minimize $L(\hat{g}_n, g_{\hat{\beta}})$ respectively correspond to $\hat{f}_n = \hat{f}_n^{\hat{\theta}^n}$ with $\hat{f}_n^{\theta}(u) = [\hat{g}_n(u + \theta) + \hat{g}_n(-u + \theta)]/2$ and $\hat{\theta}^n = \arg\min_\theta H(\hat{f}_n^{\theta})$. One may then notice that $\hat{f}_n^{\theta}$ is a kernel estimateor based on the symmetrized residuals $Y_i - \theta, -Y_i + \theta$, and $\hat{\theta}^n$ minimizes its entropy.

This presents some similarities with the Stone-Bickel approach (1975, 1982), which relies on an approximation of the score function, that is, the derivative of $\bar{H}_n(\theta)$, given by (2), with respect to $\theta$. The construction is in two stages: first, an estimator $\hat{\theta}_1^n$ asymptotically locally sufficient (in the sense of Le Cam) is constructed; secondly, an approximated score function is used to perform a Newton-Raphson step from $\hat{\theta}_1^n$. Stone (1975) shows that the construction is adaptive for the location model with symmetric errors; Bickel (1982) and then Manski (1984) extend the result to other models, including non linear regression, see also Andrews (1989). Although the construction of $\hat{H}_n(\theta)$ may rely on a similar kernel estimator, there are some key differences between the Stone-Bickel approach and the minimum-entropy method presented here. First, the estimation criterion $\hat{H}_n$, see (3,4), is minimized through a *series of* minimisation steps, using an optimisation algorithm. Secondly, all the data are treated similarly, whereas, for technical reasons, the developments in Bickel (1982), Manski (1984) rely on sample splitting: $m$ observations are used to construct a preliminary parameter estimate $\hat{\theta}^m$ to form residuals and a score function estimate; the $n-m$ remaining observations are used for the Newton-Raphson step from $\hat{\theta}_1^n$, with the requirement that $m \to \infty$ and $m/n \to 0$ as $n \to \infty$. One may notice that the numerical results presented in (Manski, 1984) show that this sample splitting degrades the performance of the estimator. The estimator proposed by Andrews (1989) does not rely on sample splitting and is defined to minimize an estimation criterion rather than to be a one-step estimator. However, the criterion (likelihood) is also constructed from a $\sqrt{n}$-consistent preliminary estimate $\hat{\theta}_1^n$, used to form a kernel estimate of $f$ and hence of the likelihood function. On the contrary, $\hat{H}_n$ does not depend on any preliminary estimate.

One motivation for minimizing the entropy of the distribution of the residuals is that it allows a lot of flexibility: many methods are available to estimate the entropy $\hat{H}_n(\theta)$, and kernel estimation is only one possibility. One may refer to Beirlant, Dudewicz, Györfi, and Van der Meulen (1997) for a survey which includes plug-in, sample spacing and nearest neighbour methods. Studying the application of these methods to semi-parametric estimation by entropy minimisation is quite challenging.

Define $\hat{\theta}^n = \arg\min_{\theta\in\Theta} \hat{H}_n(\theta)$, with $\Theta$ a compact subset of $\mathbb{R}^p$ and $\hat{H}_n(\theta)$ some estimator of the entropy of the distribution of the symmetrized residuals in the model (1). We assume that $\bar{\theta} \in \text{int}(\Theta)$ and that $\hat{H}_n(\theta)$ is twice continuously differentiable with respect to $\theta \in \Theta$. Convergence in probability when $n\to\infty$ will be denoted $\xrightarrow{\text{p}}$ ($\overset{\theta,\text{p}}{\rightsquigarrow}$ will be used when the convergence is uniform with respect to $\theta$), and convergence in distribution will be denoted $\xrightarrow{\text{d}}$; $\nabla F(\theta)$ and $\nabla^2 F(\theta)$ denote the first and second order derivatives of the function $F$ with respect to $\theta$. The adaptivity of $\hat{\theta}^n$ can be proved by following the steps below (leaving aside some usual measurability conditions, see, *e.g.*, Lemmas 2 and 3 of Jennrich (1969)):

A) show that $\hat{H}_n(\theta)\overset{\theta,\text{p}}{\rightsquigarrow}H(\theta)$, with $\hat{H}_n(\theta)$ continuous in $\theta$ for any $n$ and $H(\bar{\theta}) < H(\theta)$ for any $\theta \neq \bar{\theta}$;
B) show that $\nabla^2\hat{H}_n(\theta)\overset{\theta,\text{p}}{\rightsquigarrow}\nabla^2 H(\theta)$, with $\nabla^2 H(\bar{\theta})$ positive definite ($\succ 0$);
C) decompose $\nabla\hat{H}_n(\bar{\theta})$ into $\nabla\bar{H}_n(\bar{\theta}) + \Delta_n(\bar{\theta})$, with $\sqrt{n}\nabla\bar{H}_n(\bar{\theta})\xrightarrow{\text{d}}\mathcal{N}(\mathbf{0},\mathbf{M}_1)$ and $\sqrt{n}\Delta_n(\bar{\theta})\xrightarrow{\text{p}}\mathbf{0}$ as $n\to\infty$.

A proves that $\hat{\theta}^n\xrightarrow{\text{p}}\bar{\theta}$. A and B imply that $\nabla^2\hat{H}_n(\hat{\theta}^n)\xrightarrow{\text{p}}\mathbf{M}_2 = \nabla^2 H(\bar{\theta}) \succ 0$. Consider the following Taylor development of $\nabla\hat{H}_n(\theta)$ at $\theta = \hat{\theta}^n$, similar to that used in (Jennrich, 1969) for LS estimation: $\nabla\hat{H}_n(\hat{\theta}^n) = \mathbf{0} = \nabla\hat{H}_n(\bar{\theta}) + (\hat{\theta}^n - \bar{\theta})^\top\nabla^2 H[\alpha_n\hat{\theta}^n + (1-\alpha_n)\bar{\theta}]$, for some $\alpha_n \in [0,1]$. C then implies $\sqrt{n}(\hat{\theta}^n - \bar{\theta})\xrightarrow{\text{d}}\mathcal{N}(\mathbf{0},\mathbf{M}_2^{-1}\mathbf{M}_1\mathbf{M}_2^{-1})$ and adaptivity is proved provided that $\mathbf{M}_2^{-1}\mathbf{M}_1\mathbf{M}_2^{-1} = \mathbf{M}_F^{-1}$, the inverse of the Fisher information matrix for the model (1). Step C allows some freedom in the choice of the function $\bar{H}_n(\theta)$. However, a natural candidate is (2), for which asymptotic normality of $\sqrt{n}\nabla\bar{H}_n(\bar{\theta})$ holds under standard assumptions, with $\mathbf{M}_1 = \mathbf{M}_F$.

A most important remark at this point is that (uniform) $\sqrt{n}$-consistency of the entropy estimate $\hat{H}_n(\theta)$ is not a prerequisite for $\sqrt{n}$-consistency of $\hat{\theta}^n$ (we only need $\sqrt{n}$-consistency of $\Delta_n(\bar{\theta})$). It is important because $\sqrt{n}$-consistency does not seem to be a widespread property among entropy estimation methods, see Beirlant, Dudewicz, Györfi, and Van der Meulen (1997).

## 3 Adaptive Estimation in the Location Model

The residuals are given by $e_i(\theta) = Y_i - \theta$. Consider the kernel estimators

$$k^\theta_{n,i}(u) = \frac{1}{(n-1)h_n}\sum_{j=1,j\neq i}^{n} K\left[\frac{u-e_j(\theta)}{h_n}\right],\ i = 1,\ldots,n,$$

where the kernel $K$ is a p.d.f. symmetric about 0 that satisfies $\int |u|K(u)du < \infty$, $K$ and its first two derivatives being continuous and of bounded variation, see Schuster (1969) (the density of the standard normal satisfies these conditions). We assume that $f$ has unbounded support, $f$ and its derivatives $f^{(s)}$ are bounded for $s = 1, 2, 3$, $H(f) < \infty$ and $f$ has a finite Fisher information for location, $i(f) = \int [f'(x)]^2/f(x)dx < \infty$. Consider the entropy estimate given by

$$\hat{H}_n = -\frac{1}{n}\sum_{i=1}^{n} \log\{f^{\theta}_{n,i}[e_i(\theta)]\}U_n[e_i(\theta)] \tag{4}$$

where $f^{\theta}_{n,i}(u) = (1/2)[k^{\theta}_{n,i}(u) + k^{\theta}_{n,i}(-u)]$ and $U_n(x) = u(|x|/A_n - 1)$, with $u(z) = 1$ for $z \leq 0$, $0$ for $z \geq 1$ and $u(z)$ varying smoothly in between, $u'(0) = u'(1) = 0$, and $\max_z |u'(z)| = d_1 < \infty$, $\max_z |u''(z)| = d_2 < \infty$. $\hat{H}_n(\theta)$ is then continuous (and twice continuously differentiable) in $\theta$ for any $n$. As in Dmitriev and Tarasenko (1973), we assume that exists a (strictly increasing) function $B(x)$ such that for all $x$, $B(x) \geq \sup_{|y|\leq x} 1/f(y)$. Define $B_n = B(2A_n + L)$. Using (Dmitriev and Tarasenko, 1973, Theorem 4) and (Newey, 1991, Corollary 3.1), one can then show that $\hat{H}_n(\theta) \overset{\theta,\mathrm{p}}{\rightsquigarrow} H(\theta)$ as $n \to \infty$, provided that $A_n$ (and thus $B_n$) increases slowly enough, and the bandwidth $h_n$ of the kernel estimator decreases slowly enough ($B_n = n^{\alpha}$, $h_n = 1/[n^{\alpha}\log n]$ with $\alpha < 1/3$ is suitable). Here, $H(\theta)$ is the entropy of the true distribution of the symmetrized residuals for $\theta$, $H(\theta) = -\int \log[\pi^{\theta}(x)]\pi^{\theta}(x)dx$ with $\pi^{\theta}(x) = (1/2)[f(x+\theta-\bar{\theta}) + f(x-\theta+\bar{\theta})]$, which is minimum at $\theta = \bar{\theta}$. This proves point A of Section 2, and thus the consistency of $\hat{\theta}^n$ that minimizes $\hat{H}_n$; the details will be presented elsewhere. Similarly, with slightly stronger conditions on $f$ one can prove point B, that is, $\nabla^2\hat{H}_n(\theta) \overset{\theta,\mathrm{p}}{\rightsquigarrow} \nabla^2 H(\theta)$, with $\nabla^2 H(\bar{\theta}) = i(f)$, *e.g.*, when $B_n = n^{\alpha}$, $h_n = 1/[n^{\alpha}\log n]$, $\alpha < 1/7$. The adaptivity of $\hat{\theta}^n$, that is step C, would then follow from

$$\frac{2}{\sqrt{n}}\sum_{i=1}^{n} \frac{(k^{\bar{\theta}}_{n,i})'(\varepsilon_i)}{k^{\bar{\theta}}_{n,i}(\varepsilon_i) + k^{\bar{\theta}}_{n,i}(-\varepsilon_i)} U_n(\varepsilon_i) \overset{\mathrm{d}}{\to} \mathcal{N}(0, i(f))\,.$$

The conditions required on the functions $f$, $K$ and $u$ for this to hold are currently under investigation. One may notice that a difficulty which is not present in the Stone-Bickel approach is due to the fact that $(k^{\bar{\theta}}_{n,i})'(-x) \neq -(k^{\bar{\theta}}_{n,i})'(x)$.

*Example 1.* In this example and Example 2 we take $A_n = \infty$ in (3) and $U_n(x) \equiv 1$ in (4); the bandwidth $h$ of the kernel estimator is chosen by the double kernel method of Berlinet and Devroye (1994) and based on residuals obtained from a robust $M$-estimator. Table 1 gives the empirical value $\hat{C}_n$ of $\mathrm{E}\{\hat{v}_n\hat{v}_n{}^\top\}$, with $\hat{v}_n = \sqrt{n}(\hat{\theta}^n - \bar{\theta})$, obtained from 100 repetitions of the estimation procedure, for different choices of the estimator $\hat{\theta}$ and p.d.f. $f$. The number of observations is $n = 100$.

**Table 1.** Value of $\hat{C}_n$ in the location model for the LS, Minimum Hellinger Distance (Beran, 1978) and Minimum Entropy estimators ((3) for $\mathrm{ME}_1$ and (4) for $\mathrm{ME}_2$). The errors are standard normal, bi-exponential ($f(x) = (1/\sqrt{2})\exp(-\sqrt{2}|x|)$), and Student's $t_\nu$ with $\nu = 3, 5$ and 10 degrees of freedom.

| $f$ | $\mathcal{N}(0,1)$ | exp | $t_3$ | $t_5$ | $t_{10}$ |
|---|---|---|---|---|---|
| $\mathbf{M}_F^{-1}$ | 1 | 0.5 | 0.5 | 0.8 | 0.9455 |
| LS | 1.09 | 0.94 | 1.13 | 0.96 | 1.03 |
| MHD | 1.12 | 0.72 | 0.50 | 0.86 | 1.0 |
| $\mathrm{ME}_1$ | 1.12 | 0.71 | 0.48 | 0.83 | 0.99 |
| $\mathrm{ME}_2$ | 1.19 | 0.74 | 0.57 | 0.84 | 0.98 |

# 4 Adaptive Estimation in Nonlinear Regression

We return to the nonlinear regression model (1), and assume that the design consists of replications at fixed points: the design measure $\xi$ has a finite number of support points $X^1, \ldots, X^m$, $X^j$ receiving the weight $\xi^j$. Therefore, for a total of $n$ observations, $n_j = n\xi^j$ are made at $X = X^j$.

Consider first a 2-stage estimation method, with $m$ minimum entropy estimations at the first step and one (weighted) LS estimation at the second. For each $X^j$, $j = 1, \ldots, m$, let $Y_{j_i}$ denote the observations made at $X = X^j$, $i = 1, \ldots, n_j$, and $\hat{\eta}^j$ denote the estimated response at $X = X^j$ obtained by an adaptive estimator: we have $\sqrt{n_j}[\hat{\eta}^j - \eta(\bar{\theta}, X^j)] \xrightarrow{\mathrm{d}} \mathcal{N}(0, i^{-1}(f))$ as $n_j \to \infty$. Having solved $m$ such location problems, we form an LS estimation problem by considering the estimated responses $\hat{\eta}^j$ as pseudo-observations, and minimize $J_n^{LS}(\theta) = \sum_{j=1}^m \xi^j [\eta(\theta, X^j) - \bar{\eta}^j]^2$. Assume that $\bar{\theta} \in \mathrm{int}(\Theta)$, with $\Theta$ a compact subset of $\mathbb{R}^p$, that $\eta(\theta, X^j)$ is twice continuously differentiable with respect to $\theta \in \Theta$ for any $j$, that the Fisher information matrix $\mathbf{M}_F(\bar{\theta}) = i(f) \sum_{j=1}^m \xi^j \nabla\eta(\bar{\theta}, X^j)[\nabla\eta(\bar{\theta}, X^j)]^\top$ is definite positive and that the identifiabilty condition $\{[\eta(\theta, X^j) = \eta(\bar{\theta}, X^j)\,, j = 1, \ldots, m] \Rightarrow \theta = \bar{\theta}\}$ is satisfied. One can then show that the 2-stage $LS$ estimator $\hat{\theta}_{LS}^n$ that minimizes $J_n^{LS}(\theta)$ satisfies $\sqrt{n}(\hat{\theta}_{LS}^n - \bar{\theta}) \xrightarrow{\mathrm{d}} \mathcal{N}(\mathbf{0}, \mathbf{M}_F^{-1}(\bar{\theta}))$.

Although $\hat{\theta}_{LS}^n$ is asymptotically efficient, one can expect a one-stage adaptive estimator to exhibit a better finite sample behaviour. Using a justification similar to that given in Section 2 for the location model, we suggest the following procedure: (i) form kernel estimators $\hat{f}^{j,\theta}$ of the distribution of (symmetrized) residuals *for each design point $X^j$ separately* and compute their respective entropies $H(\hat{f}^{j,\theta})$, (ii) compute

$$\hat{\theta}^n = \arg\min_{\theta \in \Theta} E_\xi \left(\hat{H}_n(\theta, X)\right) \quad \text{with } \hat{H}_n(\theta, X^j) = H(\hat{f}^{j,\theta})\,,\ j = 1, \ldots, m. \quad (5)$$

The adaptivity of the method would then follow from adaptivity in the location model.

This approach does not extend to more general designs (whereas, for instance, randomized designs are considered in Bickel (1982) and in Manski (1984) the empirical distribution of design points converges a.s. at rate $\sqrt{n}$ to some distribution with non singular variance; see also the conditions imposed on the design in Jennrich (1969)). The difficulty comes from the fact that estimating the entropy of the symmetrized residuals at each design point $X$ is not possible without replications of observations. One possibility, which is used in (PT-2001a,b), is to mix all (symmetrized) residuals together and estimate the entropy $\hat{H}_n(\theta)$ of their distribution by (3) or (4). Replace $\xi$ by $\xi_n$ in (5), where $\xi_n$ is empirical measure of the design points $X_i$. Let $U$ be a random variable with distribution conditional on $X = X^j$ given by $\hat{f}^{j,\theta}$ of (5). Then, $\mathrm{E}_{\xi_n}\{\hat{H}_n(\theta, X)\} = \mathcal{H}(U|X)$ the conditional entropy of $U$ given $X$. From a standard result in information theory, $\mathcal{H}(U|X) \leq \mathcal{H}(U) = \hat{H}_n(\theta)$. The entropy $\hat{H}_n(\theta)$ obtained by mixing up all residuals, which can be constructed for any design, is thus an upper bound on the criterion $\mathrm{E}_{\xi_n}\{\hat{H}_n(\theta, X)\}$ minimized in (5). Studying the adaptivity of the corresponding estimator will form the subject of further work. Some preliminary numerical results are presented below.

**Table 2.** Values $(T_n, D_n)$ of the trace $(\times 10^{-3})$ and determinant $(\times 10^{-2})$ of $\hat{C}_n$ in Example 2 for different $f$ and estimators (with $q$ outliers for the last three lines).

| $f$ | $\mathcal{N}(0,1)$ | exp | $t_3$ | $t_5$ | $t_{10}$ |
|---|---|---|---|---|---|
| optimum ($\hat{C}_n = \mathbf{M}_F^{-1}$) | (6.2, 0.8) | (3.1, 0.2) | (3.1, 0.2) | (4.9, 0.5) | (5.8, 0.75) |
| LS | (8.8, 1.2) | (13.6, 3.6) | (9.1, 2.0) | (9.7, 2.2) | (9.7, 2.1) |
| MHD | (9.1, 1.4) | (3.8, 0.5) | (6.4, 0.6) | (7.9, 1.4) | (7.1, 1.7) |
| ME | (9.2, 1.25) | (3.8, 0.4) | (4.9, 0.4) | (7.8, 1.2) | (6.8, 1.35) |
| $q$ (nb. outliers, $f$ = exp) | 20 | 60 | 80 | | |
| LS | (35.0, 4.2) | (128.4, 7.3) | (20.3, 12.4) | | |
| MHD | (5.3, 1.4) | (17.9, 3.2) | (2.9, 2.5) | | |
| ME | (4.9, 0.8) | (17.9, 2.2) | (2.9, 1.6) | | |

*Example 2.* Consider the regression model $\eta(\theta, x) = \theta_1 \exp(-\theta_2 x)$, with $\bar{\theta} = (100, 2)^\top$, the design measure is supported at $X^j = 1 + (j-1)/9$, $j = 1, \ldots, 10$, with $\xi^j = 1/10$ for all $j$. The results are in Table 2, with the same notations as in Example 1. We use 100 repetitions with 100 observations for each. ME uses (3), both MHD (see Beran (1978)) and ME mix all residuals together. In the last three lines of the table we introduce $q$ outliers $\varepsilon_k$ ($\mathcal{N}(2, 10)$, randomly allocated among the $X^j$'s) in addition to the $n$ observations ($\hat{C}_n$ is still computed for $n = 100$, $f$ is the bi-exponential).

## References


D.W.K. Andrews. Asymptotics for semiparametric econometric models: III Testing and examples. Cowles Foundation Discussion Paper No. 910, 1989.

J.M. Begun, W.J. Hall, W.-M. Huang, and J.A. Wellner. Information and asymptotic efficiency in parametric-non parametric models. *Ann. Statist.*, 11(2):432–452, 1983.

J. Beirlant, E.J. Dudewicz, L. Györfi, and E.C. Van der Meulen. Nonparametric entropy estimation; an overview. *Intern. J. Math. Stat. Sci.*, 6(1): 17–39, 1997.

R. Beran. Asymptotically efficient rank estimates in location models. *Ann. Statist.*, 2:63–74, 1974.

R. Beran. An efficient and robust adaptive estimator of location. *Ann. Statist.*, 6(2):292–313, 1978.

A. Berlinet and L. Devroye. A comparison of kernel density estimates. *Publications de l'institut de statistique de l'Université de Paris*, 38:3–59, 1994.

P.J. Bickel. On adaptive estimation. *Ann. Statist.*, 10:647–671, 1982.

Yu.G. Dmitriev and F.P. Tarasenko. On the estimation of functionals of the probability density and its derivatives. *Theory of Probability and its Applications*, 18(3):628–633, 1973.

R.I. Jennrich. Asymptotic properties of nonlinear least squares estimation. *Annals of Math. Stat.*, 40:633–643, 1969.

C.F. Manski. Adaptive estimation of nonlinear regression models. *Econometric Reviews*, 3(2):145–194, 1984.

W.K. Newey. Uniform convergence in probability and stochastic equicontinuity. *Econometrica*, 9(4):1161–1167, 1991.

L. Pronzato and É. Thierry. Entropy minimization for parameter estimation problems with unknown distribution of the output noise. In *Proc. ICASSP'2001*, Salt Lake City, May 2001.

L. Pronzato and É. Thierry. A minimum-entropy estimator for regression problems with unknown distribution of observation errors. In A. Mohammad-Djafari, editor, *Bayesian Inference and Maximum Entropy Methods in Science and Engineering, Proc. 20th Int. Workshop, (Gif-sur-Yvette, 2000)*, pages 169–180, New York, 2001. Am. Inst. of Physics.

E.F. Schuster. Estimation of a probability density function and its derivatives. *Ann. Math. Statist.*, 40:1187–1195, 1969.

C. Stein. Efficient nonparametric testing and estimation. In *Proceedings of the Third Berkeley Symposium on Mathematical Statistics and Probability, 1954–1955, vol. I*, pages 187–195, Berkeley and Los Angeles, 1956. University of California Press.

C.J. Stone. Adaptive maximum likelihood estimators of a location parameter. *Ann. Statist.*, 3(2):267–284, 1975.

# Optimal Designs for Contingent Response Models

H.S. Rabie[1] and N. Flournoy[2]

[1] Department of Statistics
146 Middlebush Building
University of Missouri-Columbia
Columbia, MO 65211-4100
hsr2mf@mizzou.edu
[2] flournoyn@missouri.edu

**Summary.** We study $D$- and $c$-optimal designs for the contingent response models of Li, Durham, and Flournoy (1995). In the contingent response model there are two types of failure. We call one failure type *toxicity* and the other *disease failure.* No toxicity and no disease failure is a *success* or *cure*. We assume disease failures are *contingent* on toxicity in that they can only be observed in the absence of toxicity. We also assume the probability of toxicity increases with the dose, and the probability of disease failure given no toxicity decreases with dose. Interest is in finding $c$-optimal designs for estimating the dose that maximizes the cure probability.

**Key words:** Dose response, $D$-optimal, $c$-optimal, continuation ratio model, phase II clinical trials, stress tests, ternary responses.

## 1 Introduction

We study $D$- and $c$-optimal designs for the contingent response models of Li, Durham, and Flournoy (1995). In the contingent response model there are two types of failure. We call one failure type *toxicity* and the other *disease failure*. No toxicity and no disease failure is a *success* or *cure*. We assume disease failures are *contingent* on toxicity in that they can only be observed in the absence of toxicity. We also assume the probability of toxicity increases with the dose, and the probability of disease failure given no toxicity decreases with dose. Interest is in finding $c$-optimal designs for estimating the dose that maximizes the cure probability. We call this dose *the optimal dose* and denote it by $\nu$.

Examples of data well fit by a contingent response model arise in many areas of study, with phase II clinical trials being an obvious one. In many phase II trials, a toxicity failure is fatal or so severe as to stop the trial. Then efficacy results are obtained only in the absence of toxicity failures. Hayes, Edie, and Durham describe testing the compressive strength of fibers. A fiber may fail after it is stressed under tension to a predetermined level.

Only if the fiber does not break under initial tension is a recoil test initiated. If the initial stress level is sufficiently high (but not high enough to lead to a failure), the fiber may then fail due to compressive stresses generated as the stored strain energy is recovered. The goal is to find the stress level that maximizes the probability of a recoil success without tensile failure. Other examples are described in Fan and Chaloner (2003).

In Section 2, we define the contingent response model, give examples, and show the continuation ratio model Agresti (1990) to be a special case. In Section 3, we define Fisher's information matrix for a location-scale family of contingent response models. A location-scale family of optimal designs is constructed, in the spirit of Ford, Torsney, and Wu (1992) by defining a single design called the *canonical optimal design* for the family. Then all other designs in the family are generated by transforming the canonical optimal design in a prescribed way. In Section 4, we establish a general location-scale family of $D$-optimal designs, and we find some $D$-optimal designs for the canonical positive/negative extreme value models which is defined in Section 2. In Section 5, we established a location-scale family of $c$-optimal designs for the positive/negative extreme value model, and find $c$-optimal designs for some canonical models. Tables of optimal designs can be obtained from the author. Complete proofs of the Lemma and Theorems 1-4 can be found in Rabie (under preparation).

## 2 The Contingent Response Model

Let

$$Y_{1j} = \begin{cases} 1 & \text{if the } j\text{th subject has a toxic response} \\ 0 & \text{else} \end{cases}$$

$$Y_{2j} = \begin{cases} 1 & \text{if the } j\text{th subject has disease failure} \\ 0 & \text{else} \end{cases}$$

for $j = 1, \ldots, N$. Only three outcomes are possible, namely, $\{Y_{1j} = 0, Y_{2j} = 0,\}, \{Y_{1j} = 0, Y_{2j} = 1\}$, and $\{Y_{1j} = 1\}$. We consider a location-scale family of parametric models: $P\{Y_{1j} = 1 \mid x\} = F(\alpha_1 + \beta_1 x) = F_x$; and $P\{Y_{2j} = 0 \mid Y_{1j} = 0, x\} = G(\alpha_2 + \beta_2 x) = G_x$ with $\bar{F}_x = 1 - F_x$ and $\bar{G}_x = 1 - G_x$; $x$ is log dose. Note the probability of success is

$$H_x = P\{Y_{1j} = 0, Y_{2j} = 0 \mid x\} = \bar{F}_x G_x; \tag{1}$$

the probability of toxicity is $F_x$; and the probability of disease failure is $P(Y_{1j} = 0, Y_{2j} = 1) = \bar{F}_x \bar{G}_x$.

The optimal dose is the maximum of $H_x$ which in some cases may be found by setting the derivative of (1) equal to zero, that is,

$$H'(\cdot) = \bar{F}(\cdot)G'(\cdot) - G(\cdot)F'(\cdot) = 0, \tag{2}$$

where $F'$ and $G'$ are the derivatives of $F$ and $G$. When (2) does not have a closed form solution, numerical methods will yield a practical solution. Conditions for $H(\cdot)$ to have maximum are given in Li, Durham, and Flournoy (1995). These conditions are satisfied if $F(x)$ and $G(x)$ are continuous probability distribution functions.

The *continuation ratio model* is a special case of the contingent response model in which $F_x$ and $G_x$ are logistic. To see this, we follow (Agresti, 1990, p. 319) and define $\pi_1(x) = P(\text{toxicity } midx) = F_x$; $\pi_2(x) = P(\text{cure} \mid x) = \bar{F}_x G_x$; and $\pi_3(x) = P(\text{disease failure} \mid x) = \bar{F}_x \bar{G}_x$. Continuation ratio logits are defined as $L_j = \text{logit } \rho_j(x)$, where $\rho_j(x) = \pi_j(x)/\sum_{i=j+1}^{3} \pi_i(x), j = 1, 2$. Modeling $L_j = \alpha_j + \beta_j$ is equivalent to assuming $F_x = \exp(\alpha_1 + \beta_1 x)/(1 + \exp(\alpha_1 + \beta_1 x))$; $G_x = \exp(\alpha_2 + \beta_2 x)/(1 + \exp(\alpha_2 + \beta_2 x)), \beta_1, \beta_2 > 0$. Designs for continuation ratio model have been studied by Thall and Russell (1998) and Zocchi and Atkinson (1999). For consistency, we call the continuation ratio model the logistic/logistic, or LL response model. Fan and Chaloner (2001, 2003) found optimal designs for the continuation ratio model, so this paper is a generalization of some of their work. They give the solution to (2) when $\beta_1 = \beta_2$. For $\beta_1 \neq \beta_2$, (2) does not have an explicit solution.

Li, Durham, and Flournoy (1995) give two contingent response models for which the optimal dose does have an explicit solution:

1) **The positive/negative extreme value contingent response model**
**(PNEV).**
Let $\bar{F}_x = \exp(-\exp(\alpha_1 + \beta_1 x))$ and $G_x = \exp(-\exp(-(\alpha_2 + \beta_2 x))), \beta_1, \beta_2 > 0$. Then the optimal dose for $\beta_1 \neq \beta_2$ is $\nu = [\log\{\beta_2/\beta_1\} - \alpha_1 - \alpha_2]/(\beta_1 + \beta_2)$. For $\beta_1 = \beta_2 = \beta$, the optimal dose is $\nu = -(\alpha_1 + \alpha_2)/2\beta$.

2) **The logistic/exponential contingent response model (LE).**
Let $F_x = \exp(\alpha_1 + \beta_1 x)/1 + \exp(\alpha_1 + \beta_1 x)$ and $G_x = \exp(\alpha_2 + \beta_2 x), \beta_1, \beta_2 > 0$, $x \in (-\infty, -\frac{\alpha_2}{\beta_2})$. For $\beta_1 > \beta_2$, the optimal dose is $\nu = [\log\{\beta_2/(\beta_1 - \beta_2)\} - \alpha_1]/\beta_1$. No solution to (2) exists for $\beta_1 \leq \beta_2$, but in that case the optimal dose is $x = -\alpha_2/\beta_2$.

## 3 The Design Problem

Assume we have $n_i$ independent observations taken at $x_i$, $i = 1, \ldots, K$. Let $r_i$ be the number of toxic responses, $m_i$ the number of cures and $(n_i - m_i - r_i)$ the number of disease failures at dose level $x_i$. The likelihood function is proportional to $\prod_{i=1}^{K} F_{x_i}^{r_i} (\bar{F}_{x_i} G_{x_i})^{m_i} (\bar{F}_{x_i} \bar{G}_{x_i})^{n_i - r_i - m_i}$.

The exact design problem is to find $K$, $\{x_i\}$, and $\{n_i\}$ to maximize an optimality criterion. We find approximate continuous designs (see Kiefer (1974)) in which $\xi_i$ replaces $n_i/N$ for fixed $N = \sum_i^K n_i$ with $0 \leq \lambda_i \leq 1$. So the problem becomes finding $K$ and the measure $\boldsymbol{\xi} = \{(x_i, \lambda_i), i = 1, \cdots, K\}$. Exact designs are found by integer approximation of continuous designs.

For the $D$-optimal criterion, we maximize the log determinant of Fisher's information matrix. For the $c$-optimal criterion, we find the design that minimizes $\dot{\mathbf{g}}^{\mathbf{T}}(\boldsymbol{\Theta})\mathbf{M}^{-1}(\boldsymbol{\xi}, \boldsymbol{\Theta})\dot{\mathbf{g}}(\boldsymbol{\Theta})$, where $\dot{\mathbf{g}}$ is the gradient vector of $\nu = \mathbf{g}(\boldsymbol{\Theta}), \mathbf{M}(\boldsymbol{\xi}, \boldsymbol{\Theta}) = \sum_i^K \lambda_i \mathbf{I}(x_i, \boldsymbol{\Theta})$ and $\mathbf{I}(x_i, \boldsymbol{\Theta})$ is the information of a single subject treated at $x_i$. $\mathbf{M}$ is the average per observation information matrix. $N\mathbf{M}(\boldsymbol{\xi}, \boldsymbol{\Theta})$ is Fisher's information for the whole experiment. Define $v(x) = F_x'^2/(F_x \bar{F}_x)$; $w(x) = \bar{F}_x G_x'^2/(G_x \bar{G}_x)$. For the contingent response model with $\boldsymbol{\Theta} = (\alpha_1, \beta_1, \alpha_2, \beta_2)$, $\mathbf{I}(\mathbf{x}, \boldsymbol{\Theta})$ is given by Lemma 1.

**Lemma 1.** *If* $\beta_1 \neq \beta_2$, *then* $\boldsymbol{\Theta} = (\alpha_1, \beta_1, \alpha_2, \beta_2)$ *and* $\mathbf{I}_{\mathbf{4}\times\mathbf{4}}(x, \boldsymbol{\Theta}) = \mathrm{diag}(\boldsymbol{A}, \boldsymbol{B})$, *where* $\mathbf{A}_{\mathbf{2}\times\mathbf{2}} = v(x, \boldsymbol{\Theta}) \begin{pmatrix} 1 & x \\ x & x^2 \end{pmatrix}$; $\mathbf{B}_{\mathbf{2}\times\mathbf{2}} = w(x, \boldsymbol{\Theta}) \begin{pmatrix} 1 & x \\ x & x^2 \end{pmatrix}$.

*If* $\beta_1 = \beta_2 = \beta$, *then* $\boldsymbol{\Theta} = (\alpha_1, \beta, \alpha_2)$ *and*

$$I_{3\times3}(x, \boldsymbol{\Theta}) = v(x, \boldsymbol{\Theta}) \begin{pmatrix} 1 & x & 0 \\ x & x^2 & 0 \\ 0 & 0 & 0 \end{pmatrix} + w(x, \boldsymbol{\Theta}) \begin{pmatrix} 0 & 0 & 0 \\ 0 & x^2 & x \\ 0 & x & 1 \end{pmatrix}. \tag{3}$$

The proof of this Lemma involves deriving the Fisher information matrix, with an algebraic rearrangement of terms.

## 4 Locally $D$-optimal Designs

The $D$-optimality criterion, used when we are interested in estimating the model parameters, is the maximum of the determinant of Fisher's Information. Since there is no closed form solution for our $D$- and $c$-optimal designs, the NPSOL Gill, Murray, Saunders, and Wright (1998) algorithm was used to find designs that maximize the optimality criterion; then the General Equivalence theorem was used to verify global optimality (see Silvey (1980)). Thus for each optimal design report in this section and the next, the NPSOL algorithm was run assuming $2, 3, 4, \ldots$ point designs, each with multiple starting values covering a wide range of possible design configurations. When a potential optimal candidate design was identified, directional derivatives were graphed to see that they appeared non-positive everywhere and zero at the candidate design points. In addition, we checked algebraically that the directional derivatives evaluated at the optimal design point were zero. Tables of canonical optimal designs for the NPEV, LE and LL response models are available from the authors, together with their supporting documentation.

### 4.1 Unequal Slopes $\beta_1 \neq \beta_2; \boldsymbol{\Theta} = (\alpha_1, \beta_1, \alpha_2, \beta_2); \boldsymbol{\theta} = (\alpha_2, \beta_2, \mu, r)$

Consider the contingent response model with distribution functions $F_x = F(\alpha_1 + \beta_1 x)$ and $G_x = G(\alpha_2 + \beta_2 x)$. Reparameterize $\boldsymbol{\Theta} = (\alpha_1, \beta_1, \alpha_2, \beta_2)$ to

$\boldsymbol{\theta} = (\alpha_2, \beta_2, \mu, r)$, where $\mu = \alpha_1 - r\alpha_2$ and $r = \beta_1/\beta_2$. When $\boldsymbol{\theta} = (0, 1, \mu, r)$ we say we have the *canonical* $(\mu, r)$ *model*. Let $\boldsymbol{\xi}^*$ be the $D$-optimal design for the canonical $(\mu, r)$ model. Theorem 1 tells us that the optimal designs for all other models with parameters $(\alpha_2, \beta_2, \mu, r)$ can be generated by a linear transformation of the canonical optimal designs.

**Theorem 1.** *If the design* $\boldsymbol{\lambda}_0^* = (\lambda_1^*, \lambda_2^*, \ldots, \lambda_K^*)$ *at* $(x_1^*, x_2^*, \ldots, x_K^*)$ *is locally* $D$*-optimal for* $\boldsymbol{\theta}_0 = (0, 1, \mu, r)$*, then* $\boldsymbol{\lambda}^* = \boldsymbol{\lambda}_0^*$ *at* $(\frac{x_1^* - \alpha_2}{\beta_2}, \frac{x_2^* - \alpha_2}{\beta_2}, \ldots, \frac{x_K^* - \alpha_2}{\beta_2})$ *is locally* $D$*-optimal for* $\boldsymbol{\theta} = (\alpha_2, \beta_2, \mu, r)$.

*Proof.* By Lemma 1, $\det \mathbf{M}(\boldsymbol{\lambda}, \boldsymbol{\theta}) = \det \sum_i^K \lambda_i \mathbf{A}(x_i, \boldsymbol{\theta}) \times \det \sum_i^K \lambda_i \mathbf{B}(x_i, \boldsymbol{\theta})$. Expanding terms we obtain $\det \mathbf{A}(\lambda, \boldsymbol{\theta}) = \det \mathbf{A}(\lambda_0, \boldsymbol{\theta}_0)/\beta_2^2$ and $\det \mathbf{B}(\lambda, \boldsymbol{\theta}) = \det \mathbf{B}(\lambda_0, \boldsymbol{\theta}_0)/\beta_2^2$. Thus $\det \mathbf{M}(\boldsymbol{\lambda}, \boldsymbol{\theta}) = \det \mathbf{M}(\boldsymbol{\lambda}_0, \boldsymbol{\theta}_0)/\beta_2^4$. Now consider the one-to-one onto transformation $\tau(\boldsymbol{\xi}_0) = \boldsymbol{\xi}$ defined on the designs: $\boldsymbol{\xi}_0 = (\xi_i, x_i)$; $\boldsymbol{\xi} = (\xi_i, (x_i - \alpha_2)/\beta_2)$, $i = 1, \ldots, l$. Then $\tau^{-1}$ is well-defined and

$$
\begin{aligned}
\boldsymbol{\xi}^* &= \arg\max_{\boldsymbol{\xi}} \det(\mathbf{M}(\boldsymbol{\xi}, \boldsymbol{\theta})) && \text{definition of } \boldsymbol{\xi}^* \\
&= \arg\max_{\boldsymbol{\xi}} \beta_2^4 \det((\boldsymbol{\xi}, \boldsymbol{\theta})) && \beta_2^4 \text{ is constant} \\
&= \arg\max_{\boldsymbol{\xi}} \det((\boldsymbol{\xi}_0, \boldsymbol{\theta}_0)) && \text{from above} \\
&= \arg\max_{\tau(\boldsymbol{\xi}_0)} \det((\boldsymbol{\xi}_0, \boldsymbol{\theta}_0)) && \tau(\boldsymbol{\xi}_0) = \boldsymbol{\xi} \\
&= \arg\max_{\tau^{-1}\tau(\boldsymbol{\xi}_0)} \det((\boldsymbol{\xi}_0, \boldsymbol{\theta}_0)) && \tau \text{ is 1-1 onto}
\end{aligned}
$$

It follows that

$$
\boldsymbol{\xi}_0^* = \arg\max_{\boldsymbol{\xi}_0} \det((\boldsymbol{\xi}_0, \boldsymbol{\theta}_0)) \qquad \text{definition of } \boldsymbol{\xi}_0^* \text{ and } \tau^{-1}(\boldsymbol{\xi}) = \boldsymbol{\xi}_0.
$$

Maximizing $\det \mathbf{M}(\boldsymbol{\xi}, \boldsymbol{\theta})$ is the same as maximizing $\log\det \mathbf{M}(\boldsymbol{\xi}, \boldsymbol{\theta})$ over all possible designs $\boldsymbol{\xi}$. Therefore, if $\boldsymbol{\xi}_0^*$ is locally $D$-optimal for $\boldsymbol{\theta}_0$, then $\boldsymbol{\xi}^*$ is locally $D$-optimal for $\boldsymbol{\theta}$. □

Locally $D$-optimal designs for several canonical $(\mu, r)$ PNEV models consist of two, three, four points for small, moderate and large negative values of $\mu$, respectively. For positive values of $\mu$, the optimal designs consist of two and three design points depending on the value of $r$. For the many parameters values we studied, the number of optimal design points for PNEV model is the same as the number of optimal designs points for the LL model found in Fan and Chaloner (2003).

The $D$-optimal design for single negative extreme value function with $\alpha_2 = 0, \beta_2 = 1$ consists of two points with equal weights at $(x_1 = 1.3377, x_2 = -0.9796)$. The optimal design for the single positive extreme value function with $\alpha_1 = 0, \beta_1 = 1$ consists of two design points with equal weights at $(-x_1, -x_2)$. It is interesting that the optimal design points for these $G_x$ and

$F_x$ are the same as the optimal design points $\bar{G}_x$ and $\bar{F}_x$, respectively. When $F_x$ and $\bar{G}_x$ become quite separate, optimal designs for PNEV tend toward having four equally weighted points which are the optimal design points for $F_{\mu+rx}$ and $\bar{G}_x$ combined. This is consistent with findings in Fan and Chaloner (2003) for the LL model. As $\mu \to -\infty$, we conjecture that the optimal design for the canonical $(\mu, r)$ PNEV model will have four equally weighted points at $[x_1, x_2, (-x_1 - \mu)/r, (-x_2 - \mu)/r]$.

### 4.2 Equal Slopes $\beta_1 = \beta_2 = \beta$; $\boldsymbol{\Theta} = (\alpha_1, \beta, \alpha_2)$; $\boldsymbol{\theta} = (\alpha_2, \beta, \mu)$

Consider the contingent response model with distribution functions $F_x = F(\alpha_1 + \beta x)$ and $G_x = G(\alpha_2 + \beta x)$. Reparameterize $\boldsymbol{\Theta} = (\alpha_1, \beta, \alpha_2)$ to $\boldsymbol{\theta} = (\alpha_2, \beta, \mu)$, where $\mu = \alpha_1 - \alpha_2$. When $\boldsymbol{\theta} = (0, 1, \mu)$, we say we have a *canonical* $(\mu)$ *model.* Let $\boldsymbol{\xi}^*$ be the $D$-optimal design for the canonical $(\mu)$ model. Theorem 2 tells us that the optimal designs for all other $(\mu)$ models can be generated by a linear transformation of the canonical optimal designs.

**Theorem 2.** *If the design* $\boldsymbol{\lambda}_0^* = (\lambda_1^*, \lambda_2^*, \ldots, \lambda_K^*)$ *at* $(x_1^*, x_2^*, \ldots, x_K^*)$ *is locally* $D$*-optimal for* $\boldsymbol{\theta}_0 = (0, 1, \mu)$*, then* $\boldsymbol{\lambda}^* = \boldsymbol{\lambda}_0^*$ *at* $(\frac{x_1^* - \alpha_2}{\beta}, \frac{x_2^* - \alpha_2}{\beta}, \ldots, \frac{x_K^* - \alpha_2}{\beta})$ *is locally* $D$*-optimal for* $\boldsymbol{\theta} = (\alpha_2, \beta, \mu)$.

*Proof.* It can be shown algebraically that $\det(\mathbf{M}(\boldsymbol{\theta}, \boldsymbol{\lambda})) = \det(\mathbf{M}(\boldsymbol{\lambda}_0, \boldsymbol{\theta}_0)/\beta^2$. The rest of the proof is analogous to that of Theorem 1. □

$D$-optimal designs for several canonical $(\mu)$ PNEV models are found to consist of two, three and four points for small, moderate and large negative values of $\mu$, respectively. For positive values of $\mu$, the optimal designs consist of two points. For large negative values of $\mu$, the optimal designs are not the optimal designs for $\bar{G}_x$ and $F_{\mu+rx}$ combined. However, they are still four point designs located approximately at [-0.8537, 1.0773, (0.8537 - $\mu$), (-1.0773 -$\mu$)] with approximate weights (0.2900, 0.2100, 0.2900, 0.2100). Having four design points is consistent with the results of Fan and Chaloner (2003) for the LL model but it is different in having unequal weights.

## 5 Locally $c$-optimal Designs

Our interest is to find the $c$-optimal design for estimating the optimal dose $\nu = g(\boldsymbol{\Theta})$. For non-linear models the $c$-optimal criterion is to minimize

$$\varphi_c(\boldsymbol{\Theta}, \boldsymbol{\lambda}) = \dot{\mathbf{g}}^{\mathrm{T}}(\boldsymbol{\Theta})\mathbf{M}^{-1}(\boldsymbol{\xi}, \boldsymbol{\Theta})\dot{\mathbf{g}}(\boldsymbol{\Theta}),$$

where $\dot{\mathbf{g}}$ is the gradient vector of $g(\boldsymbol{\Theta})$. Potential optimal designs for the PNEV model were found using the NPSOL algorithm. Optimality was verified by establishing graphically that the directional derivatives are non-positive everywhere and finding algebraically that the directional derivatives evaluated at the optimal design points were equal to zero. Singular optimal designs occur, in which case, Silvey's theorem (see Silvey (1978)) was used to verify optimality. Some comments regarding the application of Silvey's theorem are provided in the proof of Theorem 5.

## 5.1 Unequal Slopes $\beta_1 \neq \beta_2; \Theta = (\alpha_1, \beta_1, \alpha_2, \beta_2); \theta = (\alpha_2, \beta_2, \mu, r)$

We find $c$-optimal designs for the optimal dose for PNEV model, namely, $\nu = g(\Theta) = (\log(\beta_2/\beta_1) - \alpha_1 - \alpha_2)/(\beta_1 + \beta_2)$. The gradient vector is $\partial g(\Theta)/\partial\Theta_i =$

$$\left(\frac{-1}{\beta_1+\beta_2}; \frac{-1}{\beta_1(\beta_1+\beta_2)} - \frac{\log(\beta_2/\beta_1)-\alpha_1-\alpha_2}{(\beta_1+\beta_2)^2}; \frac{-1}{\beta_1+\beta_2}; \frac{1}{\beta_2(\beta_1+\beta_2)} - \frac{\log(\beta_2/\beta_1)-\alpha_1-\alpha_2}{(\beta_1+\beta_2)^2}\right)^T$$

We reparameterize as in (4.1). It follows from Theorem 3 that once the locally $c$-optimal designs are found for a canonical $(\mu, r)$ model, the locally $c$-optimal designs for all other models in the family can be found from the canonical design.

**Theorem 3.** *If the design* $\boldsymbol{\lambda}_0^* = (\lambda_1^*, \lambda_2^*, \ldots, \lambda_K^*)$ *at* $(x_1^*, x_2^*, \ldots, x_K^*)$ *is locally c-optimal for* $\boldsymbol{\theta}_0 = (0, 1, \mu, r)$*, then* $\boldsymbol{\lambda}^* = \boldsymbol{\lambda}_0^*$ *at* $x = (\frac{x^*-\alpha_2}{\beta_2}, \frac{x_2^*-\alpha_2}{\beta_2}, \ldots, \frac{x_K^*-\alpha_2}{\beta_2})$ *is locally c-optimal for* $\boldsymbol{\theta} = (\alpha_2, \beta_2, \mu, r)$.

*Proof.* It can be shown that $\varphi_c(\boldsymbol{\theta}, \boldsymbol{\lambda}) = (\frac{1}{\beta_2^2})\varphi_c(\boldsymbol{\theta}_0, \boldsymbol{\lambda}_0)$ using Maple software, and simplifying. The rest of the proof is analogous to that of Theorem 1. □

We found $c$-optimal designs for the canonical $(\mu, r)$ PNEV models. The optimal designs consist of two and three points. For large positive values of $\mu$ the probability of success is negligible and not of interest.

## 5.2 Equal Slopes $\beta_1 = \beta_2 = \beta; \Theta = (\alpha_2, \beta, \alpha_1); \theta = (\alpha_2, \beta, \mu)$

If $\beta_1 = \beta_2 = \beta$, the solution to (2) is $\nu = g(\Theta) = -(\alpha_1 + \alpha_2)/2\beta$ with gradient $\partial g(\Theta)/\partial\Theta_i = \left(-\frac{1}{2}\beta, \frac{\alpha_1+\alpha_2}{2\beta^2}, -\frac{1}{2}\beta\right)^T$. We reparameterize as in (4.2). It follows from Theorem 4 that once the locally $c$-optimal designs are found for the canonical $(\mu)$ model, the locally $c$-optimal designs for all other models in the family can be found from the canonical $(\mu)$ design.

**Theorem 4.** *If the design* $\boldsymbol{\lambda}_0^* = (\lambda_1^*, \lambda_2^*, \ldots, \lambda_K^*)$ *at* $(x_1^*, x_2^*, \ldots, x_K^*)$ *is locally c-optimal for* $\boldsymbol{\theta}_0 = (0, 1, \mu)$*, then* $\boldsymbol{\lambda}^* = \boldsymbol{\lambda}_0^*$ *at* $(\frac{x_1^*-\alpha_2}{\beta}, \frac{x_2^*-\alpha_2}{\beta}, \ldots, \frac{x_K^*-\alpha_2}{\beta})$ *is locally D-optimal for* $\boldsymbol{\theta} = (\alpha_2, \beta, \mu)$.

The proof of this theorem is analogous to that of Theorem 3.

The $c$-optimal designs for some canonical $(\mu)$ PNEV models are found to consist of one and two points. When $F_x$ and $\bar{G}_x$ are quite separate, the optimal design points are close to $(-0.465, 0.465 - \mu)$ with equal weights. These points are approximately the 20%-percentiles for $G_x$ and $F_{\mu+rx}$.

Theorem 5 gives canonical $c$-optimal designs for $\boldsymbol{\theta_0} = (0,1,\mu) = (0,1,0)$ and $(0,1,1)$. These $c$-optimal designs have only one point.

**Theorem 5.** *For $\boldsymbol{\theta} = (\alpha_2, \beta, \mu) = (0,1,0)$ and $(0,1,1)$, the c-optimal design is an one point design putting mass 1 at the optimal dose $x = 0$ and $x = -1/2$, respectively.*

*Proof.* For these parameters, the information matrix is singular. Because $\varphi_c$ is concave, we can verify that the design is $c$-optimal by finding a matrix $\mathbf{H}$ such that $(\mathbf{M} + \mathbf{HH^T})^{-1}$ exists and is a generalized inverse of $\mathbf{M} = \mathbf{M}(\boldsymbol{\lambda}, \boldsymbol{\theta})$ such that $F_\varphi(\mathbf{M} + \mathbf{HH^T}, \mathbf{I}(x, \boldsymbol{\theta})) \leq 0$ for all $x$ and the sup $F_\varphi(\mathbf{M} + \mathbf{HH^T}, \mathbf{I}(x, \boldsymbol{\theta})) = 0$; $F_\varphi$ is the directional derivative of $\varphi$ as defined in Silvey (1978). The Fisher's information for PNEV when $\boldsymbol{\theta} = (\alpha_2, \beta, \mu) = (0,1,0)$ and $(\alpha_2, \beta, \mu)$, respectively, is $\mathbf{M}(\boldsymbol{\theta}, \boldsymbol{\lambda}) =$

$$\begin{pmatrix} 0.5819767 & 0 & 0 \\ 0 & 0 & 0 \\ 0 & 0 & 0.2140973 \end{pmatrix} \text{ and } \begin{pmatrix} 0.6471598 & -0.3235799 & 0 \\ -0.3235799 & 0.1929014 & -0.062223 \\ 0 & -0.062223 & 0.124446 \end{pmatrix}.$$

For $\boldsymbol{\theta} = (\alpha_2, \beta, \mu) = (0,1,0)$ and $(0,1,1)$ take $\mathbf{H} = \left(\frac{100}{59}\sqrt{3}, -\frac{2}{\sqrt{3}}, 0\right)^T$ and $\mathbf{H} = \left(\frac{100}{49}\sqrt{3}, -\frac{2}{\sqrt{3}}, 0\right)^T$, respectively. Then $\mathbf{G} = (\mathbf{M} + \mathbf{HH^T})^{-1} =$

$$\begin{pmatrix} 1.718282 & 1.456171 & 0 \\ 1.456171 & 1.984043 & 0 \\ 0 & 0 & 4.670774 \end{pmatrix} \text{ and } \begin{pmatrix} 7.222648 & 8.135674 & 4.067837 \\ 8.135674 & 9.832958 & 4.916479 \\ 4.067837 & 4.916479 & 10.493853 \end{pmatrix},$$

respectively. To find $\mathbf{H}$, we searched iteratively, starting with the value given in Fan and Chaloner (2003) for the analogous LL response model. Iterating over candidate values of $\mathbf{H}$, we used NPSOL to generate candidate optimal designs for which we graphically established that directional derivatives are non-positive and algebraically verified the sup $F_\varphi(\mathbf{M} + \mathbf{HH^T}, \mathbf{I}(x, \boldsymbol{\theta})) = 0$. □

## 6 Conclusion

In this paper, we give Fisher's information matrix for the contingent response model. We showed that the continuation ratio model is a special case.

In Sections 4.1 and 4.2 we found numerically that the $D$-optimal designs for PNEV consist of four points when $\mu$ assumes large negative values; when $\beta_1 = \beta_2$ the optimal points are not equally weighted. We conjecture that the *limiting optimal designs* in Fan and Chaloner (2003) can be used to prove these numerical results.

Under the assumptions of Section 4 we showed that once locally $D$-optimal designs are found numerically for a canonical design family, all other designs in the family are found by a linear transformation of the canonical designs. This is true for any contingent response model.

The proofs of Theorems 3 and 4 for $c$-optimal designs rely on the form of the optimal dose $\nu = g(\boldsymbol{\Theta})$ in the PNEV. Analogous results were obtained in Fan and Chaloner (2001, 2003) for the LL model. We conjecture the general findings hold for all contingent response models for which there exists a point with maximum success probability, but proofs must be done model by model.

## References

A. Agresti. *Categorical Data Analysis.* Wiley Series in Probability and Mathematical Statistics: Applied Probability and Statistics. John Wiley & Sons Inc., New York, 1990.

S.K. Fan and K. Chaloner. Optimal designs for a continuation-ratio model. In A.C. Atkinson, P. Hackl, and W.G. Müller, editors, *mODa 6—Advances in Model-Oriented Design and Analysis, (Puchberg/Schneeberg, June 25-29, 2001)*, Contributions to Statistics, pages 77–85. Physica-Verlag, Heidelberg, 2001.

S.K. Fan and K. Chaloner. Optimal designs for a trinomial response. preprint, 2003.

I. Ford, B. Torsney, and C.F.J. Wu. The use of a canonical form in the construction of locally optimal designs for nonlinear problems. *J. Roy. Statist. Soc. Ser. B*, 54(2):569–583, 1992.

P.E. Gill, W. Murray, M.A. Saunders, and M.H. Wright. *User's Guide for NPSOL 5:0 A Fortran package for Nonlinear Programming*, volume 86-1 of *Technical Report SOL.* Stanford University Department of Operations Research Systems Optimization Laboratory, Stanford, CA, 1998.

G.J. Hayes, D.D. Edie, and S.D. Durham. Statistical analysis of recoil test data for carbon fibers.

J. Kiefer. General equivalence theory for optimum designs (approximate theory). *Ann. Statist.*, 2:849–879, 1974.

Z. Li, S.D. Durham, and N. Flournoy. An adaptive design for maximization of a contingent binary response. In *Adaptive designs (South Hadley, MA, 1992)*, volume 25 of *IMS Lecture Notes Monogr. Ser.*, pages 179–196. Inst. Math. Statist., Hayward, CA, 1995.

H.S. Rabie. *Optimal Designs for the Contingent Response Model.* PhD thesis, Department of Statistics, University of Missouri-Columbia, under preparation.

S.D. Silvey. Optimal design measures with singular information matrices. *Biometrika*, 65(3):553–559, 1978.

S.D. Silvey. *Optimal design.* Chapman & Hall, London, 1980. An introduction to the theory for parameter estimation, Monographs on Applied Probability and Statistics.

P.F. Thall and K. Russell. A strategy for dose-finding and safety monitoring based on efficacy and adverse outcomes in phase I / II clinical trials. *Biometrics*, 54:251–264, 1998.

S.S. Zocchi and A.C. Atkinson. Optimum experimental designs for multinomial logistic models. *Biometrics*, 55(2):437–444, 1999.

# Bayesian $D$-Optimal Designs for Generalized Linear Models with a Varying Dispersion Parameter

E. Rodrigues Pinto[1] and A. Ponce de Leon[2]

[1] COPPE/UFRJ-Department of Product Engineering, Universidade Federal do Rio de Janeiro - Brasil edmilsonr@hotmail.com.br
[2] UERJ-Institute of Social Medicine, Universidade do Estado do Rio de Janeiro - Brasil ponce@ims.uerj.br

**Summary.** In this article we extend the results of Atkinson and Cook (1995) concerning $D$-optimum designs for heteroscedastic regression models. They introduced a variety of $D$-optimality criteria, under the hypothesis of Gaussian errors, when the mean response and the variance are parametric functions of possibly different sets of explanatory variables. The aim of this paper is to extend such ideas under other distributional assumptions rather than Gaussian. We regard an unknown likelihood but suppose the relationship between the mean response and the variance is known. Bayesian $D$-optimum designs for models pertaining to the exponential family are then proposed. This methodology is only feasible by applying the concept of extended quasi-likelihood, first proposed in Nelder and Pregibon (1987). A potential area of application of the results proposed here lies in experimental design in off-line quality control.

**Key words:** Bayesian $D$-optimum designs, heteroscedastic models, extended quasi-likelihood, Taguchi methods.

## 1 Introduction

Inspired by the methodology introduced by Taguchi, there has been an increasing interest in the last few years in models that not only can describe the mean of the response as usually, but also the response variance. Such an interest is particularly strong in applied statistics research in the industry (Vining and Schaub (1996)).

Taguchi introduced a new quality philosophy, producing a unique and powerful discipline for quality improvement that differs from traditional practices. In many fields of industrial experimentation Taguchi's philosophy translates into finding ways for achieving and controlling conditions under which the value of the expected response remains near a specified level, with minimum variability. Therefore, it is fair to say that Taguchi's methodology leads to the joint modelling of mean and variance, with a linear predictor for the response mean and another for the response variance.

Following the publication of Taguchi's ideas, several methodologies, supposedly statistically efficient, have been proposed in the literature so as to formalize such ideas. For instance, in order to reduce the number of experimental runs, fractional factorial designs are used as alternative proceedings to Taguchi's methods, however they can also produce a large number of runs.

In this context, a question that naturally arises is as follows. Is it possible to choose the points of an experimental design region, so that functions of the mean as well as variance functions are estimated optimally in some sense?

Assuming a Gaussian response, Atkinson and Cook (1995) showed how to obtain $D$-optimal designs for regression models with heteroscedastic errors. They regarded a class of models in which the error variance and the mean response are parametric functions of possibly different sets of explanatory variables. Under the assumption of a known likelihood, they developed extensions of optimal design theory that lead to optimum designs when the parameters of both the mean and the variance functions, or the parameters of either function, are of main interest.

In the same field of research, Vining and Schaub (1996) proposed a semi-Bayesian approach also for heteroscedastic Gaussian models, in which the optimality criterion is to maximize the determinant of the expected information matrix. More recently, Tack, Goos, and Vandebroek (2002) also showed, for heteroscedastic linear models, that any prior knowledge about the sign of the variance function parameters leads to considerably more efficient designs than the classical ones based on homoscedastic error assumptions.

In this article we extend the work of Atkinson and Cook into the more general framework of GLM (Generalized Linear Models). The likelihood is assumed unknown, but we suppose there is a relationship between the mean and the variance and that it is known. In this context and regarding the purpose of making statistical inference a new criterion of parameter estimation is required, namely the extended quasi-likelihood function.

The concept of extended quasi-likelihood was first proposed in Nelder and Pregibon (1987). Later, this concept was used in Nelder and Lee (1991) to propose methods based on the theory and framework of GLM as an alternative to Taguchi's methods, the latter having been criticized in the statistical literature. They introduced the joint modelling of mean and dispersion with two linked GLM, one for the mean response and the other for the response dispersion.

## 2 The Quasi-Likelihood

The quasi-likelihood function $Q$ or more conveniently the log quasi-likelihood function is defined as a function whose derivative with respect to the mean, $\mu$, is equal to that of the logarithm of the likelihood function, provided each observation, $y$, had a distribution in the exponential family.

Formally the quasi-likelihood function is defined as $Q(\mu, y) = \int_y^\mu \frac{y-t}{\varphi V(t)} dt$ or $\frac{\partial Q(\mu;y)}{\partial \mu} = \frac{y-\mu}{\varphi V(\mu)}$, where $\varphi$ is a dispersion parameter and $V(\cdot)$ is a known variance function.

The use of $Q$ as a criterion function for parameter estimation allows the class of GLM to be extended to models defined only by the properties of the first two moments. The quasi-likelihood $Q$ will be a true likelihood if there is a distribution of the GLM type where $\mathrm{Var}\,(Y) = \varphi V(\mu)$.

Quasi-likelihoods allow two sorts of extensions of GLM. In the first, GLM with fixed $\varphi = 1$ can be extended to allow a varying $\varphi$. For example, log linear models with Poisson errors for which $\mathrm{Var}\,(Y) = \mu$ can be extended to allow overdispersion, that is $\mathrm{Var}\,(Y) = \varphi\mu$. In the second extension, $V(\mu)$ may take a form which does not correspond to that of a standard GLM, for example by taking $V(\mu) = \mu^\alpha$ for different values of $\alpha$.

Quasi-likelihood yields the same estimating equations as standard GLM, resulting in maximum quasi-likelihood estimates instead of maximum likelihood estimates. It also gives rise to other usual quantities and tools developed under GLM such as the deviance function and both deviance and Pearson residuals. In brief, quasi-likelihood can be thought of as a way of defining nearly exponential distributions. Quasi-likelihoods are essential for modelling overdispersed counts or proportions (see McCullagh and Nelder (1989)).

## 3 The Extended Quasi-Likelihood

In GLM the deviance function is used to measure the discrepancy of a fit or goodness of fit that is useful to compare nested models, however it cannot be used to compare models with different variance functions or dispersion model specifications. A suitable criterion to be used for fitting the joint model for mean and dispersion is the EQL (Extended Quasi-Likelihood), denoted by $Q^+$. The function $Q^+$, for a single observation $y_i$ with mean $\mu_i$ and variance $\varphi_i V(\mu_i)$, is defined as follows.

$$Q^+(y_i, \mu_i) = -\frac{1}{2}\frac{d_i}{\varphi_i} - \frac{1}{2}\log\left\{2\pi\varphi_i V(y_i)\right\}, \tag{1}$$

where $d_i = -2\int_{y_i}^{\mu_i} \frac{y_i - u}{V(u)} du$ is the deviance individual component corresponding to the $i$th observation.

The extended quasi-deviance is given by

$$-2Q^+ = \sum_i \frac{d_i}{\varphi_i} + \sum_i \log\left(2\pi\varphi_i V(y_i)\right). \tag{2}$$

Note that $Q^+$ and $Q$ need not presuppose knowledge of the full distribution of $Y$, but only the relationship between the mean response and the response variance. The estimates of $\beta$ obtained by maximizing $Q^+$ are the same as those obtained by maximizing $Q$, that is they are maximum quasi-likelihood estimates. This follows because $Q^+$ is a linear function of $Q$ with coefficients independent of $\beta$.

If there is a distribution of the exponential family with a given variance function, it turns out that the EQL is the Saddle-point approximation to that distribution. The approximation is exact for the normal and inverse Gaussian distributions, differs for gamma distribution by a function that depends only on the shape parameter, and for discrete distributions by replacement of all the factorials by their Stirling approximations (see Jørgensen (1987)).

The quasi-likelihood function requires knowledge of the variance function up to a multiplicative constant. However, in using $Q^+$ this condition can be relaxed. In certain applications, $\varphi$ may not be a constant and thus there might be interest in modelling $\varphi$ as a function of a set of covariates. In this case, the extended quasi-likelihood function provides an objective criterion upon which this modelling can be carried out. Here, we suppose that the variance function pertains to the one family of functions indexed by an unknown parameter $\lambda$, that is

$$-2Q_\lambda^+ = \sum_i \frac{d_{\lambda i}}{\varphi_i} + \sum_i \log\left(2\pi\varphi_i V_\lambda(y_i)\right), \tag{3}$$

where $d_\lambda(y;\mu) = -2\int_y^\mu \frac{y-u}{V_\lambda(u)}du$.

A useful family of variance functions arises by considering powers of $\mu$, $V_\lambda(\mu) = \mu^\lambda$ (see Nelder and Pregibon (1987)). The most common values of $\lambda$ are $0, 1, 2, 3$. These correspond to variance functions associated with the normal, Poisson, gamma and inverse Gaussian distributions, respectively. For this family of variance functions, the deviance function individual contribution is given by

$$d_\lambda(y;\mu) = \begin{cases} 2\{y\log(y/\mu) - (y-\mu)\} & \lambda = 1 \\ 2\{y/\mu - \log(y/\mu) - 1\} & \lambda = 2 \\ \frac{2\{y^{2-\lambda} - (2-\lambda)y\mu^{1-\lambda} + (1-\lambda)\mu^{2-\lambda}\}}{(1-\lambda)(2-\lambda)} & \text{otherwise} \end{cases}.$$

Another type of variance function, that is mainly used for binary data models, is the family $V_\lambda(\mu) = \mu^\lambda(1-\mu)^\lambda$. For a fixed $\lambda$, the quasi-likelihood estimate of $\beta$ can be obtained.

## 4 The Model

Let $y_i$ be the $i$th response of interest, $f_i(x_i)$ be the $p \times 1$ vector representing the $i$th setting of covariates presumed to influence the expected response, and $g_i(z_i)$ be the $q \times 1$ vector representing the $i$th setting of covariates presumed to influence the variance. Moreover $z_i$ is allowed to contain some or all of the components of vector $x_i$ as well as other covariates.

Suppose that the distribution of $Y$ is unknown. However, it is known that $\mathrm{Var}(Y_i) = \varphi_i V(\mu_i)$, where $\mathrm{E}(Y_i) = \mu_i$. Let $k$ denote the link function for the mean, that is $\eta_i = k(\mu_i) = {f_i}^T\beta$, where $\beta$ is a $p \times 1$ vector of unknown parameters. Let $h$ denote the link function for the dispersion model, usually taken to be the natural logarithm, that is $\tau_i = h(\varphi_i) = {g_i}^T\gamma$, where $\gamma$ is a $q \times 1$ vector of unknown parameters.

When the distribution of $Y$ is supposed unknown and $\varphi$ is allowed to vary, the extended quasi-likelihood is used for parameter estimation in such a model.

Denote $X^T = (f_1(x_1), \ldots, f_n(x_n))$ and $Z^T = (g_1(z_1), \ldots, g_n(z_n))$, where ${f_i}^T(x_i) = (f_{i1}(x_i), \ldots, f_{ip}(x_i))$ and ${g_i}^T(z_i) = (g_{i1}(z_i), \ldots, g_{iq}(z_i))$, for $i = 1, \ldots, n$ and where $f_{i1}(x_i) = g_{i1}(z_i) = 1$ for all $i$.

Then, based on the extended quasi-likelihood as the estimation criterion function, the Fisher information matrix is given as

$$I(X, Z|\beta, \gamma) = \begin{bmatrix} X^TWPX & 0 \\ 0 & Z^TVPZ \end{bmatrix},$$

where $W = \mathrm{diag}(w_i)$, $V = \mathrm{diag}(v_i)$, $P = \mathrm{diag}(p_i)$, $w_i = \left(\frac{\partial \mu_i}{\partial \eta_i}\right)^2 \frac{1}{\varphi_i V(\mu_i)}$, $v_i = \left(\frac{\partial \varphi_i}{\partial \tau_i}\right)^2 \frac{1}{2{\varphi_i}^2}$, and $p_i$ is the weight for the $i$th design point. The notation $\mathrm{diag}(\cdot)$ represents the diagonal matrix.

## 5 The Equivalence Theorem

Following Atkinson and Cook (1995), we consider a generic design problem with a $t \times 1$ parameter vector $\theta$ and a design vector $u$ that is confined to a compact design region $\chi$. We assume that the information per observation $I(u|\theta)$ is the sum of $r$ rank-1 matrices, $I(u|\theta) = \sum_{j=1}^{r} h_j {h_j}^T$, where $h_j = h_j(u, \theta)$ is a function of the design vector $u$ and the unknown parameter vector $\theta$.

Here $r = 2$, thus the information matrix for the $i$th observation can be written as $I_i(u_i|\theta) = \sum_{j=1}^{2} h_{ij} {h_{ij}}^T$, where ${h_{i1}}^T = \left({f_i}^T(x_i)(w_i)^{\frac{1}{2}}, 0^T\right)$ and ${h_{i2}}^T = \left(0^T, (v_i)^{\frac{1}{2}} {g_i}^T(z_i)\right)$. Here, $u_i = (f_i(x_i), g_i(z_i))$ and $\theta = (\beta, \gamma)$.

The total information $I(\theta)$ in the experiment is represented by the average with respect to the design measure $\xi$ on $\chi$ of the information per observation,

$$I(\theta|\xi) = \int_{\chi} I(u|\theta) d\xi(u). \tag{4}$$

Then, the $D$-optimal design problem is to find the design measure from a specified class of design measures $\Xi$, that maximizes $|I(\theta|\xi)|$, the determinant of the total information.

For the models described in this paper, the total information matrix $I(\theta|\xi)$ depends on the value of the parameter $\theta$. To avoid such dependence, the approach we use here is to obtain Bayesian designs as described, for example in Chaloner and Larntz (1989) for nonlinear models or in Atkinson and Donev (1992).

The Bayesian criterion function, which is based on approximating the pre-posterior loss is $\mathrm{E}_\theta[\log|I(\theta|\xi)|]$, where $\mathrm{E}_\theta$ refers to the expectation with respect to a prior distribution for $\theta$.

Again, following Atkinson and Cook, let us denote

$$d_j(u,\theta,\xi) = {h_j}^T(u,\theta)I^{-1}(\theta|\xi)h_j(u,\theta) \tag{5}$$

and let $\Xi$ denote the class of all probability measures on $\chi$. Then, a General Equivalence Theorem holds, that is the three conditions on the optimum design measure $\xi^*$ stated below are equivalent.

a) The design $\xi^*$ maximizes $\mathrm{E}_\theta[\log|I(\theta|\xi)|]$ over $\Xi$
b) $\min_\Xi \max_\chi E_\theta \sum_j d_j(u,\theta,\xi) = \max_\xi E_\theta \sum_j d_j(u,\theta,\xi^*)$
c) $\max_\xi E_\theta \sum_j d_j(u,\theta,\xi^*) = t$,

where $t$ is the number of parameters in the model. Like in other model contexts, condition c) provides a way to check whether a given design is in fact Bayesian $D$-optimal.

Here, we can write the standardized variance for the point $i$, corresponding to the $i$th row of the matrices $X$ and $Z$ by $d_i = \sum_{j=1}^{2} d_{ij}$, where $d_{ij} = {h_{ij}}^T(u_i,\theta)I^{-1}(\theta|\xi)h_{ij}(u_i,\theta)$, that is

$$d_i = w_i {f_i}^T(x_i)(X^TWX)^{-1} f_i(x_i) + v_i {g_i}^T(z_i)(Z^TVZ)^{-1} g_i(z_i). \tag{6}$$

## 6 Cake Mix Example

This illustration is inspired in a problem concerning the development of a new cake mix for the consumer market discussed in Atkinson and Donev (1992). In the original example the product needs to be robust enough to resist incorrect cooking conditions, represented by the environmental factors of oven temperature $x_4$ and baking time $x_5$. The three design factors, which are under the manufacturer's control, are the amounts of flour, sugar and egg, denoted by $x_1$, $x_2$ and $x_3$ respectively. The experimental design consisted of a $2^3$ factorial with a central point in the design factors crossed with a $2^2$ factorial plus a central point in the environmental factors. The response was a taste index for which the larger the value the better the mixture.

In order to illustrate the theory, we consider the same problem as in Atkinson and Donev, that is the problem of preparing a cake mix robust enough to resist incorrect cooking conditions, but now with a new taste response strategy and a new criterion as the response.

Consider a panel consisting of a large number of cake experts. Each expert will eat 9 pieces of cake, from the nine trials. After tasting each piece of cake, the expert will say whether that mixture is approved or not. It is presumed that results of different tastes are independent in the sense that after a judge tries a piece of cake this will not influence the judgement of other pieces. Furthermore it is presumed that the judges are somewhat rigourous. After all the experts have tried pieces from all mixtures, the results for each mixture will be tallied.

Thus the model for the response mean is assumed to be Poisson (or Poisson-like) with a logarithmic link function and the model for the response variance is assumed to be gamma also with a logarithmic link. We consider that factors $x_2$ and $x_3$ affect the mean, such that $\eta = k(\mu) = 2.199 + 0.456x_3 - 0.629x_2x_3$ and that only $x_1$ affects the response variance such that $\tau(\varphi) = -0.744x_1$. Now we can proceed to obtain optimal designs.

We regard different scenarios in the search for an optimal design, namely Local-Local indicating a local $D$-optimal design for the parameters that affect the mean as well as the parameters that affect the variance, Bayesian-Local indicating a Bayesian $D$-optimal design with a prior distribution for the parameters that affect the mean and a fixed (local) value for the parameters affecting the variance, Local-Bayesian, and Bayesian-Bayesian. The latter are defined analogously. We also consider the efficiency of the optimal design, $\xi^*$, with respect to the design obtained using a factorial $2^3$-design with a centre point, $\xi_1$. The efficiency of $\xi^*$ was considered as $\text{eff}(\xi^*, \xi_1) = \frac{\psi(I(\theta|\xi^*))}{\psi(I(\theta|\xi_1))}$, where $\psi(I(\theta|\xi))$ is the criterion function value for the design $\xi$.

**Local-Local** : we consider the parameters values for $\beta$ and $\gamma$ equal to $\beta = (2.199, 0.456, -0.629)$ and $\gamma = (-0.744)$. The optimal design obtained was:

$$\xi^*_{LL} = \left\{ \begin{matrix} (1.0, 1.0, -1.0) & (1.0, 1.0, 1.0) & (1.0, -1.0, 1.0) \\ 0.33 & 0.33 & 0.34 \end{matrix} \right\}$$

with the criterion function value $\psi(I(\theta|\xi^*_{LL})) = 8.698$ and with $\text{eff}(\xi^*_{LL}, \xi_1) = 1.0001$.

For the next cases, we regard discrete priors for the parameters $\beta$ and $\gamma$. The prior's values were selected so as to form a cube around the local values for $\beta$ and $\gamma$. The prior to $\beta$ ($\gamma$) was obtained adding and subtracting a constant $\alpha$ ($\delta$) for the local values of $\beta$ ($\gamma$), resulting in eight (two) total points, plus the local beta (gamma) value. The nine (three) points of the prior were assigned the same weights. In addition we considered a prior for gamma with more than three points. The values considered for $\alpha$ were $0.25, 0.5$ and $1.0$ while the values for gamma were $0.25, 0.5$ and $0.75$.

**Bayesian-Local** : for $\alpha = 0.25$ the optimal design obtained was:

$$\xi^*_{BL} = \left\{ \begin{matrix} (1.0, 1.0, -1.0) & (1.0, 1.0, 1.0) & (1.0, -1.0, 1.0) & (1.0, -1.0, -1.0) & (1.0, -0.02, 1.0) \\ 0.14 & 0.25 & 0.26 & 0.30 & 0.05 \end{matrix} \right\}$$

with the criterion function value $\psi(I(\theta|\xi^*_{BL})) = 5.225$ and with $\mathrm{eff}(\xi^*_{BL}, \xi_1) = 1.0015$. For $\alpha = 1.0$ the optimal design obtained was:

$$\xi^*_{BL} = \left\{ \begin{matrix} (1.0, 1.0, -1.0) & (1.0, 1.0, 1.0) & (1.0, -1.0, 1.0) & (1.0, -1.0, -1.0) & (1.0, 1.0, 0.54) \\ 0.18 & 0.25 & 0.28 & 0.25 & 0.04 \end{matrix} \right\}$$

with the criterion function value $\psi(I(\theta|\xi^*_{BL})) = 6.585$ and with $\mathrm{eff}(\xi^*_{BL}, \xi_1) = 1.0016$.

**Local-Bayesian** : for $\delta = 0.25$ the optimal design obtained was:

$$\xi^*_{LB} = \left\{ \begin{matrix} (1.0, 1.0, -1.0) & (1.0, 1.0, 1.0) & (1.0, -1.0, 1.0) & (1.0, -1.0, -1.0) & (1.0, 0.02, 1.0) \\ 0.14 & 0.25 & 0.22 & 0.30 & 0.09 \end{matrix} \right\}$$

with the criterion function value $\psi(I(\theta|\xi^*_{LB})) = 4.89$ and with $\mathrm{eff}(\xi^*_{LB}, \xi_1) = 1.0059$. For $\delta = 0.25$ plus the points with $\delta = 0.5$ and $\delta = 0.75$ the optimal design obtained was:

$$\xi^*_{LB} = \left\{ \begin{matrix} (1.0, 1.0, -1.0) & (1.0, 1.0, 1.0) & (1.0, -1.0, 1.0) & (1.0, -1.0, -1.0) & (1.0, -0.10, -1.0) \\ 0.14 & 0.31 & 0.25 & 0.20 & 0.10 \end{matrix} \right\}$$

with the criterion function value $\psi(I(\theta|\xi^*_{LB})) = 5.354$ and with $\mathrm{eff}(\xi^*_{LB}, \xi_1) = 1.0071$.

**Bayesian-Bayesian** : for $\alpha = 0.25$ and $\delta = 0.25$ the optimal design obtained was:

$$\xi^*_{BB} = \left\{ \begin{matrix} (1.0, 1.0, -1.0) & (1.0, 1.0, 1.0) & (1.0, -1.0, 1.0) & (1.0, -1.0, -1.0) & (1.0, -0.02, 1.0) \\ 0.15 & 0.25 & 0.26 & 0.29 & 0.05 \end{matrix} \right\}$$

with the criterion function value $\psi(I(\theta|\xi^*_{BB})) = 5.175$ and with $\mathrm{eff}(\xi^*_{BB}, \xi_1) = 1.0015$. For $\alpha = 0.5$ and $\delta = 0.25$ plus the points with $\delta = 0.5$ and $\delta = 0.75$ the optimal design obtained was:

$$\xi^*_{BB} = \left\{ \begin{matrix} (1.0, 1.0, -1.0) & (1.0, 1.0, 1.0) & (1.0, -1.0, 1.0) & (1.0, -1.0, -1.0) & (1.0, 0.0, 1.0) \\ 0.16 & 0.25 & 0.29 & 0.29 & 0.01 \end{matrix} \right\}$$

with the criterion function value $\psi(I(\theta|\xi^*_{BB})) = 5.618$ and with $\mathrm{eff}(\xi^*_{BB}, \xi_1) = 1.0002$.

In all cases, the efficiency of the optimal design, $\xi^*$, in relation to the factorial design with centre point, $\xi_1$, was approximately 1.0 and the criterion function value was always higher than the criterion function value for the factorial design.

## 7 Final Considerations

The results of this article extend the theory of optimal designs for generalized linear models, for which there is no knowledge of the likelihood but the relationship between the two first moments is known. The use of the extended quasi likelihood permitted the optimal designs for these types of models to be found.

The theory of Bayesian $D$-optimal designs allowed priors for the parameters to be provided, thus allowing several situations to be considered. Meanwhile, we still need to better explore those situations, and in future work, we intend to apply the idea of Bayesian sequential designs.

## References

A.C. Atkinson and R.D. Cook. *D*-optimum designs for heteroscedastic linear models. *J. Amer. Statist. Assoc.*, 90(429):204–212, 1995.

A.C. Atkinson and A.N. Donev. *Optimum Experimental Designs.* Clarendon Press, Oxford, 1992.

K. Chaloner and K. Larntz. Optimal Bayesian design applied to logistic regression experiments. *J. Statist. Plann. Inference*, 21(2):191–208, 1989.

B. Jørgensen. Exponential dispersion models. *J. Roy. Statist. Soc. Ser. B*, 49(2):127–162, 1987.

P. McCullagh and J.A. Nelder. *Generalized Linear Models.* Monographs on Statistics and Applied Probability. Chapman & Hall, London, 1989.

J.A. Nelder and Y. Lee. Generalized linear models for the analysis of Taguchi-type experiments. *Appl. Stoch. Models Bus. Ind.*, 7:107–120, 1991.

J.A. Nelder and D. Pregibon. An extended quasilikelihood function. *Biometrika*, 74(2):221–232, 1987.

L. Tack, P. Goos, and M. Vandebroek. Efficient Bayesian designs under heteroscedasticity. *J. Statist. Plann. Inference*, 104(2):469–483, 2002.

G. Vining and D. Schaub. Experimental designs for estimating both mean and variance functions. 28:135–147, 1996.

# $L$–optimum Designs in Multi–factor Models with Heteroscedastic Errors

C. Rodríguez, I. Ortiz, and I. Martínez

Dept. Statistics and Applied Mathematics, University of Almería, Spain
crt@ual.es

**Summary.** This paper considers the construction of $L$-optimum designs for Kronecker product and additive regression models when the errors are heteroscedastic. Sufficient conditions are given so that $L$-optimum designs for the multi-factor models can be built from $L$-optimum designs for their submodels with a single factor. Some results about $A$- and $V$-optimality are included.

**Key words:** $L$-optimum design, Kronecker product model, additive model.

## 1 Introduction

Let $f(x)^T = (f_1(x), \ldots, f_m(x))$ denote $m$ linearly independent regression functions defined on some compact subset $\mathcal{X}$ of $\mathbb{R}^k$ and let $\theta^T = (\theta_1, \ldots, \theta_m)$ denote a vector of parameters. We consider the usual linear regression model

$$Y_x = f(x)^T\theta + \varepsilon, \quad x \in \mathcal{X}, \tag{1}$$

in which for each $x \in \mathcal{X}$, a random response variable with mean $\sum_{j=1}^m f_j(x)\theta_j$ and variance $\sigma^2/\lambda(x)$ can be observed. The different observations are assumed to be uncorrelated. The efficiency function $\lambda(x)$ is a positive real-valued continuous function defined on $\mathcal{X}$. A design $\xi$ will be regarded as a probability measure on $\mathcal{X}$ and the information contained in a design $\xi$ is measured by its information matrix denoted by $M(\xi)$.

Optimum experimental designs typically minimize some convex function of the inverse of the information matrix. If the interest is to estimate the parameters in the model, a suggestive criterion is $A$-optimality, which attempts to minimize the trace of the covariance matrix. In other more general situations, the experimenter is interested in an $s$-dimensional linear aspect of the parameters in the model, $\psi(\theta) = A^T\theta$, where $A^T \in \mathbb{R}^{s\times m}$ is the associated matrix. Then, $L_A$-optimality criterion can be considered, that uses the covariance matrix of the linear aspect $\psi(\theta)$. If the interest is to predict the response $Y_x$, $V$-optimum designs minimize the weighted average (with respect to some measure $\mu$) of the variance of the prediction on $\mathcal{X}$.

Most experiments are described by multi-factor models. The construction of optimum designs is more complicated for these models than for models with a single factor. Therefore, it is interesting to obtain optimum designs for multi-factor models in terms of optimum designs for their one-dimensional components. Schwabe (1996) makes a review of optimal designs for multi-factor models with homoscedastic errors. The heteroscedastic case has been less deeply studied. Wong (1994) considers $G$-optimality and Montepiedra and Wong (2001) and Rodríguez and Ortiz (2004) calculate designs for $D$-optimality.

The aim of this paper is to construct $L$-optimum designs for Kronecker product and additive models when the errors are heteroscedastic. Sufficient conditions are found so that optimum designs for the multi-factor models can be constructed from optimum designs for the single factors following the approach used by Schwabe (1996) in the homoscedastic case.

## 2 Product Models

Consider the $n$ linear one-factor heteroscedastic models

$$\mathrm{E}(Y_{x_k}) = f_k(x_k)^T\theta_k = \sum_{i_k=1}^{p_k} f_{k,i_k}(x_k)\theta_{k,i_k}, \quad x_k \in \mathcal{X}_k \ , \tag{2}$$

with $\mathrm{Var}\,(Y_{x_k}) = \sigma_k^2/\lambda_k(x_k)$, $k = 1, \ldots, n$ and $n \geq 2$.

The resulting $n$-factor Kronecker product model with interactions is

$$\mathrm{E}(Y_{x_1,\ldots,x_n}) = \sum_{i_1=1}^{p_1} \cdots \sum_{i_n=1}^{p_n} f_{1,i_1}(x_1) \cdot \ldots \cdot f_{n,i_n}(x_n)\theta_{i_1,\ldots,i_n} \tag{3}$$

with $\mathrm{Var}\,(Y_{x_1,\ldots,x_n}) = \sigma^2/(\lambda_1(x_1) \cdot \ldots \cdot \lambda_n(x_n))$, $(x_1, \ldots, x_n) \in \mathcal{X}_1 \times \ldots \times \mathcal{X}_n$.

This product model has $p_1 \cdot \ldots \cdot p_n$ parameters and the heteroscedastic structure is determined by the efficiency function

$$\lambda(x_1, \ldots, x_n) = \lambda_1(x_1) \cdot \ldots \cdot \lambda_n(x_n). \tag{4}$$

Let $\xi_k$ be a design on $\mathcal{X}_k$, $k = 1, \ldots, n$. The *product design* on $\mathcal{X}_1 \times \ldots \times \mathcal{X}_n$ is defined by the product measure $\xi(A_1, \ldots, A_n) = \xi_1(A_1) \cdot \ldots \cdot \xi_n(A_n)$, where $A_k \subseteq \mathcal{X}_k$ is measurable w.r.t. $\xi_k$, $k = 1, \ldots, n$.

As in the homoscedastic case, the information matrix and hence, the covariance matrix for a product design factorize into their marginal counterparts. Besides, let $\psi(\theta) = A^T\theta$ be an $s$-dimensional linear aspect of $\theta$ such that its associated matrix is factorized as $A = A_1 \otimes \ldots \otimes A_n$, where $A_k \in \mathbb{R}^{p_k \times s_k}$. If $\psi_k(\theta) = A_k^T\theta_k$ is identifiable under the design $\xi_k$ in the marginal model (2), $k = 1, \ldots, n$, then $\psi(\theta)$ is identifiable under $\xi = \xi_1 \otimes \ldots \otimes \xi_n$ and the factorization $\mathrm{Cov}_\psi(\xi) = \mathrm{Cov}_{\psi_1}(\xi_1) \otimes \ldots \otimes \mathrm{Cov}_{\psi_n}(\xi_n)$ holds. These facts can be proven by replacing the regression functions $f_k(x_k)$ used in (Schwabe, 1996, Section 4), by $\sqrt{\lambda_k(x_k)}f_k(x_k)$, $k = 1, \ldots, n$.

**Theorem 1.** *If $A = A_1 \otimes \ldots \otimes A_n$, where $A_k$ is a $p_k \times s_k$ matrix with rank $s_k \leq p_k$ and $\xi_k^*$ is an $L_{A_k}$-optimum design in marginal model (2), $k = 1, \ldots, n$, then $\xi_1^* \otimes \ldots \otimes \xi_n^*$ is $L_A$-optimum in the $n$–factor model (3).*

*Proof.* As $\mathrm{tr}\{A^T[M(\xi_1 \otimes \ldots \xi_n)]^{-1}A\} = \prod_{k=1}^n \mathrm{tr}\{A_k^T M_k(\xi_k)^{-1} A_k\}$, the product of the $L_{A_k}$-optimum designs $\xi_k^*$ is $L_A$-optimum within the class of product designs. Applying the Studden theorem for $L_{A_k}$-optimality we obtain that $\xi_1^* \otimes \ldots \otimes \xi_n^*$ is an $L_A$-optimum design. □

**Corollary 1.**
1. *If $\xi_k^*$ is $A$-optimum in marginal model (2), $k = 1, \ldots, n$, then $\xi_1^* \otimes \ldots \otimes \xi_n^*$ is $A$-optimum in the product model (3).*
2. *If $c = c_1 \otimes \ldots \otimes c_n$, $c_k \in \mathbb{R}^{p_k}$ and $\xi_k^*$ is $c_k$-optimum in marginal model (2), $k = 1, \ldots, n$, then $\xi_1^* \otimes \ldots \otimes \xi_n^*$ is $c$-optimum in the product model (3).*
3. *If $\mu_k$ is a weighting measure on $\mathcal{X}_k$, $\mu = \mu_1 \otimes \ldots \otimes \mu_n$ and $\xi_k^*$ is $V$-optimum with respect to $\mu_k$ in marginal model (2), $k = 1, \ldots, n$, then $\xi_1^* \otimes \ldots \otimes \xi_n^*$ is $V$-optimum with respect to $\mu$ in the product model (3).*

*Example 1.* Consider the two-factor model $\mathrm{E}(Y_{x_1,x_2}) = \theta_0 + \theta_1 x_1 + \theta_2 x_2 + \theta_3 x_1 x_2 + \theta_4 x_2^2 + \theta_5 x_1 x_2^2$, with efficiency function $\lambda(x_1, x_2) = (1+x_1^2)(1+x_2^2)$ and $x_1, x_2$ in $[-1, 1]$. This model can be expressed as Model I $\otimes$ Model II, where models I and II are first and second degree one-factor polynomial models, respectively, with efficiency functions $\lambda_1(x_1) = 1 + x_1^2$ and $\lambda_2(x_2) = 1 + x_2^2$. The $A$-optimum design for model I, $\xi_1^*$, is equally supported at $\pm 1$ and for model II, $\xi_2^*$ puts weights 0.207, 0.586 and 0.207 at -1, 0 and 1, respectively. The product design $\xi_1^* \otimes \xi_2^*$ is $A$-optimum for the above two-factor model.

# 3 Additive Models

This section deals with the construction of optimum designs for additive models from the optimum designs for the corresponding marginal models. We work with two classes of additive models, models with constant term and orthogonal models.

## 3.1 Additive Models with Constant Term

In this subsection we consider heteroscedastic models explicitly including a constant term,

$$\mathrm{E}(Y_{x_k}) = \theta_0 + f_k(x_k)^T \theta_k = \theta_0 + \sum_{i_k=1}^{p_k} f_{k,i_k}(x_k)\theta_{k,i_k}, \quad x_k \in \mathcal{X}_k \ . \tag{5}$$

The $n$-factor additive model

$$\mathrm{E}(Y_{x_1,\ldots,x_n}) = \theta_0 + \sum_{k=1}^{n} f_k(x_k)^T \theta_k \ , \tag{6}$$

has $1+p_1+\ldots+p_n$ parameters, and the efficiency function (4) will be used.

We first show the results for an additive two-factor model. The following lemma is an extension of Lemma 5.5 in Schwabe (1996).

**Lemma 1.** *The additive model (6) with $n=2$ and $\lambda(x_1,x_2)=\lambda_1(x_1)\lambda_2(x_2)$ has the following properties:*

1. *$\theta=(\theta_0,\theta_1,\theta_2)$ is estimable under $\xi_1\otimes\xi_2$ if and only if $(\theta_0,\theta_k)$ is estimable under $\xi_k$ in the $k$th marginal model, $k=1,2$.*
2. *For every admissible design $\xi_1\otimes\xi_2$, the covariance matrix is*

$$M(\xi_1\otimes\xi_2)^{-1} = \begin{pmatrix} \mathrm{Cov}_{\theta_0}(\xi_1\otimes\xi_2) & C_1^T & C_2^T \\ C_1 & \Lambda_2^{-1}\mathrm{Cov}_{1,\theta_1}(\xi_1) & \mathbf{0} \\ C_2 & \mathbf{0} & \Lambda_1^{-1}\mathrm{Cov}_{2,\theta_2}(\xi_2) \end{pmatrix}$$

*being $C_k = -\Lambda_k^{-1}\Lambda_j^{-1}\mathrm{Cov}_{k,\theta_k}(\xi_k)\int \lambda_k f_k d\xi_k$, with $\Lambda_k=\int_{\mathcal{X}_k}\lambda_k d\xi_k$ and matrix $\mathrm{Cov}_{k,\theta_k}(\xi_k)$ is the covariance for the estimator of $\theta_k$ under $\xi_k$ in model $k$, $k,j=1,2$, $k\neq j$. Besides, the covariance matrix for the estimator of $(\theta_0,\theta_k)$ under $\xi_k$ in the marginal model $k$, is*

$$M_k(\xi_k)^{-1} = \begin{pmatrix} \mathrm{Cov}_{k,\theta_0}(\xi_k) & -\Lambda_k^{-1}\int \lambda_k f_k^T d\xi_k \mathrm{Cov}_{k,\theta_k}(\xi_k) \\ -\Lambda_k^{-1}\mathrm{Cov}_{k,\theta_k}(\xi_k)\int \lambda_k f_k d\xi_k & \mathrm{Cov}_{k,\theta_k}(\xi_k) \end{pmatrix}$$

3. *The variance for the estimator of $\theta_0$ in the additive model with respect to the product design, $\mathrm{Cov}_{\theta_0}(\xi_1\otimes\xi_2)$, is related to the variances of the estimator for $\theta_0$ under $\xi_k$ in model $k$, $k=1,2$, by $\mathrm{Cov}_{\theta_0}(\xi_1\otimes\xi_2) = \Lambda_2^{-1}\mathrm{Cov}_{1,\theta_0}(\xi_1)+\Lambda_1^{-1}\mathrm{Cov}_{2,\theta_0}(\xi_2)-\Lambda_1^{-1}\Lambda_2^{-1}$.*
4. *The covariance for the estimator of $\theta_k$ in the additive model satisfies $\mathrm{Cov}_{\theta_k}(\xi_1\otimes\xi_2)=\Lambda_j^{-1}\mathrm{Cov}_{k,\theta_k}(\xi_k)$, $k,j=1,2$, $k\neq j$.*

**Theorem 2.** *For a model given by (6), $n=2$ and $\lambda(x_1,x_2)=\lambda_1(x_1)\lambda_2(x_2)$, let $\psi(\theta)=A^T\theta\in\mathbb{R}^s$ be an $s$-dimensional linear aspect of $\theta$ such that its associated matrix $A$ is partitioned as $A^T=(A_0^T|A_1^T|A_2^T)$ where $A_0^T\in\mathbb{R}^s$ and $A_k^T\in\mathbb{R}^{s\times p_k}$, $k=1,2$. Let $\psi_k(\theta_k)=A_0^T\theta_0+A_k^T\theta_k$ be an $s$–dimensional linear aspect associated with the $k$th marginal model, if $\xi_k^*$ is $L_{\psi_k}$-optimum and*

$$\max_{x_k\in\mathcal{X}_k}\lambda_k(x_k)=\int\lambda_k d\xi_k^*,\quad k=1,2, \tag{7}$$

*then, the design $\xi_1^*\otimes\xi_2^*$ is $L_\psi$-optimum within the class of product designs.*

*Proof.* Let $\lambda_k^*=\max_{x_k\in\mathcal{X}_k}\lambda_k(x_k)$, $k=1,2$. For any designs $\xi_1$ and $\xi_2$,

$$\operatorname{tr}\{A^T M(\xi_1 \otimes \xi_2)^{-1} A\} = \frac{\operatorname{tr}\{\operatorname{Cov}_{\psi_1}(\xi_1)\}}{\Lambda_2} + \frac{\operatorname{tr}\{\operatorname{Cov}_{\psi_2}(\xi_2)\}}{\Lambda_1} - \frac{\operatorname{tr}(A_0^T A_0)}{\Lambda_1 \Lambda_2}$$

$$\geq \frac{\operatorname{tr}\{\operatorname{Cov}_{\psi_1}(\xi_1^*)\}}{\lambda_2^*} + \frac{\operatorname{tr}\{\operatorname{Cov}_{\psi_2}(\xi_2^*)\}}{\lambda_1^*} - \frac{\operatorname{tr}\{A_0^T A_0\}}{\lambda_1^* \lambda_2^*} = \operatorname{tr}\{A^T M(\xi_1^* \otimes \xi_2^*)^{-1} A\}.$$

□

**Corollary 2.** *The design $\xi_1^* \otimes \xi_2^*$ is A-optimum within the class of product designs for the additive model (6) with $n = 2$ and $\lambda(x_1, x_2) = \lambda_1(x_1)\lambda_2(x_2)$, if $\xi_k^*$ is A-optimum in the kth marginal model and conditions (7) hold.*

**Theorem 3.** *For the additive model (6) and $\lambda(x_1, \ldots, x_n) = \prod_{i=1}^n \lambda_i(x_i)$, let $\psi(\theta) = A^T\theta \in \mathbb{R}^s$ be an s-dimensional linear aspect of $\theta$ such that $A$ is partitioned as $A^T = (A_0^T | A_1^T | \ldots | A_n^T)$ where $A_0^T \in \mathbb{R}^s$ and $A_k^T \in \mathbb{R}^{s \times p_k}$, $k = 1, \ldots, n$. Let $\psi_k(\theta_k) = A_0^T\theta_0 + A_k^T\theta_k$ be an s-dimensional linear aspect associated with the kth marginal model, if $\xi_k^*$ is $L_{\psi_k}$-optimum and*

$$\prod_{1 \leq j \neq k \leq n} \max_{x_j \in \mathcal{X}_j} \lambda_j(x_j) = \prod_{1 \leq j \neq k \leq n} \int_{\mathcal{X}_j} \lambda_j d\xi_j^*, \quad k, j = 1, \ldots, n, \tag{8}$$

*then the design $\xi_1^* \otimes \ldots \otimes \xi_n^*$ is $L_\psi$-optimum within the class of product designs.*

**Corollary 3.** *The design $\xi_1^* \otimes \ldots \otimes \xi_n^*$ is A-optimum within the class of product designs for the additive model (6) and $\lambda(x_1, \ldots, x_n) = \lambda_1(x_1) \cdot \ldots \cdot \lambda_n(x_n)$, if $\xi_k^*$ is A-optimum in the kth marginal model and conditions (8) hold.*

*Example 2.* Consider the model $\mathrm{E}(Y_{x_1,x_2}) = \theta_0 + \theta_1 x_1 + \theta_2 x_2$, with efficiency function $\lambda(x_1, x_2) = (1 + x_1^2)(1 + x_2^2)$ and $x_1, x_2$ in $[-1, 1]$. This model can be expressed as the sum of two first-order uni-factor polynomial models with efficiency functions $\lambda_k(x_k) = 1 + x_k^2$, $k = 1, 2$. In each model, the $L_{\theta_k}$-optimum design for the parameter $\theta_k$, $\xi_k^*$, is equally supported at $\pm 1$ and $\max_{x_k \in [-1,1]} \lambda_k(x_k) = 2 = \int_{-1}^1 \lambda_k(x_k) d\xi_k^*$. Therefore, $\xi_1^* \otimes \xi_2^*$ is $L_\psi$-optimum in the additive model for the linear aspect defined by the matrix $A^T = \begin{pmatrix} 0\,1\,0 \\ 0\,0\,1 \end{pmatrix}$.

*Example 3.* Consider a model that is the sum of two second order uni-factor polynomial models with efficiency functions $\lambda_k(x_k) = 1 + x_k^2$, $x_k \in [-1, 1]$, $k = 1, 2$. In each model, the $L_{\theta_k}$-optimum design, $\xi_k^*$, for the parameter $\theta_k$ corresponding to $x_k^2$, is supported at -1, 0 and 1 with weights 0.207, 0.586 and 0.207, respectively. Here $\max_{x_k \in [-1,1]} \lambda_k(x_k) = 2 \neq \int_{-1}^1 \lambda_k(x_k) d\xi_k^* = 1.454$ and the product design is not $L_\psi$-optimum in the additive model for the linear aspect with $A^T = \begin{pmatrix} 0\,0\,0\,1\,0 \\ 0\,0\,0\,0\,1 \end{pmatrix}$.

### 3.2 Orthogonal Additive Model

Throughout this subsection we consider additive models where a constant term is not necessarily involved, but an orthogonality assumption on the marginal regression functions is assumed. We consider $n$ heteroscedastic models defined by (2) and the resulting additive $n$–factor model

$$\mathrm{E}(Y_{x_1,\ldots,x_n}) = \sum_{k=1}^{n} f_k(x_k)^T\theta_k \ , \tag{9}$$

with efficiency function (4). At most one of the marginal models has a constant term, and if it exists the first model is the model with intercept.

**Theorem 4.** *For the additive model (9) and $\lambda(x_1,\ldots,x_n) = \lambda_1(x_1)\ldots\lambda_n(x_n)$, let $\psi(\theta)^T = \left(\psi_1(\theta_1)^T,\ldots,\psi_n(\theta_n)^T\right)^T$ be linear aspects of $\theta$, where $\psi_k(\theta_k) = A_k^T\theta_k$, $k = 1,\ldots,n$. If $\xi_k^*$ is $L_{\psi_k}$-optimum in the kth marginal model (2),*

$$\prod_{1\le j\ne k\le n} \max_{x_j\in\mathcal{X}_j} \lambda_j(x_j) = \prod_{1\le j\ne k\le n} \int_{\mathcal{X}_j} \lambda_j d\xi_j^*, \quad k,j = 1,\ldots,n, \tag{10}$$

*and*

$$\int \lambda_k f_k d\xi_k^* = \mathbf{0}, \quad k = 2,\ldots,n, \tag{11}$$

*then the design $\xi_1^*\otimes\ldots\otimes\xi_n^*$ is $L_\psi$-optimum. In particular, if $\xi_k^*$ is A-optimum in the kth marginal model (2), $k = 1,\ldots,n$, and the above conditions hold, then $\xi_1^*\otimes\ldots\otimes\xi_n^*$ is A-optimum.*

*Proof.* If $\xi^* = \xi_1^*\otimes\ldots\otimes\xi_n^*$ with $\int\lambda_k f_k d\xi_k^* = \mathbf{0}$, $k = 2,\ldots,n$, then its information matrix is partitioned as:

$$M(\xi_1^*\otimes\ldots\otimes\xi_n^*) = diag\left( \Big( \prod_{1\le j\ne k\le n} \int_{\mathcal{X}_j} \lambda_j d\xi_j^* \Big) M_k(\xi_k^*) \right),$$

where $M_k(\xi_k^*)$ is the information matrix associated with $\xi_k^*$ in the marginal model. For every admissible design $\xi$ in the additive model (9) we obtain

$$\begin{aligned}\mathrm{tr}\{\mathrm{Cov}_\psi(\xi)\} &= \sum_{k=1}^{n}\mathrm{tr}\{\mathrm{Cov}_{\psi_k}(\xi)\} \ge \sum_{k=1}^{n} \frac{\mathrm{tr}\{\mathrm{Cov}_{k,\psi_k}(\xi_k)\}}{\prod_{1\le j\ne k\le n}\max_{x_j\in\mathcal{X}_j}\lambda_j(x_j)} \\ &\ge \sum_{k=1}^{n} \frac{\mathrm{tr}\{\mathrm{Cov}_{k,\psi_k}(\xi_k^*)\}}{\prod_{1\le j\ne k\le n}\int_{\mathcal{X}_j}\lambda_j d\xi_j^*} = \mathrm{tr}\{\mathrm{Cov}_\psi(\xi_1^*\otimes\ldots\otimes\xi_n^*)\} \ .\end{aligned}$$

Hence, $\xi_1^*\otimes\ldots\otimes\xi_n^*$ is $L_\psi$-optimum. □

## 4 Some Remarks

For product models with heteroscedastic structure, the product of $L_{A_k}$-optimum designs in the marginal models is $L_A$-optimum in the product model for $A = \bigotimes_{k=1}^{n} A_k$. Hence, the product of $A$-optimum (resp. $V$-optimum with respect to $\mu_k$) designs in the marginal models is $A$-optimum (resp. $V$-optimum with respect to $\mu = \bigotimes_{k=1}^{n} \mu_k$) in the product model. Therefore, the results for the homoscedastic case happen again without additional conditions on the factors or the weighting functions.

For additive models with constant term and product efficiency function, the product of $L_{\psi_k}$-optimum designs in the marginal models is $L_\psi$-optimum within the class of product designs in the additive model if conditions (8) hold. For orthogonal additive models conditions (10) are necessary.

This work has focused on product and additive models with product efficiency functions. In practice, these models can be used for example, if the multi–factor model has an exponential efficiency function such that the exponent is the sum of uni–factor functions, as in Kunert and Lehmkuhl (1998). Other heteroscedastic structures could be considered as the sum,

$$\lambda(x_1, \dots, x_n)^{-1} = \lambda_1(x_1)^{-1} + \dots + \lambda_n(x_n)^{-1} \ . \tag{12}$$

It is interesting to study if an optimum design for a model has a good behaviour when it is used to estimate another model with a different structure. For example, we can work with product or additive models, and with efficiency functions of the type product or sum. Following example shows this question.

*Example 4.* We consider two-factor models, such that the two marginal models are polynomial regression models of first or second degree. Each submodel has the efficiency function $\lambda(x) = 1 + x^2$, $x \in [-a, a]$, $a > 0$. We denote by **PP** and **PS** a product model with product or sum efficiency function, respectively, and by **SP** and **SS**, additive models with product or sum efficiencies, respectively. When the preceding theorems are not applicable, we have implemented the necessary software to calculate $A$-optimum designs for these models using the Gauss package, in order to obtain the optimum designs in the class of all designs. In any cases the optimum designs are not product designs.

For the sake of simplicity Table 2 only includes, for some values of $a$, $A$-efficiencies of the designs for a model when these are used to estimate other model. We denote by M1/M2 the efficiency of an $A$-optimum design for model M1 with respect to model M2. Every cell in Table 2 has five values corresponding to the efficiencies included in the cell of Table 1. We leave out efficiencies that have less interest.

For submodels of degree 1 the theorems of this paper are applicable to obtain $A$-optimum designs for models **PP** and **SP**. As we can observe from Table 2 $A$-optimum designs for all models coincide. In general, efficiencies decrease as

**Table 1.** Efficiencies contained in cells of Table 2.

| | | |
|---|---|---|
| **PP/SP** | | |
| **PP/PS** | | |
| **PP/SS** | **PS/SS** | **SP/SS** |

**Table 2.** $A$-efficiencies in multi-factor models.

| $[-a, a] \times [-a, a]$ | Degree 1 | | | Degree 2 | | |
|---|---|---|---|---|---|---|
| $a = \frac{1}{2}$ | 1.0000 | | | 0.9970 | | |
| | 1.0000 | | | 0.9984 | | |
| | 1.0000 | 1.0000 | 1.0000 | 0.9969 | 0.9992 | 1.0000 |
| $a = 1$ | 1.0000 | | | 0.9717 | | |
| | 1.0000 | | | 0.9848 | | |
| | 1.0000 | 1.0000 | 1.0000 | 0.9650 | 0.9847 | 0.9992 |
| $a = 2$ | 1.0000 | | | 0.7678 | | |
| | 1.0000 | | | 0.8976 | | |
| | 1.0000 | 1.0000 | 1.0000 | 0.8457 | 0.8528 | 0.8913 |
| $a = 5$ | 1.0000 | | | 0.5628 | | |
| | 1.0000 | | | 0.7627 | | |
| | 1.0000 | 1.0000 | 1.0000 | 0.6665 | 0.7105 | 0.8001 |
| $a = 10$ | 1.0000 | | | 0.5322 | | |
| | 1.0000 | | | 0.7190 | | |
| | 1.0000 | 1.0000 | 1.0000 | 0.6175 | 0.6669 | 0.7721 |

$a$ increases. When $a \leq 1$ the product of $A$-optimum designs is a good design for the four models studied. The design for Model **SP** is good for Model **SS**, besides, in general $A$-optimum design for Model **PP** has better efficiencies for Model **SS** than for Model **SP**.

*Acknowledgement.* The authors are grateful to the referees for their helpful comments and suggestions. This work has been partially supported by Junta de Andalucía, research group FQM244.

## References

J. Kunert and F. Lehmkuhl. The generalized $\beta$–method in Taguchi experiments. In A.C. Atkinson, L. Pronzato, and H.P. Wynn, editors, *mODa 5—Advances in Model-Oriented Data Analysis and Experimental Design (Marseilles/France, 1998)*, Contributions to Statistics, pages 223–230, Heidelberg, 1998. Physica-Verlag.

G. Montepiedra and W.K. Wong. A new design criterion when heteroscedasticity is ignored. *Ann. Inst. Statist. Math.*, (53):418–426, 2001.

C. Rodríguez and I. Ortiz. $D$-optimal designs in multi-factor models with heteroscedastic errors. *J. Statist. Plann. Inference*, (to appear), 2004.

R. Schwabe. *Optimum Designs for Multi-Factor Models.* Springer, 1996.

W.K. Wong. $G$-optimal designs for multifactor experiments with heteroscedastic errors. *J. Statist. Plann. Inference*, 40:127–133, 1994.

# Multiplicative Algorithms for Constructing Optimizing Distributions : Further Developments

B. Torsney[1] and S. Mandal[2]

[1] Department of Statistics, University of Glasgow, Glasgow, G12 8QW, United Kingdom. b.torsney@stats.gla.ac.uk
[2] Department of Statistics, University of Manitoba, Winnipeg, MB, R3T 2N2, Canada. saumen_mandal@umanitoba.ca

**Summary.** We consider a class of optimization problems in which the aim is to find an optimizing distribution. In order to determine the optimizing distribution, a class of multiplicative algorithms, indexed by a function $f(\cdot)$ which depends on the derivatives of the criterion function, is considered. The function $f(\cdot)$ is positive, strictly increasing and may depend on one or more free parameters. In an attempt to improve convergence, we consider an objective approach to choosing $f(\cdot)$ which allows the criterion function to have negative derivatives. We consider objective choices of the class of functions $f(\cdot)$ for the proposed approach. The approach enjoys considerable improvements in convergence of the algorithm.

**Key words:** multiplicative algorithms, optimality conditions, optimal distributions on spaces, optimal designs, vertex directional derivatives.

## 1 Introduction

We consider the following problem.
(P1): Maximize criterion $\varphi(p)$ over $\mathcal{P} \equiv \{p = (p_1, p_2, \ldots, p_J) : p_j \geq 0, \sum p_j = 1\}$.
The equality constraint $\sum p_j = 1$ renders the problem a nondegenerate constrained optimization problem, the full constraint region being a closed bounded convex set. Examples of this include parameter estimation, stratified sampling and optimal regression design problems. We will focus on the latter later. The simplest estimation problem is that of estimating the cell probabilities $p_j$ of a $J$-cell multinomial distribution. There is an explicit solution in this case. We wish to focus on examples for which an explicit solution is not available, for example, finding Maximum Likelihood estimates of mixing components, given data from a mixture. Before turning to algorithms we need to detail optimality conditions.

## 2 Optimality Conditions

We define optimality conditions in terms of point to point directional derivatives. Making use of differential calculus, the directional derivative $F_\varphi\{p,\ q\}$ of a criterion function $\varphi(\cdot)$ at $p$ in the direction of $q$ is defined as:

$$F_\varphi\{p,q\} = \lim_{\varepsilon \downarrow 0} \frac{\varphi\{(1-\varepsilon)p + \varepsilon q\} - \varphi(p)}{\varepsilon} = \left.\frac{d\varphi}{d\varepsilon}\right|_{\varepsilon=0^+} \tag{1}$$

The derivative of $F_\varphi\{p,\ q\}$ may exist even if $\varphi(\cdot)$ is not differentiable. If $\varphi(\cdot)$ is differentiable, then (1) becomes

$$F_\varphi\{p,q\} = (q-p)^T \frac{\partial \varphi}{\partial p} = \sum_{j-1}^{J} (q_j - p_j)\, d_j, \text{ where } d_j = \frac{\partial \varphi}{\partial p_j},\ j = 1,2,\ldots,J.$$

When $p, q \in \mathcal{P}$, they can be interpreted as distributions or sets of weights. A particular case of $q$ is then $q = e_j$, where $e_j$ is the $j$th unit vector in $\mathbb{R}^J$. Let

$$F_j = F_\varphi\{p, e_j\} = \frac{\partial \varphi}{\partial p_j} - \sum_{i=1}^{J} p_i \frac{\partial \varphi}{\partial p_i}.$$

We call $F_j$ the $j$th vertex directional derivative of $\varphi(\cdot)$ at $p$. In fact, when $\varphi(p) = \psi\{M(p)\}$ (see section 4), where $\psi(M)$ is a design criterion and $M$ is a matrix, $F_j$ is a directional derivative of $\psi(\cdot)$ in the direction of a one point design. We assume $\varphi(p)$ to be differentiable at $p^*$, where $p^*$ is a potential maximizer of $\varphi(\cdot)$. Then, first-order conditions for a local maximum are

$$F_j^* \begin{cases} = 0 & \text{for } p_j^* > 0 \\ \leq 0 & \text{for } p_j^* = 0 \end{cases}. \tag{2}$$

If $\varphi(\cdot)$ is concave on its feasible region, then the first-order stationarity condition (2) is both necessary and sufficient for optimality, a result known as the general equivalence theorem in optimal design (Whittle (1973); Kiefer (1974)).

## 3 Algorithms

Problem (P1) has a distinctive set of constraints, namely the variables $p_1, p_2, \ldots, p_J$ must be nonnegative and sum to 1. An iteration which neatly submits to these and has some suitable properties is the multiplicative algorithm:

$$p_j^{(r+1)} = \frac{p_j^{(r)} f\left(d_j^{(r)}\right)}{\sum_{i=1}^{J} p_i^{(r)} f\left(d_i^{(r)}\right)} \tag{3}$$

where $d_j^{(r)} = \left.\frac{\partial\varphi}{\partial p_j}\right|_{p=p^{(r)}}$, while $f(d)$ is positive and strictly increasing in $d$. $f(d)$ may depend on one or more free parameters. Thus in view of the conditions for (local) optimality, a solution to Problem (P1) is a fixed point of the iteration but so also are the solutions to Problem (P1) when any subset of weights is set to zero.

This type of iteration was first proposed by Torsney (1977), taking $f(d) = d^\delta$, with $\delta > 0$. This requires derivatives to be positive. Various choices of $f(\cdot)$ have been considered in the literature. These include Silvey, Titterington, and Torsney (1978), Torsney (1988), Torsney and Alahmadi (1992), Mandal and Torsney (2000), Torsney and Mandal (2001). Mandal and Torsney (2002) consider developments of the above algorithm based on a clustering approach. Fedorov (1972) and Wynn (1972) consider vertex direction algorithms which perturb one $p_j$ and change the others proportionately.

### 3.1 Properties of the Iteration (3)

Under the conditions imposed on $f(\cdot)$, iterations (3) possess the following properties considered by Torsney (1988), Torsney and Alahmadi (1992) and Mandal and Torsney (2000).

1. $p^{(r)}$ is always feasible.
2. $F_\varphi\{p^{(r)}, p^{(r+1)}\} \geq 0$ with equality when the $d_j$'s corresponding to nonzero $p_j$'s have a common value, $d$, that is $d_j = \sum p_i d_i = d$ so that

$$p_j^{(r+1)} = \frac{p_j^{(r)} f(d_j)}{\sum_{i=1}^{J} p_i^{(r)} f(d_i)} = \frac{p_j^{(r)} f(d)}{f(d) \sum_{i=1}^{J} p_i^{(r)}} = p_j^{(r)}.$$

   The inequality property can be seen by letting a positive random variable $D$ take the value $\partial\varphi/\partial p_j$ with probability $p_j$ ($p_j = p_j^{(r)}$). Then

$$F_\varphi\{p^{(r)}, p^{(r+1)}\} = \frac{\mathrm{Cov}\,\{D, f(D)\}}{\mathrm{E}\,\{f(D)\}}.$$

   The argument then is that the covariance between $D$ and $f(D)$ must be nonnegative if $f(D)$ is increasing in $D$. Thus an increase in the criterion can be obtained by a partial but possibly not a full step from $p^{(r)}$ in the direction of $p^{(r+1)}$.
3. Under the above iteration $\mathrm{supp}(p^{(r+1)}) \subseteq \mathrm{supp}(p^{(r)})$, but weights can converge to zero, where $\mathrm{supp}(p)$ denotes those points with positive weight when $p$ is a distribution on some (design) space.

*Note 1.* Statements (1) to (3) are the only analytical results we cite. In particular, we report only empirical results on convergence. Of course, if the criterion $\varphi(p)$ were (strictly) concave and iterations were monotonic, then convergence would be guaranteed. Property (2) indicates that monotonicity could be a possibility. It could be achieved by cutting short, if necessary, the step from $p^{(r)}$ to $p^{(r+1)}$; see Silvey, Titterington, and Torsney (1978). Actually, as mentioned earlier, $f(\cdot)$ may depend on one or more free parameters. Judicious choice of these may ensure monotonicity; see Silvey, Titterington, and Torsney (1978), Torsney and Alahmadi (1992). We do not pursue these avenues further. Our aim is to propose, on intuitive grounds, objective choices of $f(\cdot)$.

## 4 Optimal Distribution on Spaces

Before proceeding further we add some structure to Problem (P1). A general class of examples of Problem (P1) is:
Problem (P2): Maximize $\psi(x)$ over the convex hull of $G(\mathcal{V}) = \{G(v) : v \in \mathcal{V}\}$, where $G(\cdot)$ is a given one to one function and $\mathcal{V} = \{v_1, \ldots, v_J\}$ is a known set of vector (or matrix) vertices of fixed dimension $k$. That is, maximize $\psi(x)$ over

$$\mathcal{CH}\{\mathcal{G}(\mathcal{V})\} = \left\{ x = x(p) = \sum_{j=1}^{J} p_j \, G(v_j) \; : \; p = (p_1, p_2, \ldots, p_J) \in \mathcal{P} \right\}.$$

Note that we could alternatively state that $x(p) = \mathrm{E}_p(G(v))$, where $G(v)$ is a random variable assuming the value $G(v_j)$ with probability $p_j$. That is, solve Problem (P1) for

$$\varphi(p) \;= \psi\{x\}, \;\; x = \mathrm{E}_p(G(v)) = \sum_{j=1}^{J} p_j G(v_j).$$

We focus on optimal regression design problems in which $G(v) = v\,v^T$ and $x = M(p) = \sum p_j v_j v_j^T$. Assuming $M(p)$ is nonsingular, the matrix $M^{-1}(p)$ is proportional to the covariance matrix of the least squares estimators $\hat{\theta}$ under the linear model: $\mathrm{E}(y|x_j) = f^T(x_j)\,\theta = v_j^T\,\theta$, $\mathrm{Var}\,(y|x_j) = \sigma^2$. For good estimation we wish to choose $p_1, p_2, \ldots, p_J$ to make $M^{-1}(p)$ 'small' or $M(p)$ 'large'. There are various design criteria $\psi(M)$ for doing so, the most common being the $D$-optimal one, where $\psi(M) = \log\det(M)$ or $(1/k)\log\det(M)$. We focus on this choice and consider the standardized version. Then for $\varphi(p) = \psi\{M(p)\}$, $d_j = (1/k)\, v_j^T\, M^{-1}(p)\, v_j$.

We consider five examples considered by Silvey, Titterington, and Torsney (1978), Wu (1978). The first example originated in Wynn (1970). The examples are defined by their design loci.

Example 1: $\mathcal{V} = \mathcal{V}_1 = \{ (1,-1,-1)^T, (1,-1,1)^T, (1,1,-1)^T, (1,2,2)^T\}$

Example 2: $\mathcal{V} = \mathcal{V}_2 = \{ (1,-1,-1)^T, (1,-1,1)^T, (1,1,-1)^T, (1,2,3)^T\}$

Example 3: $\mathcal{V} = \mathcal{V}_3 = \{ (1,-1,-2)^T, (1,-1,1)^T, (1,1,-1)^T, (1,2,2)^T\}$

Example 4: $\mathcal{V} = \mathcal{V}_4 = \{ (1,1,-1,-1)^T, (1,-1,1,-1)^T, (1,-1,-1,-1)^T, (1,2,2,-1)^T, (1,1,-1,1)^T, (1,-1.5,1,1)^T, (1,-1,-1,2)^T\}$

Example 5: $\mathcal{V} = \mathcal{V}_5 = \{ (1,1,-1,-1)^T, (1,-1,1,-1)^T, (1,-1,-1,-1)^T, (1,2,2,-1)^T, (1,1,-1,1)^T, (1,-1.5,1,1)^T, (1,-1,-1,2)^T, (1,1,1.5,1)^T\}$

These correspond to linear models with a constant term since the first component of each vertex ($v_j$) is always 1. Thus in examples 1-3 if $v_j \in \mathcal{V}$ then $v_j$ is of the form $(1, x_2, x_3)$. In fact, we can assume the more realistic regression model $\mathrm{E}(y) = v^T\theta$, $v \in \mathcal{V}$, where $\mathcal{V}$ is a quadrilateral. Examples 4 and 5 have similar characterisations. Also $\mathcal{V}_5 = \mathcal{V}_4 \cup \{(1, 1, 1.5, 1)^T\}$.

To set the scene we report the performance of algorithm (3) in calculating $D$-optimal designs when $f(\cdot)$ satisfies the conditions of section 3. We first consider three choices of $f(\cdot)$, namely, $f(d) = \ln(e+\delta d)$; $f(d) = \exp(\delta d)/(1+\exp(\delta d))$ and $f(d) = a - \exp(-\delta d)$, $a > 1$. Note that the first choice needs $d > 0$. For $a$ close to 1, the last choice of $f(d)$ is close to an exponential cumulative distribution function. In Table 1, we record for n=1, 2, 3, 4 the number of iterations needed to achieve $\max_{1\le j\le J}\{F_j\} \le 10^{-n}$, where $F_j$ are the vertex directional derivatives. In all the cases we take the initial design to be $p_j^{(0)} = 1/J$, $j = 1, 2, \ldots, J$. Results are given only for the best choices of $\delta$.

Convergence of the algorithm can be slow as the results in Table 1 demonstrate. However arguably it is fast to begin with. If we compare the results for the five examples for the three choices of the functions $f(\cdot)$, $f(d) = \exp(\delta d)/(1+\exp(\delta d))$ is not good whereas $f(d) = \ln(e+\delta d)$ and $f(d) = a - \exp(-\delta d)$ are not bad. One marginally positive result is that convergence is faster under the case $f(d) = a - \exp(-\delta d)$ with $a = 1.0001$. Convergence was slower for large values of $a$. Interestingly if $a \approx 1$ and $\delta$ is small then $f(d) \approx \delta d$. For this $d$ needs to be positive. This suggests that $f(d) = d$ is an efficient choice for the $D$-optimal criterion. Certainly it is known to be monotonic for this criterion.

Below are the weights to which convergence was obtained for the choices of $f(\cdot)$ and $\delta$ considered. These are the optimal designs.

Example 1: {0.125, 0.281, 0.281, 0.313}
Example 2: {0.074, 0.291, 0.311, 0.324}
Example 3: {0.243, 0.305, 0.161, 0.291}
Example 4: {0.030, 0.012, 0.231, 0.234, 0.183, 0.208, 0.102}
Example 5: {0.030, 0.012, 0.231, 0.234, 0.183, 0.208, 0.102, 0.000}

**Table 1.** Initial iteration results: Number of iterations needed to achieve $\max_{1\le j\le J}\{F_j\} \le 10^{-n}$ for $n = 1, 2, 3, 4$.

| | $f(d) = \ln(e + \delta d)$ | | | | | $f(d) = \frac{\exp(\delta d)}{(1+\exp(\delta d))}$ | | | | | $f(d) = a - \exp(-\delta d)$ | | | | |
|---|---|---|---|---|---|---|---|---|---|---|---|---|---|---|---|
| Examples | $\delta$ | $n$ | | | | $\delta$ | $n$ | | | | $\delta$ | $n$ | | | |
| | | 1 | 2 | 3 | 4 | | 1 | 2 | 3 | 4 | | 1 | 2 | 3 | 4 |
| Ex-1 | 5.0 | 2 | 14 | 38 | 63 | 1.3 | 3 | 16 | 43 | 72 | 0.01 | 1 | 4 | 11 | 18 |
| Ex-2 | 5.0 | 3 | 24 | 65 | 116 | 1.3 | 3 | 28 | 74 | 132 | 0.01 | 1 | 7 | 20 | 35 |
| Ex-3 | 5.0 | 2 | 10 | 35 | 56 | 1.3 | 2 | 12 | 40 | 64 | 0.01 | 1 | 4 | 10 | 16 |
| Ex-4 | 6.0 | 5 | 53 | 189 | 473 | 1.3 | 6 | 61 | 216 | 540 | 0.01 | 1 | 17 | 60 | 151 |
| Ex-5 | 6.0 | 5 | 74 | 294 | 634 | 1.3 | 6 | 84 | 336 | 724 | 0.01 | 2 | 24 | 95 | 204 |

Note that the design space for Example 5 added a design point to that of Example 4, that is $\mathcal{V}_5 = \mathcal{V}_4 \cup \{(1,\ 1,\ 1.5,\ 1)^T\}$. The optimal weight ($p_8$) for this vertex ($v_8$) is zero and the other optimal weights are therefore the same as in Example 4.

## 5 Objective Choices

In the previous section we considered some choices of the function $f(\cdot)$ in iteration (3). These were to some extent arbitrary. Convergence rates varied although judicious choice of a free parameter could yield improvements.

A more objective approach is motivated by the possibility that the criterion $\varphi(p)$ might have negative derivatives. For example, the likelihood under a Bradley-Terry model for paired comparisons data has the form:

$$\varphi(p) = \prod_{i<j} \frac{p_i^{O_{ij}} p_j^{O_{ji}}}{(p_i + p_j)^{O_{ij}+O_{ji}}}, \qquad p_j \ge 0,\ \sum_j p_j = 1,$$

where $O_{ij}$ is the frequency with which product $i$ is preferred to product $j$ in $(O_{ij} + O_{ji})$ comparisons. This is a homogeneous function of degree zero ($\varphi(cp) = \varphi(p)$) and hence $\sum p_i d_i = 0$. In fact, $d_j = F_j$. This leads to the idea of replacing $d_j$ by $F_j$ in algorithm (3). Since there are always positive and negative $F_j$ ($\sum p_j F_j = 0$) this requires a function $f(F)$ which is defined for positive and negative $F$, where $F$ represents a directional derivative. Note that properties of section 3 are still satisfied provided $f(F)$ is increasing in $F$. A suitable choice of $f(\cdot)$ should be governed by the fact that at the optimum

$$F_j^* \begin{cases} = 0 & \text{for} \quad p_j^* > 0 \\ \le 0 & \text{for} \quad p_j^* = 0 \end{cases}.$$

This suggests that a suitable function is one that is 'centred' on zero and changes reasonably quickly about $F = 0$. It should also be desirable to treat positive and negative $F_j$'s symmetrically, at least when all $p_j^*$'s are positive.

A class of functions $f(x)$ with the potential to satisfy these requirements has the following structure. Suppose a function $h(x)$ is defined on $\mathbb{R}$ such that

$$h(0) = 1, \quad h(x) > 0 \quad \text{and} \quad h'(x) \; > \; 0. \tag{4}$$

Then take

$$f(x) = \begin{cases} h(x) & \text{for } x < 0 \\ 2 - h(-x) & \text{for } x > 0 \end{cases}, \tag{5}$$

that is $f(x) = (1+s) - s\,h(-sx)$, $s = \text{sign}(x)$. Clearly $f(x)$ is increasing, while for $y > 0$, $(y, f(y))$ and $(-y, f(-y))$ are reflections of each other in the point $(0,1) = (0, f(0))$, that is $f(-y) = 2 - f(y)$. Equivalently $f'(y)$ is symmetric about zero. Note that $0 < f(x) < 2$, so that $f(x)$ is bounded and $f(0) = 1$.

Possible choices of $h(x)$ with $x = F$ and $\delta$ being a positive free parameter are

$$h(x) = \begin{cases} \exp(\delta x) \\ 1/(1-\delta x) \\ (1-x)^{-\delta} \\ 2H(\delta x) \end{cases}, \tag{6}$$

where $H(\cdot)$ is a c.d.f. such that $H(0) = 1/2$.

Note that if $h(x)$ is bounded on $\mathbb{R}^+$ then a converse choice would be possible. In that case we need $h(x)$ to be

$$h^*(x) \;=\; 1 + \frac{h(x) - h(0)}{h(\infty) - h(0)}$$

so that $h^*(0) = 1$ and $h^*(\infty) = 2$. Then

$$f(x) \;=\; \begin{cases} h^*(x) & \text{for} \;\; x > 0 \\ 2 - h^*(-x) & \text{for} \;\; x < 0. \end{cases}$$

Using the functions $h(x) = \exp(\delta x)$, $h(x) = 1/(1-\delta x)$ and $h(x) = (1-x)^{-\delta}$ in (6) and considering $f(x)$ in (5) we calculate $D$-optimal designs for the 5 examples given in Section 4. Thus, we use the algorithms

$$p_j^{(r+1)} = \begin{cases} p_j^{(r)} h(F_j^{(r)}) \Big/ D & \text{for } F_j^{(r)} < 0 \\ p_j^{(r)} \left\{2 - h(-F_j^{(r)})\right\} \Big/ D & \text{for } F_j^{(r)} > 0 \end{cases}, \tag{7}$$

where $D = \sum_{i=1}^{J} p_i^{(r)} f(F_i^{(r)})$, with $f(\cdot)$ as in (5). Similar to the table in the previous section we record for $n$ = 1, 2, 3, 4, the number of iterations in

Table 2 for the above three choices of $h(x)$ respectively. We take the initial weights to be equal, that is $p_j^{(0)} = 1/J$, $j = 1, 2, \ldots, J$. Comparing the results in Table 2 to those in Table 1 it shows that convergence (to the same solution as in section 4) is improved by considering this approach. The choices of $h(\cdot)$ give almost same number of iterations within each example.

**Table 2.** Iteration results with the proposed approach : number of iterations needed to achieve $\max_{1\le j\le J}\{F_j\} \le 10^{-n}$ for $n = 1, 2, 3, 4$.

| | $h(x) = \exp(\delta x)$, $x = F$ | | | | | $h(x) = 1/(1-\delta x)$, $x = F$ | | | | | $h(x) = (1-x)^{-\delta}$, $x = F$ | | | | |
|---|---|---|---|---|---|---|---|---|---|---|---|---|---|---|---|
| Examples | $\delta$ | $n$ | | | | $\delta$ | $n$ | | | | $\delta$ | $n$ | | | |
| | | 1 | 2 | 3 | 4 | | 1 | 2 | 3 | 4 | | 1 | 2 | 3 | 4 |
| Ex-1 | 1.6 | 1 | 3 | 6 | 10 | 1.7 | 1 | 3 | 6 | 10 | 1.7 | 1 | 3 | 7 | 10 |
| Ex-2 | 1.8 | 2 | 5 | 12 | 20 | 1.7 | 1 | 5 | 12 | 20 | 1.8 | 2 | 4 | 12 | 18 |
| Ex-3 | 1.5 | 1 | 3 | 6 | 10 | 1.7 | 1 | 3 | 6 | 8 | 1.7 | 1 | 2 | 6 | 9 |
| Ex-4 | 2.0 | 1 | 8 | 30 | 74 | 2.0 | 1 | 9 | 30 | 75 | 2.0 | 1 | 9 | 30 | 75 |
| Ex-5 | 2.1 | 1 | 12 | 45 | 96 | 2.1 | 1 | 12 | 46 | 97 | 2.1 | 1 | 12 | 45 | 97 |

## 6 Conclusions

It seems reasonable to claim that we have identified versions of algorithm (3) which perform reasonably efficiently. The proposed approach improves convergence considerably by a factor of 6 or 7. The examples considered are small scale in that the design spaces are in effect small discrete sets. In practice design spaces are continuous. In Mandal and Torsney (2002) we have explored a clustering approach for large discretized versions of such spaces. This involves running a version of algorithm (3) for 5 or 10 iterations. Plots of the weights or derivatives readily exhibit features typical of derivatives of the optimum namely local maxima at support points and minimal turning points between them. The minimal turning points clearly correspond to regions of zero weight. The weights on the points in these regions are converging to zero. Assuming such zero weights we would be left with a number of disjoint clusters of points which must contain the optimal support points. Clusters are taken to have boundaries near to minimal turning points. We then switch to finding optimal conditional distributions within clusters and a marginal distribution across clusters. Thus we wish to find not one but several optimizing distributions, an extension of Problem (P1). Combining this clustering device with the approach of section 5 should lead to further improvements in convergence.

*Acknowledgement.* The authors wish to thank the referee for some valuable comments which led to some improvement over an earlier version of the manuscript. The second author is supported by a research grant from the Natural Sciences and Engineering Research Council (NSERC) of Canada.

## References

V.V. Fedorov. *Theory of Optimal Experiments.* Academic Press, New York, 1972.

J. Kiefer. General equivalence theory for optimum designs (approximate theory). *Ann. Stat.*, 2:849–879, 1974.

S. Mandal and B. Torsney. Algorithms for the construction of optimizing distributions. *Comm. Statist. Theory Methods*, 29:1219–1231, 2000.

S. Mandal and B. Torsney. Construction of optimal designs using a clustering approach. *J. Statist. Plann. Inference*, 2002.

S.D. Silvey, D.M. Titterington, and B. Torsney. An algorithm for optimal designs on a finite design space. *Comm. Statist. A—Theory Methods*, 14: 1379–1389, 1978.

B. Torsney. Contribution to discussion of 'Maximum likelihood estimation via the EM algorithm' by Dempster et al. *J. Roy. Stat. Soc. B*, 39:26–27, 1977.

B. Torsney. Computing optimizing distributions with applications in design, estimation and image processing. In Y. Dodge, V.V. Fedorov, and H.P. Wynn, editors, *Optimal Design and Analysis of Experiments*, pages 361–370, Amsterdam, 1988. Elsevier Science Publishers B.V.

B. Torsney and A.M. Alahmadi. Further development of algorithms for constructing optimizing distributions. In V.V. Fedorov, W.G. Müller, and I.N. Vuchkov, editors, *Model oriented data analysis*, pages 121–129. Physica-Verlag, 1992. Proc. 2nd IIASA Workshop in St. Kyrik, Bulgaria.

B. Torsney and S. Mandal. Construction of constrained optimal designs. In A. Atkinson, B. Bogacka, and A.A. Zhigljavsky, editors, *Optimum Design (Cardiff, 2000)*, pages 141–152. Kluwer Academic Publishers, 2001.

P. Whittle. Some general points in the theory of optimal experimental design. *J. Roy. Stat. Soc. B*, 35:123–130, 1973.

C.F.J. Wu. Some iterative procedures for generating nonsingular optimal designs. *Comm. Statist. A—Theory Methods*, 14:1399–1412, 1978.

H.P. Wynn. The sequential generation of $D$-optimal experimental designs. *Ann. Math. Stat.*, 41:1655–1664, 1970.

H.P. Wynn. Results in the theory and construction of $D$-optimum experimental designs (with discussion). *J. Roy. Stat. Soc. B*, 34:133–147,170–186, 1972.

# Locally Optimal Designs for an Extension of the Michaelis-Menten Model

C. Trandafir and J. López-Fidalgo

University of Salamanca, Departamento de Estadistica Edificio de la Merced, Plza. de la Merced, 1-4, Fac. Ciencias, Salamanca, Spain `a110619,fidalgo@usal.es`

**Summary.** In this paper we study the problem of designing experiments for nonlinear regression models in pharmacokinetics. In particular, we consider an extension of the popular Michaelis-Menten model (MMM) with the addition of a linear term. $D$-optimal designs supported at 3 points with equal weights are calculated for the MMM extension using the general equivalence theorem. This model is a special case of partially nonlinear regression. We obtain $c$-, $D$-optimal designs for estimating the three parameters of the model. Next we calculate compound designs. Some numerical examples are provided with efficiencies for estimating each of the parameters.

**Key words:** $c$-optimal design, $D$-optimal design, compound designs, Lagrange interpolation polynomial.

## 1 Introduction

The MMM is used in biology to describe the equilibrium kinetics of the enzyme system and to analyze the data from drug, neurotransmitter and hormone receptor assay (see for instance Cressie and Keightley (1981), Raaijmakers (1987), Dunn (1988)). The MMM is also used in compartmental models to model the rate of change from one compartment to another. A general theory for enzyme kinetics was first developed by Michaelis and Menten (1913).

This model predicts the velocity rate $\nu$ of formation of a product in the chemical reaction when the substrate concentration $x$ is given. The MMM is represented by:

$$\mathrm{E}[\nu] = \frac{Vx}{K+x}, \quad \mathrm{Var}\,[\nu] = \sigma^2, \quad x \in \chi = [0, \infty).$$

Let the parameter $V$ denote the maximum velocity of the enzyme process; the velocity may approach $V$ but will never reach it. The parameter $K$ is the Michaelis-Menten constant. At concentration $K$ the rate of process equals half of the maximum rate. In practice $\chi$ is truncated to a compact interval. Duggleby (1979) and Currie (1982) suggested without reason that the right end point of $\chi$ is a multiple of the nominal value of $K$. López-Fidalgo and

Wong (2002) showed that an advantage of assuming $\chi = [0, bK]$, where $b$ is a user-selected constant, is that behavioural properties of locally optimal designs Chernoff (1953) depend on $b$ only.

Chan and Endrenyi (1981) and Currie (1982) incorporated the heteroscedasticity into the MMM. Dunn (1988) considered a variation of the original MMM model, which is the basis of this paper. This extension of the Michaelis-Menten model (EMMM) can be described as follows:

$$\mathrm{E}[\mathrm{v}] = \frac{Vx}{K+x} + Fx, \quad \mathrm{Var}\,[\mathrm{v}] = \sigma^2, \quad x \in \chi = [0, bK], \tag{1}$$

where we fix $b \in [1, \infty)$ and $K > 0$. The model adds a new parameter $F$ which measures the elimination rate constant of the so-called kinetics first-order process. This model seems to be a little closer to reality but is less known. So far, only $c$-optimal designs have been computed for this model López-Fidalgo and Rodríguez-Díaz (2004). As a matter of fact, this model allows discrimination between the MMM model and the linear model. Our goal is to determine a $D$-optimal design, which in this case will be supported at three points. The model represents a special case of partially nonlinear regression. In the next section we briefly present the general theory necessary for the calculations that are to be done later in Section 3. In Section 4 quasi-optimal compound designs for three $c$-optimality criteria for estimating parameters $K$, $V$ and $F$ are computed.

## 2 Theoretical Background

### 2.1 The Nonlinear Regression Model

Consider a general nonlinear regression model of the form:

$$\eta = f(x, \theta) + \varepsilon, \; x \in \chi,$$

where $\eta$ is the response variable, $x$ is the explanatory variable, $\chi$ is the design space, $\theta = (\theta_1, \ldots, \theta_k)^T$ is a $k \times 1$ vector of unknown parameters and $f(x, \theta)$ is a nonlinear function of $\theta$. If $f(x, \theta)$ is differentiable with respect to $\theta$, the gradient is defined as the $k \times 1$ vector $\nabla f(x, \theta)$ with $j$th entry

$$\{\nabla f(x, \theta)\}_j = \frac{\partial f(x, \theta)}{\partial \theta_j}, \; j = 1, \ldots, k.$$

For an approximate design $\xi$ with $n$ support points $x_1, \ldots, x_n$ and weights $\xi(x_i)$, $i = 1, \ldots, n$, the $k \times k$ information matrix is

$$M(\xi) = \sum_{i=1}^{n} \nabla f(x_i, \theta) \nabla f(x_i, \theta)^T \xi(x_i).$$

For independent observations, $M^{-1}(\xi)$ is asymptotically proportional to the covariance matrix of the maximum likelihood estimator of $\theta$. The generalized variance function $d(x,\xi)$ is defined as

$$d(x,\xi) = \nabla f(x,\theta)^T M^{-1}(\xi) \nabla f(x,\theta). \tag{2}$$

This is a measure of the variability of a prediction at a point $x$.

### 2.2 The Design Criteria

**Definition 1.** *The D-optimality criterion minimizes the volume of the confidence ellipsoid of the parameters and is defined by*

$$\Phi_D[M(\xi)] = \det M^{-1/k}(\xi). \tag{3}$$

**Theorem 1 (General Equivalence Theorem of Kiefer and Wolfowitz (1960)).** *The following three statements are equivalent:*

1. $\xi^\star$ *maximizes* $\det M(\xi)$,
2. $\xi^\star$ *minimizes* $\max_x d(x,\xi^\star)$,
3. $\max_x d(x,\xi^\star) = k$.

For more details on $D$-optimal design we refer to the monographs Fedorov (1972), Silvey (1980), Pázman (1986), Atkinson and Donev (1992) and Pukelsheim (1993).

## 3 Extension of the Michaelis-Menten Model

In this case we will search for an experimental design supported at three points

$$\xi = \begin{Bmatrix} sK & tK & bK \\ p & q & (1-p-q) \end{Bmatrix},$$

where $p$, $q \in [0,1]$ are the weights of the support points. Let $\theta^T = (K, V, F)$ be the parameter vector and let

$$\nabla f(x,\theta) = \left( -\frac{Vx}{(K+x)^2}, \frac{x}{K+x}, x \right)$$

be evaluated at the nominal values of $\theta$. The nominal values usually represent the best guess for the parameter vector $\theta$ at the beginning of the experiment. On the other hand, we can write

$$\nabla f(x,\theta) = \frac{x}{(K+x)^2} B(\theta) g(x),$$

where

$$B(\theta) = \begin{pmatrix} -V & 0 & 0 \\ K & 1 & 0 \\ K^2 & 2K & 1 \end{pmatrix} \quad \text{and} \quad g(x)^T = (1, x, x^2).$$

Thus, this model may be considered as a classical polynomial regression model (see Dette (1992), Dette and Studden (1995), Huang et al. (1995) and Ortiz and Rodríguez (1998)).

The Fisher information matrix of the design $\xi$ is

$$M(\xi) = \sum_{i=1}^{n} \nabla f(x_i, \theta) \nabla f(x_i, \theta)^T \xi(x_i) = B(\theta) X^T D X B(\theta)^T,$$

where

$$X = \begin{pmatrix} 1 & sK & (sK)^2 \\ 1 & tK & (tK)^2 \\ 1 & bK & (bK)^2 \end{pmatrix}, \quad D = \text{diag}\left( \frac{(sK)^2}{(K+sK)^4}, \frac{(tK)^2}{(K+tK)^4}, \frac{(bK)^2}{(K+bK)^4} \right).$$

The aim is to obtain $s$, $t$, $p$, $q$ such that the determinant of $M(\xi)$ is maximum. Therefore, we can write

$$\det M(\xi) = pq(1-p-q) \frac{V^2 b^2}{(1+b)^4} \frac{(b-s)^2 (b-t)^2 (s-t)^2 s^2 t^2}{(1+s)^4 (1+t)^4}$$

and the maximization problem splits into two different problems:

1. To determine the values of $p$, $q$ which maximize $pq(1-p-q)$ under the constraints $p$, $q \geq 0$, $p + q \leq 1$.
2. To determine the values of $s$, $t \geq 0$ such that

$$\frac{(b-s)(b-t)(s-t)st}{(1+s)^2(1+t)^2} \quad \text{is maximum.}$$

For the first problem one easily obtains that $p = q = \frac{1}{3}$, therefore we conclude that if there exists a locally $D$-optimal design with 3 points in its support then it has equal weights at each point. The solution of the second problem is given by the unique pair of values (up to symmetry) $s = \frac{c+d}{\sqrt{2}}$ and $t = \frac{c-d}{\sqrt{2}}$, which are both positive, where $c$, $d$ are solutions of the system

$$\begin{cases} -3b(5+b) + \sqrt{2}(15-b)c + 2(b+8)c^2 + 2\sqrt{2}c^3 & = 0 \\ 6b - d^2 - 3c(2\sqrt{2} + c) & = 0. \end{cases}$$

Now we want to prove that this design is actually $D$-optimal. It is enough to prove that $d(x, \xi^\star) \leq 3$, $x \in [0, bK]$. For this model

$$d(x, \xi^\star) = 3 \sum_{i=1}^{3} \frac{(K + x_i)^4}{x_i^2} \frac{x^2}{(K+x)^4} \ell_i^2(x),$$

where the $x_i$ are the previously determined support points and $\ell_i(x) = \prod_{j \neq i}(x - x_j)/(x_i - x_j)$ is the fundamental Lagrange interpolation polynomial corresponding to the support points. By the General Equivalence Theorem of Kiefer and Wolfowitz (1960) we have $d'(x_i, \xi^\star) = 0$ for $i = 1, 2$. Since $(s, t)$ is the unique maximum point of $d(x, \xi^\star)$, it follows that $d'(x, \xi^\star)$ is a polynomial of degree 6, $\lim_{x \to \infty} d(x, \xi^\star) = \infty$, $d(0, \xi^\star) = 0$ and $d(bK, \xi^\star) = 3$ then $d(x, \xi^\star) \leq 3, x \in [0, bK]$. In fact, $\max_{x \in [0, bK]} d(x, \xi^\star) = 3$. This completes the proof.

# 4 Compound Optimal Designs for 3 Parameters

## 4.1 $c$-optimal Designs

An elegant method for finding an optimal design to estimate a linear combination of the parameters, say $c^\top \theta$, is given in Elfving (1952). Using Elfving's argument, for nominal values $K = V = 1$ (cf. Dunn (1988)) we compute the $c_K$, $c_V$, $c_F$-optimal designs for estimating each of the three parameters

$$\xi^*_{c_K} = \begin{Bmatrix} 0.31 & 2.37 & 5 \\ 0.55 & 0.33 & 0.12 \end{Bmatrix}, \quad \xi^*_{c_V} = \begin{Bmatrix} 0.31 & 2.37 & 5 \\ 0.37 & 0.44 & 0.19 \end{Bmatrix}, \quad \xi^*_{c_F} = \begin{Bmatrix} 0.31 & 2.37 & 5 \\ 0.28 & 0.43 & 0.29 \end{Bmatrix}.$$

These optimal designs do not depend on the value of $V$ and the values of the criterion function are respectively

$$\Phi_K(\xi^*_{c_K}) = 445.803 \, \frac{K^2}{V^2}, \quad \Phi_V(\xi^*_{c_V}) = 212.165, \quad \Phi_F(\xi^*_{c_F}) = \frac{4.16629}{K^2}.$$

For the calculation of the efficiencies we use the following formula:

$$\text{eff}_{c_{\theta_i}}(\xi^*_{\theta_j}) = \frac{\Phi_{c_{\theta_i}}(\xi^*_{\theta_i})}{\Phi_{c_{\theta_i}}(\xi^*_{\theta_j})}.$$

The efficiencies of these designs for estimating the other parameters depend only on $b$. The efficiencies for b=5 are given in Table 1. It can be seen that the efficiencies for estimating each of the parameters are over 79%.

**Table 1.** Efficiencies for estimating the parameters $V$, $K$ and $F$ when $b = 5$.

| | $\xi^*_{c_K}$ | $\xi^*_{c_V}$ | $\xi^*_{c_F}$ |
|---|---|---|---|
| $\text{eff}_{c_K}$ | 1 | 0.877 | 0.793 |
| $\text{eff}_{c_V}$ | 0.905 | 1 | 0.942 |
| $\text{eff}_{c_F}$ | 0.956 | 0.932 | 1 |

## 4.2 $c$-efficiencies of the $D$-optimal Design

To assess the performance of the $D$-optimal design for estimating each of the parameters, we calculate the variances of the estimates for $b = 5$

$$\Phi_K(\xi_D) = 599.103\frac{K^2}{V^2}, \; \Phi_V(\xi_V) = 256.756, \; \Phi_F(\xi_F) = \frac{4.715}{K^2}.$$

Therefore the efficiencies of the $D$-optimal design for estimating $K$, $V$ and $F$ are:

$$\text{eff}_K(\xi_D) = 0.744, \; \text{eff}_V(\xi_D) = 0.826, \; \text{eff}_F(\xi_D) = 0.884.$$

For $b = 5$ the efficiency for estimating $K$ is about 75%, the efficiency for estimating $V$ is about 83% and the efficiency for estimating $F$ is more than 85%. The $c_K$-optimal design has about 91% (96%) efficiency for estimating parameters $V$ ($F$), the $c_V$-optimal has about 88% (93%) for parameters $K$ ($F$) and the $c_F$-optimal has about 80% (94%) for parameters $K$ ($V$).

## 4.3 Compound Optimal Designs

Very often some parameters are more interesting than others. This implies that it is desirable to have designs with greater efficiency for estimating the important parameters. To find such designs consider the standardized compound criterion given by

$$\begin{aligned}\Phi_{\lambda,\mu}(\xi) &= \lambda\frac{\Phi_K(\xi)}{\Phi_K(\xi_K^*)} + \mu\frac{\Phi_V(\xi)}{\Phi_V(\xi_V^*)} + (1-\lambda-\mu)\frac{\Phi_F(\xi)}{\Phi_F(\xi_F^*)} \\ &= \frac{\lambda}{\text{eff}_K(\xi)} + \frac{\mu}{\text{eff}_V(\xi)} + \frac{1-\lambda-\mu}{\text{eff}_F(\xi)} \\ &= \lambda\frac{\Phi_K(\xi)}{445.803} + \mu\frac{\Phi_V(\xi)}{212.165} + (1-\lambda-\mu)\frac{\Phi_F(\xi)}{4.166}\end{aligned}$$

where $\lambda, \mu \geq 0$, $\lambda + \mu \leq 1$ are user-selected constants. The criterion is a convex combination of the three convex criteria and so the criterion is essentially similar to the single-objective design problem. To tackle the case of different magnitudes of variances we standardize them dividing by their optimal values. A design that minimizes $\Phi_{\lambda,\mu}(\xi)$, for fixed $\lambda$ and $\mu$, is called a compound optimal design. The values of $\lambda$, $\mu$ and $1-\lambda-\mu$ represent the weights to be assigned to each of the criteria.

We will search for designs among those of the form:

$$\xi = \left\{ \begin{array}{ccc} sK & tK & bK \\ p & q & 1-p-q \end{array} \right\}.$$

For fixed $b$, $\lambda$ and $\mu$, we seek values of $s$, $t$, $p$, $q$ minimizing $\Phi_{\lambda,\mu}(\xi)$. From the equivalence theorem the design obtained is optimal among all designs, if and only if

**Table 2.** Quasi-optimal compound designs and their efficiencies for estimating the parameters $K$, $V$ and $F$ for different values of $\lambda$ and $\mu$.

| $\lambda$ | $\mu$ | 0.1 | 0.2 | 0.3 | 0.4 | 0.5 | 0.6 | 0.7 | 0.8 | 0.9 |
|---|---|---|---|---|---|---|---|---|---|---|
| 0.1 | $\text{eff}_K$ | 79.9 | 81.3 | 82.6 | 84.0 | 85.3 | 86.6 | 87.8 | 89.0 | 90.2 |
| | $\text{eff}_V$ | 97.1 | 97.8 | 98.3 | 98.8 | 99.2 | 99.6 | 99.8 | 99.9 | 99.8 |
| | $\text{eff}_F$ | 99.2 | 98.8 | 98.3 | 97.6 | 96.8 | 95.9 | 94.79 | 93.4 | 91.7 |
| 0.2 | $\text{eff}_K$ | 84.1 | 85.4 | 86.6 | 87.8 | 88.9 | 90.1 | 91.2 | 92.3 | |
| | $\text{eff}_V$ | 98.2 | 98.6 | 99.0 | 99.3 | 99.5 | 99.6 | 99.6 | 99.4 | |
| | $\text{eff}_F$ | 97.8 | 97.2 | 96.5 | 95.7 | 94.6 | 93.4 | 91.9 | 90.0 | |
| 0.3 | $\text{eff}_K$ | 87.7 | 88.8 | 89.9 | 91.0 | 92.1 | 93.1 | 94.1 | | |
| | $\text{eff}_V$ | 98.5 | 98.8 | 99.0 | 99.2 | 99.2 | 99.1 | 98.7 | | |
| | $\text{eff}_F$ | 96.1 | 95.3 | 94.3 | 93.2 | 91.9 | 90.3 | 88.3 | | |
| 0.4 | $\text{eff}_K$ | 90.8 | 91.8 | 92.8 | 93.8 | 94.7 | 95.6 | | | |
| | $\text{eff}_V$ | 98.4 | 98.5 | 98.6 | 98.5 | 98.3 | 97.8 | | | |
| | $\text{eff}_F$ | 93.92 | 92.9 | 91.7 | 90.31 | 88.6 | 86.4 | | | |
| 0.5 | $\text{eff}_K$ | 93.4 | 94.3 | 95.2 | 96.1 | 96.9 | | | | |
| | $\text{eff}_V$ | 97.8 | 97.8 | 97.7 | 97.3 | 96.8 | | | | |
| | $\text{eff}_F$ | 91.4 | 90.1 | 88.6 | 86.7 | 84.4 | | | | |
| 0.6 | $\text{eff}_K$ | 95.6 | 96.5 | 97.3 | 98.0 | | | | | |
| | $\text{eff}_V$ | 96.8 | 96.6 | 96.2 | 95.5 | | | | | |
| | $\text{eff}_F$ | 88.4 | 86.8 | 84.8 | 82.3 | | | | | |
| 0.7 | $\text{eff}_K$ | 97.5 | 98.2 | 98.8 | | | | | | |
| | $\text{eff}_V$ | 95.4 | 94.9 | 94.0 | | | | | | |
| | $\text{eff}_F$ | 84.9 | 82.8 | 80.1 | | | | | | |
| 0.8 | $\text{eff}_K$ | 99.0 | 99.5 | | | | | | | |
| | $\text{eff}_V$ | 93.4 | 92.3 | | | | | | | |
| | $\text{eff}_F$ | 80.7 | 77.7 | | | | | | | |
| 0.9 | $\text{eff}_K$ | 99.9 | | | | | | | | |
| | $\text{eff}_V$ | 90.4 | | | | | | | | |
| | $\text{eff}_F$ | 75.1 | | | | | | | | |

$$\Psi(x,\xi) = f^T(x)\nabla\Phi(\xi)f(x) - trM(\xi)\nabla\Phi(\xi) \geq 0, \; x \in \chi \tag{4}$$

with equality at the support points. This requires that these optimal values satisfy

$$\begin{cases} \Psi(sK,\xi) = 0, \quad \Psi(tK,\xi) = 0, \; \Psi(bK,\xi) = 0 \\ \left(\frac{\partial\Psi}{\partial x}\right)_{x=sK} = 0, \; \left(\frac{\partial\Psi}{\partial x}\right)_{x=tK} = 0. \end{cases}$$

The last two of these equations can be simplified using the fact that

$$\nabla\Phi(\xi) = -M^{-1}(\xi)\,\text{diag}\left(\frac{\lambda}{\Phi_K(\xi_K)}, \frac{\mu}{\Phi_V(\xi_V)}, \frac{1-\lambda-\mu}{\Phi_F(\xi_F)}\right) M^{-1}(\xi).$$

Using the equivalence theorem equation (4) one can verify that the computed designs are quasi-optimal. Table 2 shows the efficiencies of these designs, for

$b = 5$ and selected values of $\lambda$ and $\mu$ for estimating $K, V$ and $F$. We can see that the efficiencies are quite high for different values of $\lambda$ and $\mu$. Also we can notice the high efficiencies for $V$ in all the cases. Thus, if $\lambda$ and $\mu$ satisfy approximately the inequality $1 \leq 2\lambda + \mu \leq 1.2$ we obtain a design with good efficiencies for estimating each of the parameters $K$, $V$ and $F$ simultaneously.

## References

A. C. Atkinson and A. N. Donev. *Optimum experimental design.* Oxford University Press, New York, 1992.

F.Y. Chan and L. Endrenyi. Optimal design of experiments for estimation of precise hyperbolic kinetic and binding parameters. *J. Theor. Biol.*, 90 (2):241–263, 1981.

H. Chernoff. Locally optimal design for estimating parameters. *Ann. Math. Stat.*, 24:586–602, 1953.

N.A.C. Cressie and D.D. Keightley. Analysing data from hormone-receptor assay. *Biometrics*, 37(2):235–250, 1981.

D.J. Currie. Estimating Michaelis-Menten parameters: bias, variance and experimental design. *Biometrics*, 38(4):907–919, 1982.

H. Dette. Optimal designs for a class of polynomials of odd or even degree. *Ann. Stat.*, 20(1):238–259, 1992.

H. Dette and W. J. Studden. Optimal designs for polynomial regression when the degree is not known. *Statistica Sinica*, 5(2):459–473, 1995.

R.G. Duggleby. Experimental designs for estimating the kinetic parameter for enzyme- catalysed reactions. *J. Theor. Biol.*, 81(4):671–684, 1979.

G. Dunn. Optimal designs for drug, neurotransmitter and hormone receptor assays. *Stat. Med.*, 7:805–815, 1988.

G. Elfving. Optimum allocation in linear regression theory. *Ann. Math. Stat.*, 23:255–262, 1952.

V.V. Fedorov. *Theory of optimal experiments.* Academic Press, New York, 1972.

M.N.L. Huang, F.C. Chang, and W.K. Wong. *D*-optimal designs for polynomial regression without an intercept. *Statistica Sinica*, 5(2):441–458, 1995.

J. Kiefer and J. Wolfowitz. The equivalence of two extremum problems. *Canad.J.Math.*, 12:363–366, 1960.

J. López-Fidalgo and J.M. Rodríguez-Díaz. Elfving's method for $m$–dimensional models. *Metrika*, 2004.

J. López-Fidalgo and W.K. Wong. Design issues for the Michaelis-Menten model. *J. Theor. Biol.*, 215(1):1–11, 2002.

L. Michaelis and M.L. Menten. Kinetics for intertase action. *Biochemische Zeitung*, 49:333–369, 1913.

I. Ortiz and C. Rodríguez. *D*-optimal designs for weighted polynomial regression without any initial terms. In A.C. Atkinson, L. Pronzato, and

H.P. Wynn, editors, *MODA 5 - Advances in Model -Oriented Data analysis and Experimental design (Marseilles, June 22-26, 1998)*, pages 67–74. Physica-Verlag, 1998.

A. Pázman. *Foundation of optimum experimental design.* Reidel, Dordrecht, 1986.

F. Pukelsheim. *Optimal design of experiments.* J. Wiley and Son, New York, 1993.

J.G.W. Raaijmakers. Statistical analysis of the Michaelis-Menten equation. *Biometrics*, 43(4):793–803, 1987.

S.D. Silvey. *Optimal design.* Chapman and Hall, London, 1980.

# Asymptotic Properties of Urn Designs for Three-arm Clinical Trials

Y. Tymofyeyev[1], W.F. Rosenberger[2], and F. Hu[3]

[1] Department of Mathematics and Statistics, University of Maryland, Baltimore County, 1000 Hilltop Circle, Baltimore, Maryland 21250, U.S.A. ytymof1@math.umbc.edu

[2] Department of Mathematics and Statistics, University of Maryland, Baltimore County, 1000 Hilltop Circle, Baltimore, Maryland 21250, U.S.A. billr@math.umbc.edu

[3] Department of Statistics, University of Virginia, P.O. Box 400135, Charlottesville, Virginia 22904, U.S.A. fh6e@virginia.edu

**Summary.** The generalized Friedman's urn design for response-adaptive clinical trials generates allocation proportions of treatments assignments that, under certain conditions, asymptotically follow a normal distribution. It is important to know the asymptotic variance-covariance matrix of this distribution in practice, as one can compute the power of the test that compares treatments with binary response. In this paper we develop the explicit form of the asymptotic variance-covariance matrix of allocation proportions as well as the asymptotic variance-covariance matrix for the urn composition for a particular urn model in three-treatment comparisons.

**Key words:** urn design, clinical trial, allocation.

## 1 Introduction

One type of response-adaptive randomization procedure, where sequentially accumulated information is used to change randomization probabilities in a clinical trial is the generalized Friedman's urn (GFU) model (cf. Athreya and Karlin (1968)), which sometimes is referred to as the generalized Pòlya's urn (GPU) model in the literature. This design skews patients allocation assigning more patients to the best performing treatment during a clinical trial. Wei (1979) originally proposed to apply the GFU in the context of clinical trials. One can refer to Rosenberger (1996) for a review of the subject. We describe the GFU design below.

At each stage of the binary response clinical trial, one patient is randomly assigned to one possible treatment in the following way. There are $K$ types of particles representing each type of the treatments. Let the vector $\mathbf{Y}_0 = (Y_{01}, \ldots, Y_{0K})$ specify the original urn composition, here $Y_{0k}$ denotes the number of particles of type $k$, $(k = 1, \ldots, K)$. At stage $i$, $(i = 1, \ldots, n)$, the $i$th patient arrives, the particle is randomly drawn from the urn and replaced. If it is the particle of type $k$, then the patient is assigned to the $k$th treatment. The response to the treatment of patient $i$ is the binary random variable $\xi_i$, which might be death or cure. As soon as $\xi_i$ is observed, we update the urn by adding $d_{kj}(\xi_i)$ particles of type $j$, $j = 1, \ldots, K$, to the urn. The urn composition at the general stage $n$ is denoted by the vector $\mathbf{Y}_n = (Y_{n1}, \ldots, Y_{nK})$. We also define the vector $\mathbf{N}_n = (N_{n1}, \ldots, N_{nK})$, where the component $N_{nk}$ stands for number of times the particle of type $k$ was chosen during the first $n$ stages in the trial. We use $\mathbf{D}_i(\ell) = \{d_{kj}(\xi_i = \ell)\}_{k,j=1}^K$ to specify the *adding rule*, when outcome $\ell$ is observed ($\ell = 1, 2$). Then $\mathbf{H}_i = \mathrm{E}(\mathbf{D}_i)$, $(i = 1, \ldots, n)$ is called the *generating matrix*. The asymptotic properties of $\mathbf{Y}_n$ and $\mathbf{N}_n$ are the primary subjects of our consideration.

The simplest setup is when the $\mathbf{D}_i$ depend only on the last treatment and its outcome, then the $\mathbf{D}_i$ are independent and identically distributed random variables with $\mathbf{H}_i = \mathbf{H}$, for all $i$. In this case, the model is said to be *homogeneous*. Let the vector $\mathbf{v} = (v_1, \ldots, v_K)$ be the left eigenvector that corresponds to the largest eigenvalue of $\mathbf{H}$. Athreya and Karlin (1968) under conditions $|P\{d_{kj} = 0 \text{ for all } j\} = 0$ for all $i$ and $\mathbf{H}$ is positive regular proved that

$$\frac{N_{nk}}{\sum_{j=1}^K N_{nj}} \to v_k \quad \text{and} \quad \frac{Y_{nk}}{\sum_{j=1}^K Y_{nj}} \to v_k,$$

almost surely, where $k = 1, \ldots, K$. There are applications where the adding rule $\mathbf{D}_i$ depends on the outcome history of the previous stages (see Andersen, Faries, and Tamura (1994) and Bai, Hu, and Shen (2002) for details). In such situations $\mathbf{H}_i$ becomes the conditional expectation of $\mathbf{D}_i$.

With respect to the availability of outcomes of the treatments, the simplest case is to assume immediate response (the value $\xi_i$ is observed before the $(i+1)$th patient arrives), but typically clinical trials may not have such feature. If outcomes of the previously randomized patients become available during the recruitment period, then it is still possible to use the GFU design for clinical trial. Asymptotic properties of GFU designs with delayed response are considered in Bai, Hu, and Rosenberger (2002).

In this work we consider the special case of a homogeneous GFU model when we have $K = 3$ treatments. We investigate a generating matrix of the form:

$$\mathbf{H} = \begin{bmatrix} 1 - q_a & q_a/2 & q_a/2 \\ q_b/2 & 1 - q_b & q_b/2 \\ q_c/2 & q_c/2 & 1 - q_c \end{bmatrix}, \qquad (1)$$

where $q_a, q_b$ and $q_c$ are the probabilities of "failure" for treatments A, B and C. In other words, in case of success on treatment $k$, one particle of type $k$ is added, and in case of failure 1/2 particle of two other types are added into the urn, where "particle" is conceptual and can be fractional when generated by computer. So without lost of generality, we assume that $\beta = 1$ particle is added at each stage, otherwise one can consider the sequence $\mathbf{Y}_n/\beta$. In section 2 we will perform a spectral analysis of $\mathbf{H}$ that will be employed in the derivation of asymptotic variances of $\mathbf{Y}_n$ and $\mathbf{N}_n$ in Sections 3 and 4.

We are primarily interested in the asymptotic variance-covariance matrix for $\mathbf{N}_n$, because based on it one can evaluate power of the test that performs treatment comparisons following a response-adaptive randomization procedure. We refer to the work of Hu and Rosenberger (2003) that presents the theoretical template for such evaluation.

## 2 Spectral Analysis of Generating Matrix

As we will see below an asymptotic properties of $\mathbf{Y}_n$ and $\mathbf{N}_n$ are determined by the spectral representation of $\mathbf{H}$. Omitting the tedious algebraic computation, $\mathbf{H}$ has the following eigenvalues $\lambda_0$, $\lambda_1$ and $\lambda_2$,

$$\begin{bmatrix} \lambda_0 \\ \lambda_1 \\ \lambda_2 \end{bmatrix} = \begin{bmatrix} 1 \\ 1 - \frac{1}{2}q_a - \frac{1}{2}q_b - \frac{1}{2}q_c + \frac{1}{2}D \\ 1 - \frac{1}{2}q_a - \frac{1}{2}q_b - \frac{1}{2}q_c - \frac{1}{2}D \end{bmatrix}, \tag{2}$$

where

$$D = \sqrt{q_a^2 + q_b^2 + q_c^2 - q_a q_b - q_a q_c - q_b q_c}. \tag{3}$$

The corresponding right eigenvectors of $\mathbf{H}$ form the columns of the matrix $\mathbf{T}$,

$$\mathbf{T} = [\mathbf{1}, \mathbf{t}_1, \mathbf{t}_2] = \begin{bmatrix} 1 & \frac{q_a(q_c - q_a + D)}{q_c(q_a - q_b)} & \frac{q_a(q_c - q_a - D)}{q_c(q_a - q_b)} \\ 1 & \frac{q_b(q_c - q_b + D)}{q_c(q_a - q_b)} & \frac{q_b(q_c - q_b - D)}{q_c(q_a - q_b)} \\ 1 & 1 & 1 \end{bmatrix},$$

where $\mathbf{1} = (1,1,1)'$ corresponds to the largest eigenvalue 1, and eigenvectors $\mathbf{t}_1$, $\mathbf{t}_2$ correspond to eigenvalues $\lambda_1, \lambda_2$, respectively. We write

$$(\mathbf{T}^{-1})^* = (\mathbf{v}^*, (\boldsymbol{\ell}_1)^*, (\boldsymbol{\ell}_2)^*),$$

where $\mathbf{v}, \boldsymbol{\ell}_1$ and $\boldsymbol{\ell}_2$ are the row vectors and $^*$ stands for conjugate transpose. The left eigenvector of $\mathbf{H}$ that corresponds to the largest eigenvalue 1 is

$$\mathbf{v} = \left[ \frac{q_b q_c}{q_a q_b + q_a q_c + q_b q_c} \; \frac{q_a q_c}{q_a q_b + q_a q_c + q_b q_c} \; \frac{q_a q_b}{q_a q_b + q_a q_c + q_b q_c} \right]. \tag{4}$$

The vectors $\boldsymbol{\ell}_1$ and $\boldsymbol{\ell}_2$ are given by

$$\boldsymbol{\ell}_1^* = \begin{bmatrix} -\frac{1}{2}\frac{q_c(q_b^2 - q_a q_c + q_b D)}{D(q_a q_b + q_a q_c + q_b q_c)} \\ -\frac{1}{2}\frac{q_c(q_a^2 - q_b q_c + q_a D)}{D(q_a q_b + q_a q_c + q_b q_c)} \\ \frac{1}{2}\frac{q_a^2 + q_b^2 + (q_a + q_b)(D - q_c)}{D(q_a q_b + q_a q_c + q_b q_c)} \end{bmatrix}$$

and

$$\boldsymbol{\ell}_2^* = \begin{bmatrix} \frac{1}{2}\frac{q_c(q_b^2 - q_a q_c - q_b D)}{D(q_a q_b + q_a q_c + q_b q_c)} \\ \frac{1}{2}\frac{q_c(q_a^2) q_b q_c - q_a D)}{D(q_a q_b + q_a q_c + q_b q_c)} \\ \frac{1}{2}\frac{-q_a^2 - q_b^2 + (q_a + q_b)(D + q_c)}{D(q_a q_b + q_a q_c + q_b q_c)} \end{bmatrix}.$$

Furthermore, the asymptotic distributions for vectors $\mathbf{Y}_n$ and $\mathbf{N}_n$ will be stated under assumption that $\text{Re}(\lambda_1 \leq 1/2)$, where $\text{Re}(\cdot)$ is the real part of a number. Figure 1 visualizes the set of $(q_a, q_b, q_c)$ that satisfies this condition. So the region of interest is the part of the cube that is separated by the plotted surface, where $q_a, q_b, q_c$ are large.

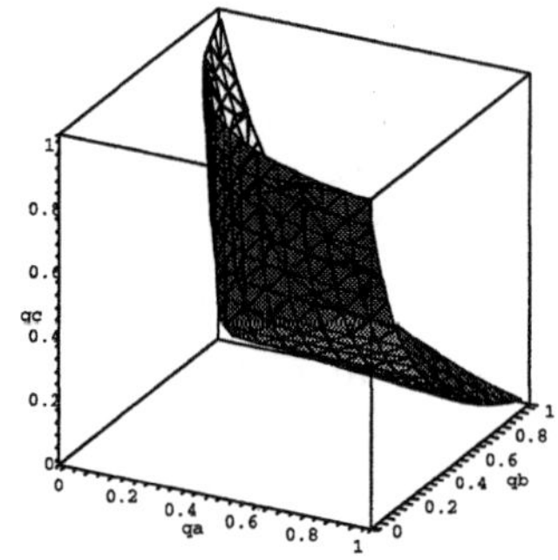


**Fig. 1.** $(q_a, q_b, q_c)$ such that $\text{Re}(\lambda_1) < 1/2$.

## 3 Asymptotic Distribution of $\mathbf{Y}_n$

While our principal motivation is finding the asymptotic variance-covariance structure of $\mathbf{N}_n$, we first obtain the variance-covariance structure of $\mathbf{Y}_n$ as we will use some results of this derivation when we consider $\mathbf{N}_n$. In this section we will employ the results obtained by Bai, Hu, and Rosenberger (2002) to investigate the asymptotic properties of the GFU design with generating matrix (1). In that paper the very general setup, where the number of treatments is arbitrary and responses on them are obtained with delay was considered (there were some non stringent assumption on delay mechanism). Now, we apply the theorem that is valid for general generating matrix, but we restrict ourselves to the case when $\mathbf{H}$ assumes eigenvalue decomposition, thus $D$ defined by (3) is not equal to zero. To account for delays in responses we define $\mu_{j\ell m}$ as the probability that the response $\ell$ ($\ell = 1, 2$) of the patient on treatment $j$ becomes available right after $m$ more patients are enrolled into randomization. So we have $\sum_{\ell,m} \mu_{j\ell m} = 1$ for $j = 1, \ldots, K$. From Bai, Hu, and Rosenberger (2002) in general we have the following result. Let

$$\lambda = \max\{\text{Re}(\lambda_1), \ldots, \text{Re}(\lambda_s)\},$$

where $1, \lambda_1, \ldots, \lambda_s$ are the eigenvalues of $\mathbf{H}$. For $\lambda < 1/2$, $n^{1/2}(\mathbf{Y}_n / \sum_{j=1}^{K} Y_{jn} - \mathbf{v})$ converges in law to a normal distribution $N(0, \mathbf{\Sigma}^Y)$, and the variance-covariance matrix $\mathbf{\Sigma}^Y$ is given below:

$$\mathbf{\Sigma}^Y = (\mathbf{T}^{-1})^* \begin{bmatrix} 0 & 0 & 0 \\ 0 & \sigma_{11} & \sigma_{12} \\ 0 & \sigma_{21} & \sigma_{22} \end{bmatrix} \mathbf{T}^{-1},$$

where the elements are the following

$$\sigma_{ij} = \frac{1}{1 - \lambda_i - \bar{\lambda}_j} \mathbf{t}_i^* \mathbf{R} \mathbf{t}_j,$$

$$\mathbf{R} = \sum_{\ell=1}^{2} \mathbf{D}(\ell) \operatorname{diag}\left( v_1 \sum_{m=0}^{\infty} \mu_{1\ell m}, v_2 \sum_{m=0}^{\infty} \mu_{2\ell m}, v_3 \sum_{m=0}^{\infty} \mu_{3\ell m} \right) \mathbf{D}(l) - \mathbf{v}'\mathbf{v}, \quad (5)$$

and diag($\mathbf{a}$) is the notation for diagonal matrix formed by vector $\mathbf{a}$. In our case $\sum_{m=0}^{\infty} \mu_{i1m} = p_i$, $\sum_{m=0}^{\infty} \mu_{i2m} = q_i$, $(i = 1, 2, 3)$, $\mathbf{D}(1) = I$ and

$$\mathbf{D}(2) = \begin{bmatrix} 0 & 1/2 & 1/2 \\ 1/2 & 0 & 1/2 \\ 1/2 & 1/2 & 0 \end{bmatrix}.$$

After some algebraic manipulation the first term in (5) reduces to

$$\frac{q_a q_b q_c}{4S} \begin{bmatrix} v_1 p_a + \frac{1}{4}(v_2 q_b + v_3 q_c) & \frac{1}{4} v_3 q_c & \frac{1}{4} v_2 q_b \\ \frac{1}{4} v_3 q_c & v_2 p_b + \frac{1}{4}(v_1 q_a + v_3 q_c) & \frac{1}{4} v_1 q_a \\ \frac{1}{4} v_2 q_b & \frac{1}{4} v_1 q_a & v_3 p_c + \frac{1}{4}(v_1 q_a + v_2 q_b) \end{bmatrix}.$$

By substituting the expressions for $v_i$, $i = 1, 2, 3$ into the above formula, we write

$$\mathbf{R} = \frac{q_a q_b q_c}{4S} = \begin{bmatrix} -2 + 4/q_a & 1 & 1 \\ 1 & -2 + 4/q_b & 1 \\ 1 & 1 & -2 + 4/q_c \end{bmatrix} - \mathbf{v}^*\mathbf{v}. \quad (6)$$

Now the expression for $\mathbf{\Sigma}^Y$ can be simplified to

$$\mathbf{\Sigma}^Y = \frac{1}{(1 - \lambda_1 - \bar{\lambda}_1)} \frac{(\mathbf{t}_1 \boldsymbol{\ell}_1)^* \mathbf{R} (\mathbf{t}_1 \boldsymbol{\ell}_1)}{(\boldsymbol{\ell}_1 \mathbf{t}_1)^2} + \frac{1}{(1 - \lambda_2 - \bar{\lambda}_2)} \frac{(\mathbf{t}_2 \boldsymbol{\ell}_2)^* \mathbf{R} (\mathbf{t}_2 \boldsymbol{\ell}_2)}{(\boldsymbol{\ell}_1 \mathbf{t}_1)^2} +$$

$$\frac{1}{(1 - \lambda_1 - \bar{\lambda}_2)} \frac{(\mathbf{t}_1 \boldsymbol{\ell}_1)^* \mathbf{R} (\mathbf{t}_2 \boldsymbol{\ell}_2) + (\mathbf{t}_1 \boldsymbol{\ell}_2)^* \mathbf{R} (\mathbf{t}_2 \boldsymbol{\ell}_1)}{(\boldsymbol{\ell}_1 \mathbf{t}_1)(\boldsymbol{\ell}_2 \mathbf{t}_2)}.$$

Finally, the asymptotic variance-covariance matrix for particles allocation proportions, vector $\mathbf{Y}/\sum_{j=1}^{K} Y_{jn}$ is given by

$$\Sigma^Y = \frac{q_a q_b q_c}{4(1 - 2\lambda_1)(1 - 2\lambda_2)(q_a q_b + q_a q_c + q_b q_c)^2} \begin{bmatrix} \sigma_{11}^Y & \sigma_{12}^Y & \sigma_{13}^Y \\ \sigma_{12}^Y & \sigma_{22}^Y & \sigma_{23}^Y \\ \sigma_{13}^Y & \sigma_{23}^Y & \sigma_{33}^Y \end{bmatrix}, \quad (7)$$

where

$$
\begin{aligned}
\sigma_{11}^Y &= -(3q_b^2q_a - 2q_aq_b + 6q_aq_bq_c + 3q_aq_c^2 \\
&\quad -2q_aq_c - 8q_b^2 + 3q_b^2q_c + 3q_bq_c^2 + 4q_b - 10q_bq_c + 4q_c - 8q_c^2), \\
\sigma_{12}^Y &= 3q_aq_bq_c + 3q_aq_b - 5q_aq_c + 3q_aq_c^2 - 5q_bq_c + 3q_bq_c^2 + 4q_c - 8q_c^2, \\
\sigma_{13}^Y &= 3q_aq_b^2 - 5q_aq_b + 3q_aq_bq_c + 3q_cq_a - 8q_b^2 + 3q_b^2q_c - 5q_bq_c + 4q_b, \\
\sigma_{22}^Y &= -(3q_a^2q_b + 3q_a^2q_c - 8q_a^2 + 6q_aq_bq_c - 2q_aq_b + 4q_a + 3q_aq_c^2 \\
&\quad -10q_aq_c - 2q_bq_c + 3q_bq_c^2 - 8q_c^2 + 4q_c), \\
\sigma_{23}^Y &= 3q_a^2q_b + 3q_a^2q_c - 8q_a^2 - 5q_aq_b + 3q_aq_bq_c - 5q_cq_a + 4q_a + 3q_bq_c, \\
\sigma_{33}^Y &= -(3q_a^2q_b - 8q_a^2 + 3q_a^2q_c + 3q_aq_b^2 + 6q_aq_bq_c \\
&\quad -10q_aq_b + 4q_a - 2q_cq_a - 8q_b^2 + 3q_b^2q_c + 4q_b - 2q_bq_c).
\end{aligned}
$$

We compare matrices computed from (7) with the corresponding variance-covariance matrix obtained by simulation with 10000 replications for three treatment clinical trial that involves 200 patients for the generating matrix in (1). The results are reported in Table 1, where the comparison is done for several combinations of $q_a, q_b$ and $q_c$ that satisfy $\lambda < 1/2$.

**Table 1.** Simulated vs. theoretical covariance matrices for $\mathbf{Y}_n / \sum_{j=1}^K Y_{jn}$.

| $(q_a, q_b, q_c)$ | theoretical | simulated |
|---|---|---|
| (0.95, 0.90, 0.85) | $\begin{pmatrix} 0.0373 & -0.0159 & -0.0214 \\ -0.0159 & 0.0429 & -0.0269 \\ -0.0214 & -0.0269 & 0.0484 \end{pmatrix}$ | $\begin{pmatrix} 0.0374 & -0.0161 & -0.0212 \\ -0.0161 & 0.0421 & -0.0260 \\ -0.0212 & -0.0260 & 0.0472 \end{pmatrix}$ |
| (0.90, 0.80, 0.70) | $\begin{pmatrix} 0.0491 & -0.0161 & -0.0330 \\ -0.0161 & 0.0661 & -0.0500 \\ -0.0330 & -0.0500 & 0.0830 \end{pmatrix}$ | $\begin{pmatrix} 0.0494 & -0.0167 & -0.0326 \\ -0.0167 & 0.0655 & -0.0488 \\ -0.0326 & -0.0488 & 0.0814 \end{pmatrix}$ |
| (0.80, 0.70, 0.60) | $\begin{pmatrix} 0.0730 & -0.0224 & -0.0506 \\ -0.0224 & 0.1011 & -0.0787 \\ -0.0506 & -0.0787 & 0.1292 \end{pmatrix}$ | $\begin{pmatrix} 0.0710 & -0.0218 & -0.0492 \\ -0.0218 & 0.0976 & -0.0758 \\ -0.0492 & -0.0758 & 0.1250 \end{pmatrix}$ |

## 4 Asymptotic Distribution of $\mathbf{N}_n$

The development of this section is based on results by Bai and Hu (2003). In that work they derived the asymptotic normality of $\mathbf{N}_n$ and gave the expression for asymptotic variance matrix in its general form. In particular, we will use the following result which assumes that there exists the usual eigenvalue decomposition of $\mathbf{H}$.

Suppose that $\lambda < 1/2$ and $\mathbf{H}$ has the representation $\mathbf{T}^{-1}\mathbf{H}\mathbf{T} = \mathbf{J}$, where $\mathbf{J} = \text{diag}(1, \lambda_1, \ldots, \lambda_{K-1})$ is the diagonal matrix of eigenvalues of $\mathbf{H}$ and as before $\mathbf{T} = (\mathbf{1}, \mathbf{t}_1, \ldots, \mathbf{t}_{K-1})$. Denote

$$a_{ij} = \mathbf{t}_{i-1}^*(\text{diag}(\mathbf{v}) - \mathbf{v}^*\mathbf{v})\mathbf{t}_{j-1}, b_{ij} = \lambda_{j-1}(1-\lambda_{j-1})^{-1}\mathbf{t}_{i-1}^*(\text{diag}(\mathbf{v}) - \mathbf{v}^*\mathbf{v})\mathbf{t}_{j-1}$$

and

$$c_{ij} = \left((1-\bar{\lambda}_{i-1})^{-1} + (1-\lambda_{j-1})^{-1}\right)(1-\bar{\lambda}_{i-1}-\lambda_{j-1})^{-1}\mathbf{t}^*_{i-1}\mathbf{R}\mathbf{t}_{j-1},$$

where $\mathbf{R}$ is defined by (5) and $i,j = 2,\ldots,K$. Then $n^{-1/2}(\mathbf{N}_n - E\mathbf{N}_n)$ is asymptotically normal with mean vector $\mathbf{0}$ and the variance-covariance matrix $\mathbf{\Sigma}^N = (\mathbf{T}^{-1})^*\tilde{\mathbf{\Sigma}}\mathbf{T}^{-1}$, where $\tilde{\mathbf{\Sigma}}$ has the following form:

$$\tilde{\sigma}_{11} = \tilde{\sigma}_{1j} = \tilde{\sigma}_{i1} = 0, \text{ and } \tilde{\sigma}_{ij} = a_{ij} + b_{ij} + \bar{b}_{ji} + c_{ij}$$

for $i,j = 2,\ldots,K$.

By using (2), (4) and (6), the closed form of the asymptotic variance-covariance matrix for case $K=3$ can be computed as

$$\mathbf{\Sigma}^N = \frac{3}{4}\frac{q_a q_b q_c}{(q_a q_b + q_a q_c + q_b q_c)^2 \prod_{i=1}^{2}(1-\lambda_i)(1-2\lambda_i)} \begin{bmatrix} \sigma_{11}^N & \sigma_{12}^N & \sigma_{13}^N \\ \sigma_{12}^N & \sigma_{22}^N & \sigma_{23}^N \\ \sigma_{13}^N & \sigma_{23}^N & \sigma_{33}^N \end{bmatrix}, \quad (8)$$

where

$$\begin{aligned}
\sigma_{11}^N &= -(3q_a q_b^2 + 6q_a q_b q_c - 2q_a q_b - 2q_a q_c + 3q_a q_c^2 \\
&\quad +3q_c q_b^2 - 8q_b^2 + 4q_b - 10q_b q_c + 3q_b q_c^2 + 4q_c - 8q_c^2), \\
\sigma_{12}^N &= 3q_a q_b q_c + 3q_a q_b - 5q_a q_c + 3q_a q_c^2 - 5q_b q_c + 3q_b q_c^2 + 4q_c - 8q_c^2, \\
\sigma_{13}^N &= 3q_a q_b^2 - 5q_a q_b + 3q_a q_b q_c + 3q_a q_c - 8q_b^2 + 3q_b^2 q_c + 4q_b - 5q_b q_c, \\
\sigma_{22}^N &= -(3q_a^2 q_b + 3q_a^2 q_c - 8q_a^2 - 2q_a q_b + 6q_a q_b q_c + 3q_a q_c^2 \\
&\quad -10q_a q_c + 4q_a - 2q_b q_c + 3q_b q_c^2 + 4q_c - 8q_c^2), \\
\sigma_{23}^N &= 3q_a^2 q_b - 8q_a^2 + 3q_a^2 q_c - 5q_a q_b + 3q_a q_b q_c - 5q_a q_c + 4q_a + 3q_b q_c, \\
\sigma_{33}^N &= -(3q_a^2 q_b + 3q_a^2 q_c - 8q_a^2 + 3q_a q_b^2 - 10q_a q_b \\
&\quad +6q_a q_b q_c - 2q_a q_c + 4q_a + 3q_b^2 q_c - 8q_b^2 - 2q_b q_c + 4q_b).
\end{aligned}$$

Again, we observe agreement between theoretical matrices computed by means of (8) and corresponding simulated variance-covariance matrices for $\mathbf{N}_n$ in Table 2. As before, $n = 200$, and the number of replications for the simulation is 10000.

## 5 Discussion

A primary objective in a clinical trial is to test hypotheses about treatments effects. Hu and Rosenberger (2003) give the general template for comparison of response-adaptive randomization procedures. The comparison addresses the power of the corresponding test and expected failure rates of patients. For a specific procedure it requires knowledge of target allocation proportions and its asymptotic distributions. In particular, the power is a function of the asymptotic variances and covariances of allocation proportions when the asymptotic distribution is multivariate normal. The variance-covariance matrix of the vector of assignment proportions obtained in this paper allows us to evaluate the design without relying on simulation. We also have shown by simulation that our asymptotic formulas are quite accurate for $n = 200$.

**Table 2.** Simulated vs. theoretical covariances matrices for $\mathbf{N}/n$.

| $(q_a, q_b, q_c)$ | theoretical $\mathbf{\Sigma}^N$ | simulated $\mathbf{\Sigma}^N$ |
|---|---|---|
| (0.95 ,0.90, 0.85) | $\begin{pmatrix} 0.1493 & -0.0636 & -0.0857 \\ -0.0636 & 0.1714 & -0.1078 \\ -0.0857 & -0.1078 & 0.1935 \end{pmatrix}$ | $\begin{pmatrix} 0.1468 & -0.0637 & -0.0830 \\ -0.0637 & 0.1675 & -0.1038 \\ -0.0830 & -0.1038 & 0.1869 \end{pmatrix}$ |
| (0.90, 0.80, 0.70) | $\begin{pmatrix} 0.1965 & -0.0644 & -0.1321 \\ -0.0644 & 0.2643 & -0.1999 \\ -0.1321 & -0.1999 & 0.3321 \end{pmatrix}$ | $\begin{pmatrix} 0.1879 & -0.0670 & -0.1208 \\ -0.0670 & 0.2499 & -0.1828 \\ -0.1208 & -0.1828 & 0.3037 \end{pmatrix}$ |
| (0.80, 0.70, 0.60) | $\begin{pmatrix} 0.2920 & -0.0898 & -0.2022 \\ -0.0898 & 0.4045 & -0.3147 \\ -0.2022 & -0.3147 & 0.5170 \end{pmatrix}$ | $\begin{pmatrix} 0.2760 & -0.0973 & -0.1787 \\ -0.0973 & 0.3696 & -0.2723 \\ -0.1787 & -0.2723 & 0.4511 \end{pmatrix}$ |

*Acknowledgement.* The authors were supported by grant DMS-0204232 from the National Science Foundation.

## References

J. Andersen, D. Faries, and R.N. Tamura. Randomized play-the-winner design for multi-arm clinical trials. *Comm. Statist. Theory Methods*, 23:309–323, 1994.

K.B. Athreya and S. Karlin. Embedding of urn schemes into continuous time Markov branching processes and related limit theorems. *Ann. Math. Statist.*, 39:1801–1817, 1968.

Z.D. Bai and F. Hu. Asymptotics in randomized urn model. Submitted, 2003.

Z.D. Bai, F. Hu, and W.F. Rosenberger. Asymptotic properties of adaptive designs for clinical trials with delayed response. *Ann. Statist.*, 30:122–139, 2002.

Z.D. Bai, F. Hu, and L. Shen. An adaptive design for multi-arm clinical trials. *J. Multivariate Anal.*, 81:1–18, 2002.

F. Hu and W.F. Rosenberger. Optimality, variability, power: Evaluating response-adaptive randomization procedures for treatment comparisons. *J. Amer. Statist. Assoc.*, 98:671–678, 2003.

W.F. Rosenberger. New directions in a adaptive design. *Statist. Sci.*, 11: 137–149, 1996.

L.J. Wei. The generalized Pòlya's urn design for sequential medical trials. *Ann. Statist.*, 7:291–296, 1979.

# $T$-Optimum Designs for Multiresponse Dynamic Heteroscedastic Models

D. Uciński[1] and B. Bogacka[2]

[1] Institute of Control and Computation Engineering, University of Zielona Góra, ul. Podgórna 50, 65–246 Zielona Góra, Poland, D.Ucinski@issi.uz.zgora.pl
[2] School of Mathematical Sciences, Queen Mary, University of London, Mile End Road, London E1 4NS, U.K., B.Bogacka@qmul.ac.uk

**Summary.** We generalize the $T$-optimality criterion for discrimination between two rival multiresponse models with observations corrupted by normally distributed noise with zero mean and a covariance matrix which depends on unknown parameters. Calculation of optimum designs amounts to solving an optimal control problem whose discretization leads to a semi-infinite programming problem. It is then solved numerically using an exchange method. As an example we calculate $T$-optimum design for discrimination between two complex chemical kinetic reactions. The design variables are the temperature profile during the process run and the initial concentration of reactants.

**Key words:** $T$-optimality, model discrimination, kinetic reactions, semi-infinite programming.

## 1 Introduction

Experimental design theory for the precise estimation of model parameters has been developed in recent years, cf. Atkinson and Donev (1992), Fedorov and Hackl (1997) and Pukelsheim (1993). However, the design problem for discrimination between competing models has had much less attention and has been developed for simple models only. Various criteria were considered by Atkinson and Fedorov (1975a,b), Burke, Duever, and Pendilis (1994), Fedorov and Khabarov (1986), Felsenstein (1992), Müller and Ponce de Leon (1996), Ponce de Leon and Atkinson (1991), and Stewart, Shon, and Box (1998).

The criterion, called $T$-optimality, introduced by Atkinson and Fedorov (1975a,b) for a single response, has an interesting statistical interpretation as the power of a test for the fit of a second model when the first one is true. We propose here an approach to construct $T$-optimum designs for discrimination between two rival multiresponse models with observations corrupted by normally distributed noise with zero mean and a covariance matrix which depends on unknown parameters. We assume that the observations are not correlated in time, but we allow correlations among different responses. Such a heteroscedastic framework was already considered in the experimental design context, cf. Atkinson and Cook (1995), Downing, Fedorov, and Leonov.

(2001), but, to the best of our knowledge, not for model discrimination. A main motivation comes from discrimination between chemical kinetic models, cf. Pilling and Seakins (1995). When the mechanism of a reaction is not fully known, several alternative models are often proposed. We then wish to conduct experiments so as to select the model that best fits the data.

As a concrete example, consider two chemical reactions: a reaction where substance $A$ changes into substance $B$, which in turn changes into substance $C$, but the first part of the reaction may be reversible, that is

$$A \underset{k_3^{(1)}}{\overset{k_1^{(1)}}{\rightleftharpoons}} B \overset{k_2^{(1)}}{\to} C, \tag{1}$$

with rate constants $k_1^{(1)}$, $k_2^{(1)}$ and $k_3^{(1)}$, and a very similar irreversible process

$$A \overset{k_1^{(2)}}{\to} B \overset{k_2^{(2)}}{\to} C, \tag{2}$$

with rates $k_1^{(2)}$ and $k_2^{(2)}$.

In many chemical systems the reverse reaction is often neglected, when it is a very slow one, resulting in the simpler consecutive model (2). Whether this is the right thing to do may be examined using a model discrimination technique. Recently, Uciński and Bogacka (2003) have found a $T$-optimum solution for discrimination between models (1) and (2), where the design consisted of two time points of support and the initial concentration of the chemical components. However, observations at many time instants can often easily be taken and it might be more effective to optimize the temperature profile during the run of the experiment for discrimination between two rival models.

To include temperature $T$ into the models (1) and (2), we assume that the rate constants $k$ depend on $T$ according to the Arrhenius relation

$$k(T) = \alpha \mathrm{e}^{-E/RT}, \tag{3}$$

where $\alpha$ is the so-called pre-exponential factor, $E$ is the activation energy and $R$ is the universal gas constant of 1.986 cal/mole K.

The changes in concentrations in the reaction (1) are described by

$$\mathcal{M}_1 : \begin{cases} \dfrac{\mathrm{d}[A]}{\mathrm{d}t} = -\alpha_1^{(1)} \mathrm{e}^{-E_1^{(1)}/RT} [A]^{\lambda_1^{(1)}} + \alpha_3^{(1)} \mathrm{e}^{-E_3^{(1)}/RT} [B]^{\lambda_3^{(1)}}, \\ \dfrac{\mathrm{d}[B]}{\mathrm{d}t} = \alpha_1^{(1)} \mathrm{e}^{-E_1^{(1)}/RT} [A]^{\lambda_1^{(1)}} - \alpha_2^{(1)} \mathrm{e}^{-E_2^{(1)}/RT} [B]^{\lambda_2^{(1)}} \\ \qquad\quad - \alpha_3^{(1)} \mathrm{e}^{-E_3^{(1)}/RT} [B]^{\lambda_3^{(1)}}, \\ \dfrac{\mathrm{d}[C]}{\mathrm{d}t} = \alpha_2^{(1)} \mathrm{e}^{-E_2^{(1)}/RT} [B]^{\lambda_2^{(1)}}, \end{cases} \tag{4}$$

where $[A]$, $[B]$ and $[C]$ are concentrations of chemical compounds $A$, $B$ and $C$ as functions of time. The $\lambda_i^{(1)}$'s denote the reaction orders. Similarly, for the reaction (2) we have

$$\mathcal{M}_2: \begin{cases} \dfrac{\mathrm{d}[A]}{\mathrm{d}t} = -\alpha_1^{(2)} \mathrm{e}^{-E_1^{(2)}/RT} [A]^{\lambda_1^{(2)}}, \\ \dfrac{\mathrm{d}[B]}{\mathrm{d}t} = \alpha_1^{(2)} \mathrm{e}^{-E_1^{(2)}/RT} [A]^{\lambda_1^{(2)}} - \alpha_2^{(2)} \mathrm{e}^{-E_2^{(2)}/RT} [B]^{\lambda_2^{(2)}}, \\ \dfrac{\mathrm{d}[C]}{\mathrm{d}t} = \alpha_2^{(2)} \mathrm{e}^{-E_2^{(2)}/RT} [B]^{\lambda_2^{(2)}}. \end{cases} \tag{5}$$

The initial conditions for both (4) and (5) are given by

$$[A](0) = a_0, \quad [B](0) = b_0, \quad [C](0) = c_0, \tag{6}$$

where $a_0$, $b_0$ and $c_0$ denote initial concentrations. Here $\alpha_i^{(\ell)}$, $E_i^{(\ell)}$ and $\lambda_i^{(\ell)}$ are estimated from experimental data. Designs for estimating parameters of $\mathcal{M}_2$ were calculated by Atkinson and Bogacka (2002) when only $[B]$ was measured, and by Atkinson (2003) when more than one response was observed.

The solution of each of these systems is an expected response of the form

$$\eta(t) = ([A](t), [B](t), [C](t))'. \tag{7}$$

$A$, $B$ and $C$ are measured in one process run at fixed times $t_i$, $i = 1, \ldots, N$, which yields the sequence of three-element observation vectors $y_i$, $i = 1, \ldots, N$. The latter are independent for different time instants, but there may be correlations between responses. The objective is to maximize the certainty of the discrimination between $\mathcal{M}_1$ and $\mathcal{M}_2$, and the decision variables are the temperature profile $T(\cdot)$ and the initial concentrations $\beta = (a_0, b_0, c_0)'$.

## 2 $T$-optimality Criterion for Heteroscedastic Models

We assume that the statistical model of observations is of the form

$$y_i = \eta(t_i) + \varepsilon_i, \quad i = 1, \ldots, N, \tag{8}$$

where $\eta(t_i)$ is a vector of $m$ expected response functions of time $t$. As for the disturbances $\varepsilon_i$, we assume that they are independent for different times

$$\varepsilon_i \sim \mathcal{N}(0, V(t_i)), \quad i = 1, \ldots, N, \tag{9}$$

where $V(t_i) \in \mathbb{R}^{m \times m}$ is a symmetric positive definite matrix. We consider two competing *heteroscedastic* models

$$\begin{aligned} \mathcal{M}_1: \quad & \mathrm{E}[y_i] = \eta_1(t_i, \vartheta_1), \quad \mathrm{Cov}(\varepsilon_i) = V_1(t_i, \vartheta_1), \quad \mathrm{Cov}(\varepsilon_i, \varepsilon_j) = 0, \quad i \neq j, \\ \mathcal{M}_2: \quad & \mathrm{E}[y_i] = \eta_2(t_i, \vartheta_2), \quad \mathrm{Cov}(\varepsilon_i) = V_2(t_i, \vartheta_2), \quad \mathrm{Cov}(\varepsilon_i, \varepsilon_j) = 0, \quad i \neq j, \end{aligned}$$

where $\vartheta_1$ and $\vartheta_2$ are vectors of unknown parameters.

The main novelty here is that the unknown parameters can be present in $\mathrm{Cov}(\varepsilon_i)$. If $\mathcal{M}_1$ is true (that is $V(t_i) = V_1(t_i, \vartheta_1^0)$, $\eta(t_i) = \eta_1(t_i, \vartheta_1^0)$ for some known $\vartheta_1^0$), discrimination between $\mathcal{M}_1$ and $\mathcal{M}_2$ can be performed based on maximizing the following generalization of the $T$-optimality criterion:

$$T_{12}(\xi) = \min_{\vartheta_2 \in \Theta_2} \Delta_{12}(\xi, \vartheta_2), \tag{10}$$

where

$$\begin{aligned}\Delta_{12}(t_i, \vartheta_2) = \sum_{i=1}^{N} \Big\{ & \Psi_{V(t_i)}[V_2^{-1}(t_i, \vartheta_2)] \\ & + [\eta(t_i) - \eta_2(t_i, \vartheta_2)]' V_2^{-1}(t_i, \vartheta_2)[\eta(t_i) - \eta_2(t_i, \vartheta_2)] \Big\},\end{aligned} \tag{11}$$

$$\Psi_V(V_2^{-1}) = \operatorname{tr}(V_2^{-1}V) - \log\det(V_2^{-1}V), \tag{12}$$

$\Theta_2$ is the set of admissible $\vartheta_2$'s, and $\xi = \{\xi_1, \dots, \xi_n\}$ is an experimental design, which may consist of time instants or other variables the response depends on. Here, temperature and initial concentrations are the design variables.

A rationale for $T_{12}$ is the following. Assume that $\mathcal{M}_1$ is true with known $\vartheta_1^0$, and that $\vartheta_2^\star$ is an argument in the definition of $T_{12}$. Discrimination between $\mathcal{M}_1$ and $\mathcal{M}_2$ can be viewed as testing the following simple hypothesis:

$$H_0 : \ \eta(t_i) = \eta_1(t_i, \vartheta_1^0), \quad V(t_i) = V_1(t_i, \vartheta_1^0), \quad i = 1, \dots, N \tag{13}$$

against the alternative

$$H_1 : \ \eta(t_i) = \eta_2(t_i, \vartheta_2^\star), \quad V(t_i) = V_2(t_i, \vartheta_2^\star), \quad i = 1, \dots, N. \tag{14}$$

Since the disturbances are Gaussian and independent, the logarithm of the likelihood ratio is

$$\begin{aligned}\ln L = & -\frac{1}{2}\sum_{i=1}^{N} \ln\det[V_2^{-1}(t_i, \vartheta_2^\star)V_1(t_i, \vartheta_1^0)] \\ & + \frac{1}{2}\sum_{i=1}^{N} [y_i - \eta_2(t_i, \vartheta_2^\star)]' V_2^{-1}(t_i, \vartheta_2^\star)[y_i - \eta_2(t_i, \vartheta_2^\star)] \\ & - \frac{1}{2}\sum_{i=1}^{N} [y_i - \eta_1(t_i, \vartheta_1^0)]' V_1^{-1}(t_i, \vartheta_1^0)[y_i - \eta_1(t_i, \vartheta_1^0)].\end{aligned} \tag{15}$$

Furthermore, assuming that $H_0$ is true, we have

$$E[y_i - \eta_1(t_i, \vartheta_1^0)] = 0.$$

Since for a vector random variable $x$ such that $\mathrm{E}[x] = \mu$, $\mathrm{Cov}(x) = V$ we have $\mathrm{E}[x'Ax] = \operatorname{tr}AV + \mu' A\mu$, we obtain

$$
\begin{aligned}
2\mathrm{E}[\ln L] &= -\sum_{i=1}^{N} \ln\det[V_2^{-1}(t_i,\vartheta_2^\star)V(t_i)] + \sum_{i=1}^{N} \mathrm{tr}[V_2^{-1}(t_i,\vartheta_2^\star)V(t_i)] \\
&\quad + \sum_{i=1}^{N} [\eta(t_i) - \eta_2(t_i,\vartheta_2^\star)]' V_2^{-1}(t_i,\vartheta_2^\star)[\eta(t_i) - \eta_2(t_i,\vartheta_2^\star)] - Nm \\
&= \Delta_{12}(\xi,\vartheta_2^\star) - Nm.
\end{aligned}
\tag{16}
$$

Thus, for constant $Nm$, $\mathrm{E}[\ln L]$ is proportional to $\Delta_{12}(\xi,\vartheta_2^\star)$.

Also, the $T$-optimality criterion (10) seems to be intuitively justified. We wish to find the appropriate form of $\eta(\,\cdot\,)$ and, at the same time, the appropriate form of $V(\,\cdot\,)$. This is achieved by conducting an experiment such that $\eta(\,\cdot\,)$ and $\eta_2(\,\cdot\,,\vartheta_2)$, as well as $V(\,\cdot\,)$ and $V_2(\,\cdot\,,\vartheta_2)$ differ as much as possible, while the value of $\vartheta_2$ is chosen to make the competing model closest to the true one. Discrimination between the expectations is accomplished via the lack-of-fit term in (10), while discrimination between the error structures is carried out by introducing the function $\Psi[\,\cdot\,]$.

**Lemma 1.** *Let* $\mathrm{PD}(m)$ *be the set of all symmetric positive-definite* $m \times m$ *matrices. Given* $A \in \mathrm{PD}(m)$ *define*

$$
\Psi_A(X) = \mathrm{tr}(XA) - \ln\det(XA), \quad \textit{for all } X \in \mathrm{PD}(m). \tag{17}
$$

*Then*

1. $\Psi_A(\,\cdot\,)$ *is strictly convex in* $\mathrm{PD}(m)$, *and*
2. $\Psi_A(X) \geq m$, *for all* $X \in \mathrm{PD}(m)$, *the equality being attained only at* $X = A^{-1}$.

Thus, the higher $\Psi_{V(t_i)}[V_2^{-1}(t_i,\vartheta_2)]$, the larger is the difference between $V_2(t_i,\vartheta_2)$ and $V(t_i)$. Moreover, when the covariance structure of both models is the same, (10) simplifies to the criterion of Uciński and Bogacka (2002).

## 3 Numerical Construction of Optimum Designs

Let $T \in L_\infty[0,t_f]$ denote the control profile to be determined so as to maximize (10). Here $t_f = 10$ was assumed, and the admissible temperatures satisfy

$$
T_{\min} \leq T(t) \leq T_{\max}, \quad t \in [0,t_f], \tag{18}
$$

where $T_{\min} = 300$ and $T_{\max} = 700$. To compute the optimal temperature numerically, it is necessary to discretize the problem. We thus introduce a finite-dimensional parametrization

$$
T(t) = \sum_{i=1}^{M} T_i b_i(t), \tag{19}
$$

where the $b_i$'s are $B$-cubic spline basis functions (here $M = 13$ such functions were taken), and $T_i$, $i = 1, \dots, M$, are to be determined as components of $\xi$. The design also includes the initial concentrations $a_0$, $b_0$ and $c_0$, where

$$0.5 \le a_0 \le 1, \quad 0.1 \le b_0 \le 0.7, \quad 0.1 \le c_0 \le 0.7, \quad a_0 + b_0 + c_0 = 1. \tag{20}$$

Consequently, the vector of the design variables $\xi$ contains 16 elements. For the true model, $\mathcal{M}_1$, the parameters

$$\vartheta_1 = (\alpha_1^{(1)}, \alpha_2^{(1)}, \alpha_3^{(1)}, E_1^{(1)}, E_2^{(1)}, E_3^{(1)}, \lambda_1^{(1)}, \lambda_2^{(1)}, \lambda_3^{(1)})$$

were chosen to be

$$\vartheta_1^0 = (0.7, 0.2, 0.1, 1000, 1000, 1000, 2.0, 2.0, 1.0).$$

As for the structure of the covariance matrix for $\mathcal{M}_1$ we assumed:

$$V(t) = 0.01\,\mathrm{diag}[([A](t, \vartheta_1^0))^2, ([B](t, \vartheta_1^0))^2, ([C](t, \vartheta_1^0))^2]$$

whereas for the alternative model $\mathcal{M}_2$ we had

$$V_2(t) = \omega\,\mathrm{diag}[([A](t, \vartheta_2))^2, ([B](t, \vartheta_2))^2, ([C](t, \vartheta_2))^2].$$

Note that $V_2$ depends on both $\omega$ (explicitly) and the parameters in the alternative model (implicitly, through the concentrations). In $\mathcal{M}_2$ we have

$$\vartheta_2 = (\alpha_1^{(2)}, \alpha_2^{(2)}, E_1^{(2)}, E_2^{(2)}, \lambda_1^{(2)}, \lambda_2^{(2)}, \omega) \in \Theta_2,$$

where $\Theta_2$ is the Cartesian product of the parameter ranges taken as

$$[.55, .85] \times [.05, .35] \times [700, 1300] \times [700, 1300] \times [1.5, 2.5] \times [1.5, 2.5] \times [.01, .015].$$

All the responses were sampled at time moments $t_i = \frac{10}{19} i$, $i = 0, \dots, 19$.

Finding a $T$-optimum design means solving a maximin problem, in which we seek $\xi$ maximizing $T_{12}(\xi)$. This task is complicated since each evaluation of $T_{12}(\xi)$ involves minimizing $\Delta_{12}(\xi, \vartheta_2)$ w.r.t. $\vartheta_2 \in \Theta_2$. However, the problem so formulated can be cast as the following semi-infinite programming one:

$$\text{Maximize } \gamma, \text{subject to } \Delta_{12}(\xi, \vartheta_2) \ge \gamma, \quad \vartheta_2 \in \Theta_2, \tag{21}$$

where $\gamma$ is some real number. The maximization above is performed with respect to $\xi \in \Xi$ ($\Xi$ is the set of admissible design variables) and $\gamma$. The term "semi-infinite programming" (SIP) derives from the property that we have finitely many decision variables (these are the elements of $\gamma$ and $\xi$) in infinitely many constraints (the number of elements in the set $\Theta_2$). Numerical approaches to solving SIP problems are characterized by Hettich and Kortanek (1993), Polak (1987), Reemtsen and Görner (1998), and Sacks (1998). In order to calculate approximations to the optimal temperature profile, we applied an exchange method, cf. Hettich and Kortanek (1993). As for the global optimization, the ARS method was adopted, cf. Walter and Pronzato (1997).

In our implementation, SAS/IML software was used. The ordinary differential equations (4) and (5) along with the corresponding sensitivity equations required to calculate the gradient $\Delta_{12}(\xi, \vartheta_2)$ were solved using the built-in `ODE` subroutine (it is an adaptive, variable order, variable step-size, stiff integrator based on implicit backward-difference methods). The local search was performed using the standard `NLPQN` subroutine which implements the algorithm used for nonlinearly constrained quasi-Newton optimization. The results obtained after some three minutes of calculations on a PC equipped with Pentium 4 CPU 2.40 GHz and running SAS Rel. 8.2 under Windows 2000 are presented in Fig. 1.

The optimum design $\xi^\star = \{T_1, \ldots, T_{13}, a_0, b_0, c_0\}$ is transformed by (19) to $\xi^\star_{\text{profile}} = \{T(t), a_0, b_0, c_0\}$, and $T(\cdot)$ is shown in Fig. 1(b), while the optimum choice of the initial concentrations $(0.5, 0.1, 0.4)$ is shown in Fig. 1(a) together with the resulting responses of the two competing models. The optimal temperature steeply decreases from its maximum to minimum possible values. These results were obtained for

$$\vartheta_2^\star = (0.83403, 0.35, 1154.67, 1057.86, 2.34329, 1.5, 0.015). \tag{22}$$

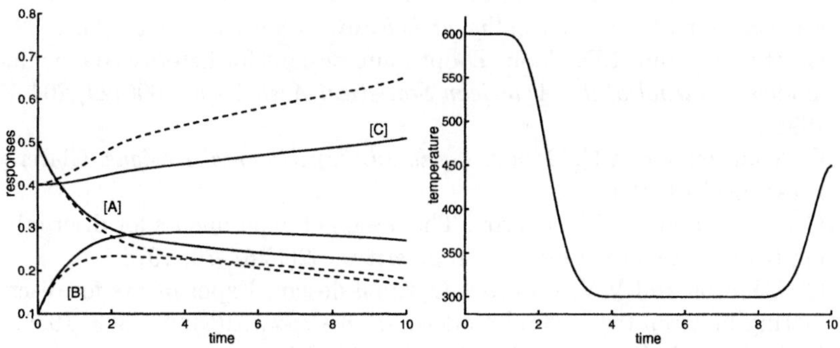


(a) (b)

**Fig. 1.** (a) Responses vs. time for the $T$-optimal temperature (continuous and dashed lines denote the responses of the true and alternative models, respectively), and (b) the $T$-optimal temperature profile.

# 4 Conclusion

Introducing temperature profile as a design variable made the discrimination between the competing models more practical as it is often the case in chemical kinetic experiments that a series of many observations of concentrations in time is obtained and it is easy to control temperature.

Further work in this direction would relax the assumption of independent observations for different time instants, leading to observed time series with some correlation structure. Also, for the proposed criterion, we can show that an equivalence condition (cf. Uciński and Bogacka (2003)) holds as the extra term in the criterion is a convex function, cf. Lemma 1. The optimum design depends on the choice of the model assumed to be true and its parameter values. Further investigation concerning this dependence is needed, as well as work on more robust methods of model discrimination.

*Acknowledgement.* This work was supported by the British Council and the State Committee for Scientific Research within the framework of a Polish-British Research Partnership Programme.

## References

A.C. Atkinson. Horwitz's rule, transforming both sides and the design of experiments for mechanistic models. *Journal of the Royal Statistical Society, Series C, Applied Statistics*, 52:261–278, 2003.

A.C. Atkinson and B. Bogacka. Compound and other optimum designs for systems of nonlinear differential equations arising in chemical kinetics. *Chemometrics and Intelligent Laboratory Systems*, 61:17–33, 2002.

A.C. Atkinson and R.D. Cook. *D*-optimum designs for heteroscedastic linear models. *Journal of the American Statistical Association*, 90(429):204–212, 1995.

A.C. Atkinson and A.N. Donev. *Optimum Experimental Designs.* Clarendon Press, Oxford, 1992.

A.C. Atkinson and V.V. Fedorov. The design of experiments for discriminating between two rival models. *Biometrika*, 62(1):57–70, 1975.

A.C. Atkinson and V.V. Fedorov. Optimal design: Experiments for discriminating between two several models. *Biometrika*, 62(2):289–303, 1975.

A.L. Burke, T.A. Duever, and A. Pendilis. Model discrimination via designed experiments: Discriminating between the terminal and penultimate models on the basis of composition data. *Macromolecules*, 27:386–399, 1994.

D. Downing, V.V. Fedorov, and S. Leonov. Extracting information from the variance function: Optimal design. In A. C. Atkinson, P. Hackl, and W. G. Müller, editors, *mODa 6 - Advances in Model-Oriented Design and Analysis (Puchberg/Schneeberg, June 25-29, 2001)*, pages 45–52. Physica-Verlag, Heidelberg, 2001.

V. V. Fedorov and P. Hackl. *Model-Oriented Design of Experiments.* Lecture Notes in Statistics. Springer-Verlag, New York, 1997.

V.V. Fedorov and V. Khabarov. Duality of optimal design for model discrimination and parameter estimation. *Biometrika*, 73(1):183–190, 1986.

K. Felsenstein. Optimal Bayesian design for discriminating among rival models. *Computational Statistics & Data Analysis*, 14:427–436, 1992.

R. Hettich and K.O. Kortanek. Semi-infinite programming: Theory, methods and applications. *SIAM Review*, 35(5):380–429, 1993.

W.G. Müller and A.C. Ponce de Leon. Discriminating between two binary data models: Sequentially designed experiments. *Journal of Statistical Computation and Simulation*, 55:87–100, 1996.

M.J. Pilling and P.W. Seakins. *Reaction Kinetics.* Oxford University Press, Oxford, 1995.

E. Polak. On the mathematical foundations of nondifferentiable optimization in engineering design. *SIAM Review*, 29(1):21–89, 1987.

A.C. Ponce de Leon and A.C. Atkinson. Optimum experimental design for discriminating between two rival models in the presence of prior information. *Biometrika*, 78(3):601–608, 1991.

F. Pukelsheim. *Optimal Design of Experiments.* Probability and Mathematical Statistics. John Wiley & Sons, New York, 1993.

R. Reemtsen and S. Görner. Numerical methods for semi-infinite programming: A survey. In R. Reemtsen and J.-J. Rückmann, editors, *Semi-Infinite Programming*, pages 195–275. Kluwer Academic Publishers, Boston, 1998.

E.W. Sacks. Semi-infinite programming in control. In R. Reemtsen and J.-J. Rückmann, editors, *Semi-Infinite Programming*, pages 389–411. Kluwer Academic Publishers, Boston, 1998.

W.E. Stewart, Y. Shon, and G.E.P. Box. Discrimination and goodness of fit of multiresponse mechanistic models. *AIChE Journal*, 44(6):1404–1412, 1998.

D. Uciński and B. Bogacka. Construction of $T$-optimum designs for multiresponse dynamic systems. In W. Härdle and B. Rönz, editors, *Compstat 2002, Proceedings in Computational Statistics 2002*, pages 267–272, Berlin, 2002. Springer.

D. Uciński and B. Bogacka. $T$-optimum design for discrimination between two multiresponse dynamic models. Submitted for publication, available at `http://www.maths.qmw.ac.uk/~bb/UcinskiBogacka1.pdf`, 2003.

É. Walter and L. Pronzato. *Identification of Parametric Models from Experimental Data.* Communications and Control Engineering. Springer-Verlag, Berlin, 1997.

# Error Transmission in Mixture Experiments

I.N. Vuchkov

University of Chemical Technology and Metallurgy, Sofia, Bulgaria
vuchkov@uctm.edu

**Summary.** In this paper we propose explicit models of mean and variance of a mixture performance characteristic for the case when the mixture amounts are set with errors in the production process. They are based on Taylor expansion up to third order terms of a mixture model expressed through the component amounts. Mixture models with errors only in component amounts and mixture models with errors in component amounts and process variables are considered. Properties of these models are studied. Quality improvement problem is defined as minimization of variance while keeping the mean value on the target. An example from rubber industry is given.

**Key words:** mixture design, component amounts.

## 1 Introduction

Mixture experiments are typical for chemical, food, metallurgical and other industries. The aim of these experiments is to find optimal component proportions that provide desired values of some product performance characteristics. Let $A_i$, $i = 1, 2, \ldots, q$ represent the amounts of the components in a mixture. The component proportions can be expressed as follows:

$$X_i = \frac{A_i}{\sum\limits_{i=1}^{q} A_i}. \tag{1}$$

They are subject to the following constraints:

$$0 \leq X_i \leq 1, i = 1, 2, \ldots, q, \quad \sum_{i=1}^{q} X_i = 1. \tag{2}$$

Let $\eta$ be the undisturbed performance characteristic of a product. Usually there are disturbances in the production process and as a result the measured value of the performance characteristic is $y = \eta + \varepsilon$. In Scheffé (1958) canonical models for mixture experiments are proposed. A second order canonical model is:

$$\eta = \sum_{i=1}^{q} \beta_i X_i + \sum_{i=1}^{q-1} \sum_{j=i+1}^{q} \beta_{ij} X_i X_j. \tag{3}$$

First order canonical model can be obtained from (3) if we put $\beta_{ij} = 0$. In this paper we consider only first and second order canonical models, but the results can easily be extended to other canonical models. In production processes errors occur due to poor measurements or violations of technological prescriptions. Let $e_i$, $i = 1, 2, \ldots, q$ be the errors in component amounts. The component proportions in a mixture experiment that are subject to errors in the component amounts are:

$$x_i = \frac{A_i + e_i}{\sum\limits_{i=1}^{q} (A_i + e_i)} \tag{4}$$

Note that $\mathrm{E}(x_i) \neq X_i$ and that for these components the constraints (2) also hold.

The errors in component amounts are transmitted to the response and cause variations in the performance characteristics, which exist even if $\varepsilon = 0$. A model-based approach has been proposed in (Vuchkov and Boyadjieva, 2001, Chapter 5)to describe error transmission for regression experiments, which are not subject to constraints (2). In the polynomial models with errors the product or process parameters are expressed as $p_i + e_i$ and mean and variance models for the performance characteristic are developed. In mixture experiments this approach cannot be used directly, because the errors occur both in the numerator and denominator of component proportions. Therefore the model is nonlinear with respect to errors and the distribution moments cannot be calculated directly.

Steiner and Hamada (1997) considered the problem of error transmission in mixture experiments. They used simulations to estimate the transmitted variance. To do that they need the relative probability of each simulated point. In this paper we propose an analytical approach that is free of simulations.

## 2 Mean and Variance Models for Mixture Experiments

In Vuchkov and Boyadjieva (2001) is used Taylor expansion for nonlinear performance characteristic $\eta$. Mean and variance models were obtained for product performance characteristics with factors that are subject to errors $e_i$ with zero expectations and variances $\sigma^2(e_i)$. These models can be used for description of mean and variance in mixture experiments. In this context and if the errors are independent and normally distributed, the models, based on Taylor expansion of $\eta$, can be written as follows:

*Mean value:*

$$\tilde{y} = \eta + \sum_{i=1}^{q} h_{ii}\sigma^2(e_i), \tag{5}$$

*Variance:*

$$\sigma^2 = \sum_{i=1}^{q} \delta_i^2 \sigma^2(e_i) + 2\sum_{i=1}^{q} h_{ii}^2 \sigma^4(e_i) + 4\sum_{i=1}^{q-1}\sum_{j=i+1}^{q} h_{ij}^2 \sigma^2(e_i)\sigma^2(e_j) +$$

$$2\sum_{i=1}^{q} \delta_i \sigma^2(e_i)\left(3\sigma^2(e_i)t_{ii}^{(i)} + \sum_{j=1,j\neq i}^{q} \sigma^2(e_j)t_{jj}^{(i)} + 2\sum_{j=1,j\neq i}^{q} \sigma^2(e_i)t_{ij}^{(i)}\right) + \sigma_\varepsilon^2, \tag{6}$$

where

$$\delta_i = \frac{1}{\sum_{i=1}^{q} A_i}\Delta_i, \tag{7}$$

$$h_{ii} = \frac{1}{\left(\sum_{i=1}^{q} A_i\right)^2} H_{ii}, \quad h_{ij} = \frac{1}{\left(\sum_{i=1}^{q} A_i\right)^2} H_{ij}, \tag{8}$$

$$t_{ii}^{(i)} = \frac{1}{\left(\sum_{i=1}^{q} A_i\right)^3} T_{ii}^{(i)}, \quad t_{ii}^{(j)} = \frac{1}{\left(\sum_{i=1}^{q} A_i\right)^3} T_{ii}^{(j)}, \quad t_{ij}^{(j)} = \frac{1}{\left(\sum_{i=1}^{q} A_i\right)^3} T_{ij}^{(j)}. \tag{9}$$

The values of $\Delta_i, H_{ii}, H_{ij}, T_{ii}^{(i)}, T_{ii}^{(j)}, T_{ij}^{(j)}$ can be written through the undisturbed component proportions as follows:

$$\Delta_i = \beta_i(1 + X_i) + \sum_{j\neq i}(\beta_j + \beta_{ij})X_j - 2\eta, \tag{10}$$

$$H_{ii} = \eta + 2\sum_{u=1}^{q-1}\sum_{v=u+1}^{q} \beta_{uv}X_uX_v - 2\sum_{u\neq i}\beta_{iu}X_u - \beta_i, \tag{11}$$

$$H_{ij} = \eta + 2\sum_{u=1}^{q-1}\sum_{v=u+1}^{q} \beta_{uv}X_uX_v - \beta_{ij}(X_i + X_j) - \sum_{u\neq i}(\beta_{iu} + \beta_{ju})X_u +$$

$$+ (\beta_{ij} - \beta_i - \beta_j)/2, \tag{12}$$

$$T_{ii}^{(i)} = \beta_i + 3(1 - X_i)\sum_{u\neq i}\beta_{iu}X_u - 3\sum_{u\neq i}\sum_{v\neq i,u}\beta_{uv}X_uX_v - \eta, \tag{13}$$

$$\begin{aligned} T_{ii}^{(j)} = T_{ij}^{(i)} =& \frac{1}{3}(2\beta_i + \beta_j) + \frac{1}{3}\beta_{ij}(3X_i + 6X_j - 9X_iX_j - 2) \\ &+ (2 - 3X_i)\sum_{u\neq i,j}\beta_{iu}X_u + + (1 - 3X_j)\sum_{u\neq i,j}\beta_{ju}X_u \\ &- 3\sum_{u\neq i,j}\sum_{v\neq i,j,u}\beta_{uv}X_uX_v - \eta, \end{aligned} \tag{14}$$

$$T_{jj}^{(i)} = T_{ij}^{(j)} \tag{15}$$

The proportion of the error in a mixture component is $e_i / \sum A_i$. Its variance is equal to

$$\sigma_i^2 = \frac{\sigma^2(e_i)}{\left(\sum A_i\right)^2}.$$

Therefore the proportion's standard deviation is

$$\sigma(e_i) = \sigma_i \sum A_i. \tag{16}$$

Substituting (7)-(16) into (5) and (6) we obtain mean and variance models expressed only through component proportions, and not the component amounts in explicit form. These models are as follows:

*Mean value:*

$$\tilde{y} = \eta + \sum_{i=1}^{q} H_{ii}\sigma_i^2, \tag{17}$$

*Variance:*

$$\begin{aligned} \sigma^2 = & \sum_{i=1}^{q} \Delta_i^2 \sigma_i^2 + 2\sum_{i=1}^{q} H_{ii}^2 \sigma_i^4 + 4\sum_{i=1}^{q-1}\sum_{j=i+1}^{q} H_{ij}^2 \sigma_i^2 \sigma_j^2 \\ & + 2\sum_{i=1}^{q} \Delta_i \sigma_i^2 \left( 3\sigma_i^2 T_{ii}^{(i)} + \sum_{j=1, j\neq i}^{q} \sigma_j^2 T_{jj}^{(i)} + 2\sum_{j=1, j\neq i}^{q} \sigma_j^2 T_{ij}^{(i)} \right) + \sigma_\varepsilon^2. \end{aligned} \tag{18}$$

These models make it possible to calculate the mean and variance values of the performance characteristic in any point of the simplex and thus explore ways to improve the product quality.

The quality improvement problem can be defined as minimization of variance (18) under the condition that the mean value of the performance characteristic is on target. Another method is to minimize the Taguchi's loss function, which for our problem can be written as follows (Taguchi (1996)):

$$L = k\left[\left(\tilde{y} - \tau\right)^2 + \sigma^2\right] \tag{19}$$

In this formula $\tau$ is the target value for the performance characteristic, while $\tilde{y}$ and $\sigma^2$ can be calculated by use of formulae (17) and (18).

## 3 Mean and Variance Models for Experiments with Mixture and Process Variables

In some mixture experiments additional independent variables are of interest. Usually they are process variables like temperature, pressure, etc.. However in some technologies all mixture components are divided into two groups. The sum of the mixture components of the first group should always be 100 weight parts (w.p.), while the components of the second group are chosen independently within some intervals. For example in the rubber industry usually several types of rubber (natural and synthetic) form a group of components that should always be equal to 100 w.p.. All other components such as soot, oil, accelerators, etc. are varied independently within some intervals, provided that the rubbers are 100 w.p.

The components of the first group are linearly dependent and are subject to constraints (2), while the components of the second group or the process variables can independently be varied within intervals

$$-1 \leq z_i \leq 1, \quad i = q+1, q+2, \ldots, m. \tag{20}$$

Regression models for this case were considered by Vuchkov, Damgaliev, and Yontchev (1981). A second order canonical model of this type is

$$\eta = \sum_{i=1}^{q} \beta_i X_i + \sum_{i=1}^{q-1} \sum_{j=i+1}^{q} \beta_{ij} X_i X_j + \sum_{i=1}^{q} \sum_{j=i+1}^{m} \gamma_{ij} X_i z_j + \sum_{i=q+1}^{m} \lambda_{ii} z_i^2. \tag{21}$$

The mean and variance models in this case are similar to (17) and (18). The only difference is that all derivatives are calculated both with respect to mixture components and process variables. We will put in brackets $x$ or $z$ to denote the derivatives and error variances with respect to each group of variables. The mean and variance models are:

*Mean value:*

$$\tilde{y} = \eta + \sum_{i=1}^{q} H_{ii}(x)\, \sigma_i^2(x) + \sum_{i=q+1}^{m} H_{ii}(z)\, \sigma_i^2(z). \tag{22}$$

*Variance:*

$$\begin{aligned}
\sigma^2 = {} & \sum_{i=1}^{q} \Delta_i^2(x)\, \sigma_i^2(x) + 2\sum_{i=1}^{q} H_{ii}^2(x)\, \sigma_i^4(x) + 4\sum_{i=1}^{q-1} \sum_{j=i+1}^{q} H_{ij}^2(x)\, \sigma_i^2(x)\, \sigma_j^2(x) \\
& + \sum_{i=q+1}^{m} \Delta_i^2(z)\, \sigma_i^2(z) + 2\sum_{i=q+1}^{m} H_{ii}^2(z)\, \sigma_i^4(z) + 4\sum_{i=q+1}^{m-1} \sum_{j=i+1}^{m} H_{ij}^2(z)\, \sigma_i^2(z)\, \sigma_j^2(z) \\
& + 2\sum_{i=1}^{q} \Delta_i(x)\, \sigma_i^2(x) \left( 3\sigma_i^2(x)\, T_{ii}^{(i)}(x) + \sum_{j=1, j\neq i}^{q} \sigma_j^2(x)\, T_{jj}^{(i)}(x) + \right. \\
& \left. 2\sum_{j=1, j\neq i}^{q} \sigma_j^2(x)\, T_{ij}^{(i)}(x) \right) + \sigma_\varepsilon^2.
\end{aligned} \tag{23}$$

The third derivatives with respect to the process variables were found to be zero ($T_{ii}^{(i)}(z) = T_{jj}^{(i)}(z) = 0$) and this is the reason why they do not appear in (23). The other derivatives can be calculated as follows:

*First derivatives with respect to mixture variables:*

$$\Delta_i(x) = \Delta_i + \sum_{j=q+1}^{m} \gamma_{ij} z_i - \sum_{\ell=1}^{q} \sum_{j=q+1}^{m} \gamma_{\ell j} X_\ell z_j,$$

where $\Delta_i$ is calculated by (10).

*First derivatives with respect to process variables:*

$$\Delta_i(z) = \sum_{\ell=1}^{q} \gamma_{\ell i} x_i + 2\lambda_{ii} z_i + \sum_{\ell=q+1, \ell \neq i}^{m} \lambda_{i\ell} z_\ell$$

*Second derivatives with respect to mixture components:*

$$H_{ii}(x) = H_{ii} + \sum_{\ell=1}^{q} \sum_{s=q+1}^{m} \gamma_{\ell s} X_\ell z_s - \sum_{s=q+1}^{m} \gamma_{is} z_s,$$

where $H_{ii}$ is given by (11) , and

$$H_{ij}(x) = H_{ij} + \sum_{\ell=1}^{q} \sum_{s=q+1}^{m} \gamma_{\ell s} X_\ell z_s - \frac{1}{2} \sum_{s=q+1}^{m} (\gamma_{is} + \gamma_{js}) z_s$$

where $H_{ij}$ is given by (12).

*Second derivatives with respect to process variables:*

$$H_{ii}(z) = \lambda_{ii} \quad \text{and } H_{ij}(z) = \frac{1}{2}\lambda_{ij}.$$

*Third derivatives with respect to the mixture components:*

$$T_{ii}^{(i)}(x) = T_{ii}^{(i)} + \sum_{\ell=1}^{q} \sum_{s=q+1}^{m} \gamma_{\ell s} X_\ell z_s - \sum_{s=q+1}^{m} \gamma_{is} z_s$$

$$T_{jj}^{(i)}(x) = T_{jj}^{(i)} + \frac{2}{3} \sum_{s=q+1}^{m} \gamma_{is} z_s + \frac{1}{3} \sum_{s=q+1}^{m} \gamma_{is} z_s - \sum_{\ell=1}^{q} \sum_{s=q+1}^{m} \gamma_{\ell s} X_\ell z_s,$$

where $T_{ii}^{(i)}$ and $T_{jj}^{(i)}$ are given by (13) and (14), respectively.

# 4 Example

Consider a formulation for a protector of truck tires. The performance characteristic of interest $y$ is the modulus of 300 % elongation of the mixture formulation. It is studied as function of three types of synthetic rubber: $X_1$ (Bulex 1500), $X_2$ (Bulex M-27) and $X_3$ (SKD). A second-order simplex lattice is used and the results of experiments are given in Table 1.

**Table 1.** Second order simplex design for rubber mixture experiment.

| No. | $X_1$ | $X_2$ | $X_3$ | $y$ |
|---|---|---|---|---|
| 1 | 1 | 0 | 0 | 100.0 |
| 2 | 0 | 1 | 0 | 69.5 |
| 3 | 0 | 0 | 1 | 65.5 |
| 4 | 0.5 | 0.5 | 0 | 77.5 |
| 5 | 0.5 | 0 | 0.5 | 71.5 |
| 6 | 0 | 0.5 | 0.5 | 82.0 |

The following canonical model was found and proved to be adequate:

$$\hat{y} = 100X_1 + 69.5X_2 + 65.6X_3 - 29X_1X_2 - 45X_1X_3 + 58X_2X_3. \qquad (24)$$

In order to study the transmitted variation we take into account that there are no errors when some of the component proportions equal 0, because in this case the corresponding component is not used in the mixture formulation. That is why all calculations were carried out for points without zero proportions. We considered a smaller simplex defined by the following vertices: $V_1$ $(0.1; 0.8; 0.1)$, $V_2$ $(0.1; 0.1; 0.8)$ and $V_3$ $(0.8; 0.1; 0.1)$. The values of biased response in the vertices and centres of the edges were calculated using (17) and the biased response was approximated by use of second order canonical model. We obtained the following model for the biased response:

$$\tilde{y} = 102.13X_1 + 68.65X_2 + 64.5X_3 - 31.61X_1X_2 - 49.05X_1X_3 + 63.22X_2X_3 \qquad (25)$$

There is some difference in the coefficients of (25) as compared with (24), which is due to the errors in the component proportions. Contours of the predicted and biased responses based on (24) and (25) are shown on Figures 1 and 2.

The variance was also calculated in the vertices and edge centres of the restricted simplex area by (18). The response variance was neglected in this model. The following canonical model approximated it:

$$\begin{aligned}\sigma^2 = {} & 97.3X_1 + 40.5X_2 + 56.5X_3 - 251.7X_1X_2 - 252.6X_1X_3 \\ & -133.7X_2X_3 + 190.2X_1X_2X_3\end{aligned}$$

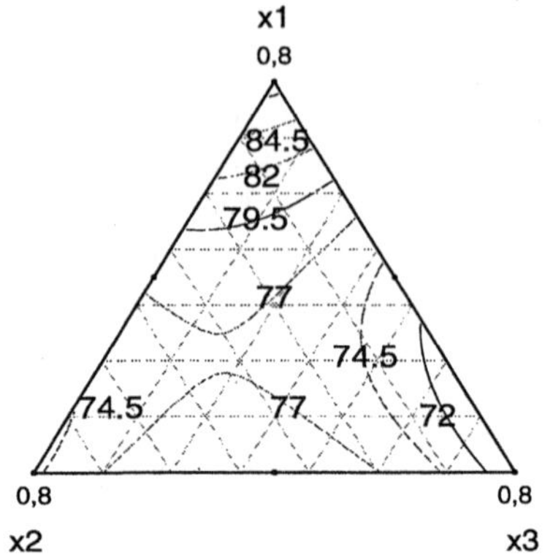


**Fig. 1.** Predicted response.

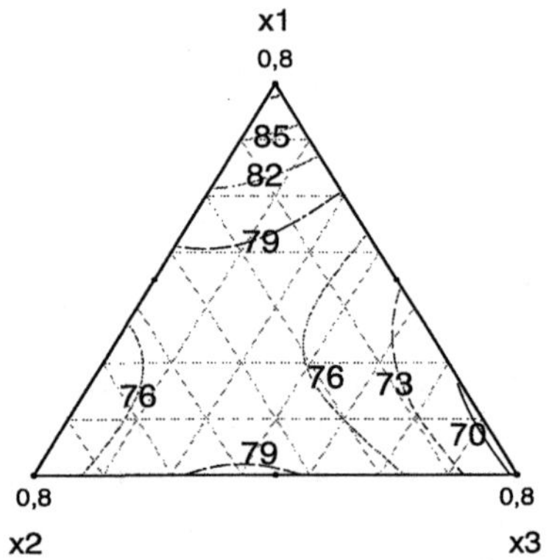


**Fig. 2.** Biased response.

Figure 3 shows the contours of variance transmitted from factors to the response. It is interesting to note that the minimal variance can be obtained for component proportions corresponding to the saddle point of the response surface, where the first derivatives of the response are equal to zero. An explanation of this property for second order polynomial models of the response is given in Vuchkov and Boyadjieva (2001). This example shows that this property holds also for second order canonical models.

The optimal component proportions depend on the target definition. For example if the modulus of 300 % elongation is required to be no more than 80, then from Figures 2 and 3 we see that the optimal proportions that make the variance equal to 0 are approximately $X_1 = 0.31, X_2 = 0.45, X_3 = 0.24$. Figure 4 shows another solution based on Taguchi's loss function with target $\tau = 80$. In this case the optimal proportions are: $X_1 = 0.26$, $X_2 = 0.48$, $X_3 = 0.26$. These two solutions are not too different.

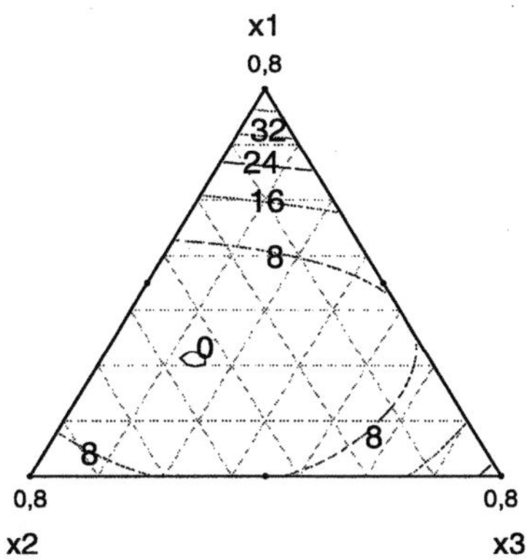


**Fig. 3.** Contour plot for variance.

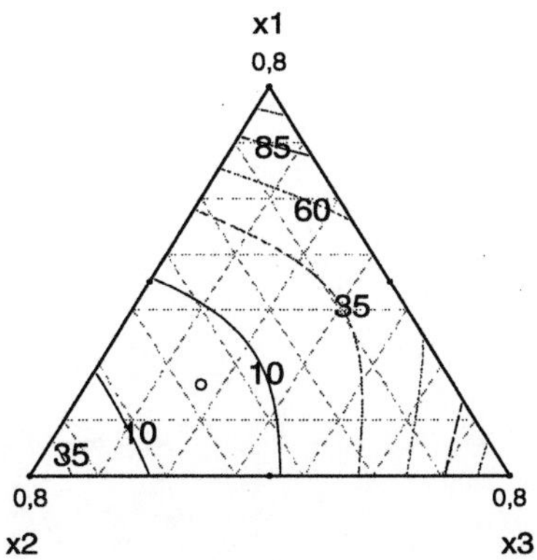


**Fig. 4.** Contour plot for $L$.

## References

H. Scheffé. Experiments with mixtures. *J. Roy. Statist. Soc. Ser. B*, 20: 344–360, 1958.

S.H. Steiner and M. Hamada. Making mixtures robust to noise and mixing measurement errors. *Journal of Quality Technology*, 29:441–450, 1997.

G. Taguchi. *Introduction to quality improvement.* APO, Tokyo, 1996.

I. N. Vuchkov, D. L. Damgaliev, and Ch. A. Yontchev. Sequentially generated second order quasi-*D*-optimal designs for experiments with mixture and process variables. *Technometrics*, 23(3):233–238, 1981.

I.N. Vuchkov and L.N. Boyadjieva. *Quality improvement with design of experiments. A response surface approach*, volume 7 of *Topics in Safety, Risk, Reliability and Quality.* Kluwer Academic Publishers, Dordrecht, 2001.

# Maximum Entropy Sampling and General Equivalence Theory

H.P. Wynn

London School of Economics, Houghton Street, London WC2A 2AE, UK
h.wynn@lse.ac.uk

**Summary.** The relationship between maximum entropy sampling (MES) and the General Equivalence Theory of Kiefer and Wolfowitz is discussed. In MES a subsample with maximum entropy is selected from a population to minimise the expected posterior entropy for the unsampled units. Taking limits of MES as the prior variation on a regression parameter tends to infinity leads to constrained $D$-optimality. The general case benefits from a spectral resolution of the covariance function.

**Key words:** entropy, maximum entropy, $D$-optimality, general equivalence theory.

## 1 Introduction

This paper draws together two somewhat distinct theories of optimal design, both of which have appeared in the MODA proceedings over the years. The first of these is maximum entropy sampling (MES) and is closely related to a type of Bayesian optimal design. The term maximum entropy sampling was introduced by Shewry and Wynn (1987), but Caseleton and Zidek (1984) had already used the same ideas in environmental monitoring. Lee (1998) and Hoffman, Lee, and Williams (2001) have taken up the topic from an algorithmic point of view. A definitive account of applications in monitoring is by Zidek, Sun, and Le (2000). The present discussion draws particularly on the ideas of Sebastiani and Wynn (2000). The second area is the General Equivalence Theory of Kiefer and Wolfowitz which is assuredly the core theory of optimal design; see for example Kiefer and Wolfowitz (1960) and Kiefer (1974).

A motivation for this paper was the desire to give a simple version of the Bayes theory at the end of a course on optimal design, to students already familiar with the main theory. In the final presentation it was stated informally that "the General Equivalence Theorem of Kiefer and Wolfowitz is the limit of Shannon's theorem as our prior ignorance increases to infinity".

The first part of the connection between the two areas, which we described in next sections, is relatively satisfactory and was mentioned in Sebastiani and Wynn (2000). One can indeed take the limit of a Bayesian/Shannon theory and obtain a $D$-optimality statement for finite samples. A deeper link lies in a version which incorporates the continuous design theory of Kiefer and Wolfowitz. One obtains in the limit the constrained optimal design theory, which was introduced first by Wynn (1982), and followed by a definitive series of works: Federov and Hackl (1997), Federov and Müller (1989), Müller (1998).

### 1.1 Maximum Entropy Sampling

Let $S = \{1, \dots N\}$ and define a population jointly distributed random variables $Y_i, i \in S$. A subset $s$ is called a *sample*: $Y_s = \{Y_i, \; i \in s \subset S\}$ and let $|s| = n$. Where useful we will consider $Y_S$ and $Y_s$ as vectors. Let us refer to the unsampled $Y_i$ as $Y_{\bar{s}} = \{Y_j, \;\; j \in \bar{s} = S \backslash s\}$. Define the entropy of a random variable $U$ with density $p_U$ in the usual way as

$$\text{Ent}(U) = -\text{E}_U(\log p_U) = -\int \log p_U p_U du$$

What we call here the "Shannon theorem" is stated as

$$\text{Ent}(Y) = \text{Ent}(Y_S) + \text{E}_{Y_s}\text{Ent}\{Y_{\bar{s}}|Y_s)\}. \tag{1}$$

See, for example (Cover and Thomas, 1991, chapter 2). Maximum Entropy Sampling simply says the following. Suppose we are free to choose the set $s \subset S$ to minimise the expected posterior entropy of the unsampled population, represented by the second term on the right-hand side of (1). Then, since the entropy of the full population on the left-hand side of (1) is fixed, it we can equivalently minimise the first term on the right-hand side, namely the (marginal) entropy of the sample.

The first and somewhat surprising connection with $D$-optimality theory is that under a suitable model MES is equivalent to a version of Bayesian $D$-optimality. For this we need the standard parametric regression model for $Y_s$. For the sample let $|s| = n < N$. Then assumed the regression model:

$$Y = X_s\theta + \varepsilon$$

and let the prior distribution on $\theta$ be $\theta \sim N(0, V)$, which is assumed to be independent of the error distribution $\varepsilon \sim N(0, \Sigma)$.

We seek to minimise by choice of $s \subset S$ the expected posterior entropy of $\theta$ namely $\text{Ent}\{\theta|Y_s)\}$. Note that in the Gaussian case entropies are given (up to constants) by the logarithm of the determinant of the relevant conditional or unconditional covariance matrix (we ignore all constants in what follows). The joint entropy of $\theta$ and the $Y_s$ can be split, again using the Shannon theorem:

$$\text{Ent}(Y_s, \theta) = \text{Ent}(\theta) + \text{E}_\theta\text{Ent}\{Y_s|\theta)\}$$

Under the independence assumptions the second term is just $\text{Ent}(\varepsilon)$, so that

$$\text{Ent}(Y_s, \theta) = \text{Ent}(\theta) + \text{Ent}(\varepsilon),$$

which does not depend on the choice of $s$. In matrix terms the result is

$$\det\left(\begin{bmatrix} X_s V X_s^T + \Sigma & X_s V \\ V X_s^T & X_s V X_s^T \end{bmatrix}\right) = \det(V)\det(\Sigma).$$

We may expand one more time by the Shannon theorem, the other way around:

$$\text{Ent}(Y_s, \theta) = \text{Ent}(Y_s) + \text{E}_{Y_s}\text{Ent}(\theta|Y_s) \tag{2}$$

Since we have just shown that left hand side of (2) does not depend on $s$, we minimise the second term on the right-hand side over $s \subset S$ by maximising the first term, namely $\text{Ent}(Y_s)$. But this is just MES.

In summary have under the condition described above the following result

**Theorem 1.** *Under a suitable Bayesian regression model Maximum Entropy Sampling for choice of a subsample s from a population S is equivalent to (i) minimisation of expected posterior entropy for the unsampled population units and (ii) Bayes minimum entropy design the for regression parameters.*

The covariance matrix of a sample is $\text{cov}(Y_s) = X_s V X_s^T + \Sigma$ and so the MES criterion is easily computed as

$$\max_{s \subset S} \det(X_s V X_s^T + \Sigma)$$

This form is nice (*e.g.*, for students!) because it avoids inverses.

Theorem 1 is not the General Equivalence Theorem, which depends on continuous design conditions. However it has the same flavour: a $D$-optimality criterion for parameters is equivalent to a minimax criterion for prediction.

## 1.2 Limit of Bayes Case

To obtain the classical case from the Bayes case we take limits in a special way. For simplicity take $\Sigma = \sigma_\varepsilon^2 I_n$ and $V = \sigma^2 I_p$, where $I_n$ and $I_p$ are the $n \times n$ and $p \times p$ identity matrices, respectively. Then in all formulae take limits as $\rho = \frac{\sigma_\varepsilon^2}{\sigma^2} \to 0$. This is the "ignorance" case alluded to above. Note that for $r \leq s$ we have for an $s \times r$ matrix $A$ the general result

$$\det(\sigma^2 A A^T + \sigma_\varepsilon I_s) = \sigma^{2r}\sigma_\varepsilon^{2(s-r)}(\det(A^T A) + O(\rho)).$$

The limit of the entropy decomposition is, then, the well-known augmentation formula:

$$\det(X_s^T X_s + X_{\bar{s}}^T X_{\bar{s}}^T) = \det(X_s^T X_s)\det(I_{N-n} + X_{\bar{s}}(X_s^T X_s)^{-1} X_{\bar{s}}^T). \tag{3}$$

We now have the exact analogue of MES. If we seek to maximise $\det(X_s^T X_s)$, $s \subset S$, then, since the left-hand side of (3) does not depend on $s$, the optimisation is equivalent to minimising $\det(I_{N-n} + X_{\bar{s}}(X_s^T X_s)^{-1} X_{\bar{s}}^T)$. The variance of the least squares estimator of predictor $\hat{Y}_i$ of $\mathrm{E}(Y_i)$ under corresponding to the $i$th row, $x_i^T$, of $X_{\bar{s}}$ is $v_i = \sigma_\varepsilon^2 x_i^T (X_s^T, X_s)^{-1} x_i$, the variance function. The best linear unbiased estimator of an observation $Y_i$ has mean squared error

$$\mathrm{E}\{(\hat{Y}_i - Y_i)^2\} = \mathrm{var}(Y_i) + \mathrm{var}((\hat{Y}_i) = \sigma_\varepsilon^2(1 + v_i).$$

The RHS of (3) (times $\sigma^2$) is just the multivariate MSE of $\hat{Y}_{\bar{s}}$.

In summary: *under suitable conditions the limit of maximum entropy sampling is D-optimality for a subsample.*

## 2 Continuous Theory

We start with a base measure $\xi_0$ on a measure space $\mathcal{X}$ to represent the full population $S$. Then the sample $s \subset S$ is modelled by a submeasure $\xi$ dominated by $\xi_0$ in the strict sense: $\xi \ll \xi_0 : \ \xi(A) \leq \xi_0(A)$ for all measurable sets $A \subseteq \mathcal{X}$. We let a fixed parameter $0 < \beta < 1$ plays the role of the sample fraction $\frac{n}{N}$ so that we set $\xi_0(\mathcal{X}) = 1, \ \xi(\mathcal{X}) = \beta$. Let $\xi_0 = \xi + \bar{\xi}$, where $\bar{\xi}(\mathcal{X}) = 1 - \beta$.

We introduce the standard set-up for optimal design theory on $\mathcal{X}$: $\mathrm{E}(Y_x) = \sum_j^p f_j(x)\theta_j = f(x)^T\theta$ and define moment matrices $M(\xi) = \int_{\mathcal{X}} f(x)f(x)^T\xi(dx)$ and the variance function: $d(x,\xi) = f(x)^T M^{-1}(\xi) f(x)$. Then we have the analogue of the entropy formula: since $M(\xi_0) = M(\xi) + M(\bar{\xi})$,

$$\det(M(\xi_0)) = \det(M(\xi)) \det\left(I + M(\bar{\xi})^{\frac{1}{2}} M(\xi)^{-1} M(\bar{\xi})^{\frac{1}{2}}\right).$$

Again the left-hand side does not depend on the choice of $\xi$, being the value for the dominating measure. The second term on the right-hand side is the measure version of the mean squared error of prediction for the unsampled population and minimising this quantity is equivalent to constrained $D$-optimality: $\max \det(M(\xi))$, under the constraints.

A key point of this formulation is that the constrained $D$-optimality problem is essentially covered in work on constrained $D$-optimality referred to in Section 1. We give a very brief description for this case.

Let $A, B \subseteq \mathcal{X}$ be two measurable sets with $A \cap B = \emptyset$ such that $\xi(A) = \bar{\xi}(B)) = \gamma \neq 0$. Let $\xi_A$ be the measure which when confined to $A$ (and all measurable subsets of $A$) is identical to $\xi$ but otherwise (on $\mathcal{X}\backslash A$) is zero. Let $\bar{\xi}$ have the same relationship to $\bar{\xi}$ and $B$. Then for small $0 < \alpha < 0$ construct the perturbation of $\xi$:

$$\xi_{\alpha,A,B} = \xi - \alpha\xi_A + \alpha\bar{\xi}_B.$$

It is straightforward to check that $\xi_{\alpha,A,B}$ satisfies all the constraints: $\xi_{\alpha,A,B} \ll \xi_0$, $\xi_{\alpha,A,B}(\mathcal{X}) = \beta$. Then a necessary condition for $\xi$ to be constrained $D$-optimal is

$$\frac{\partial}{\partial \alpha} \log \det M(\xi_{\alpha,A,B}) \mid_{\alpha \downarrow 0} \leq 0, \tag{4}$$

for all choices of $A$, $B$ subject to the stated conditions. Using a standard computation this is equivalent to

$$\int_A f(x) M^{-1}(\xi) f(x) \leq \int_B f(x) M^{-1}(\bar{\xi}) f(x) \tag{5}$$

for all choices of $A$, $B$. Now select $x_1 \in \text{supp}(\xi)$ and $x_2 \in \text{supp}(\bar{\xi})$ and take a sequence of $A \ni x_1$ and $B \ni x_2$ with diam$(A)$, diam$(B) \to 0$ and $\gamma \to 0$. This is possible for example if $f(x)$ is continuous in $\mathbb{R}^d$ and $\xi_0$ is absolutely continuous with respect to Lebesgue measure. Then (5) is equivalent to

$$f(x_1) M^{-1}(\xi) f(x_1) \leq f(x_2) M^{-1}(\bar{\xi}) f(x_2) \tag{6}$$

for all $x_1 \in \text{supp}(\xi)$, $x_2 \in \text{supp}(\bar{\xi})$. Moveover in addition we can take

$$\xi(A) = \begin{cases} 1 & A \subseteq \text{supp}(\xi) \\ 0 & A \subseteq \text{supp}(\bar{\xi}) \end{cases}.$$

A valuable description of condition is that the variance function $d(x, \xi)$ for the optimal $\xi$ *separates* $\xi$ and $\bar{\xi}$. This gives a clear geometric understanding of the support of $\xi$ in cases such as polynomials and solutions are often unique. But there are special cases such Fourier models where variance functions are flat and this is not the case. A limiting case of the constrained case, as $\beta \to 0$, is the unconstrained case. We may take $\xi_o$ as fixed and proportional to Lebesgue measure and then as $\beta \to 0$ and $\xi$ puts mass on arbitrary small intervals. This gives in the limit the classical Kiefer-Wolfowitz equivalence theorem.

## 2.1 Constrained Measures for the MES Case

The analysis of the last section can be applied the to original MES case. We take the simple case, namely the measure version of the condition $\max_s \det(\sigma X X^T + \sigma_\varepsilon I_n)$. Note that

$$\det(\sigma^2 X X^T + \sigma_\varepsilon I_n) = \det(\sigma^2 X^T X + \sigma_\varepsilon I_p) \sigma_\varepsilon^{2(n-p)}.$$

The measure analogue of this is (ignoring the scalar multiplier)

$$\det(\tau^2 M(\xi) + \sigma_\varepsilon I_p),$$

where $\tau^2 = \beta \sigma_\varepsilon^2 n$. Here, $\beta$ is inserted to reflect the condition of constrained optimisation. We then follow exactly the same argument as before with $M(\xi)$ replaced by $M_+(\xi) = \tau^2 M(\xi) + \sigma_\varepsilon I_p$. The separation condition is

$$f(x_1) M_+^{-1}(\xi) f(x_1) \leq f(x_2) M_+^{-1}(\bar{\xi}) f(x_2), \tag{7}$$

for all pairs $x_1, x_2$ with $x_1 \in \text{supp}(\xi)$ and $x_2 \in \text{supp}(\bar{\xi})$. But

$$f(x)^T(\sigma^2 X^T X + \sigma_\varepsilon^2 I_p)^{-1} f(x) = \frac{1}{\sigma^2 \sigma_\varepsilon^2}\{f(x)^T f(x) - \sigma^4 f(x)^T X^T (\sigma_\varepsilon^2 I_n + \sigma^2 X X^T)^{-1} X f(x)\}. \tag{8}$$

Comparing (7) with (8) in the Bayes/MES case, we obtain the posterior variance of $Y_x$.

We summarise this as *there is a constrained D-optimality theory which is the measure generalisation of the finite population MES, in the Bayes regression case.*

## 3 General MES and *D*-optimality

Much of the original work on MES has been motivated by applications to areas such as environmental sampling and computer experiments in which more general stochastic processes are used, such as Gaussian covariance processes. Let us briefly review how to link MES with *D*-optimality in such a case.

The joint distribution of the population has full rank $N \times N$ covariance matrix:

$$\text{cov}(Y_S) = \Gamma.$$

The spectral resolution of $\Gamma$ can be written as:

$$\Gamma = U \Lambda U^T,$$

where $U = [z_1 : \ldots : z_n]$ is the matrix of normalised eigenvectors of $\Gamma$ and $\Lambda$ the diagonal matrix of eigenvalues. This gives the "spectral resolution"

$$Y_S = U \Lambda^{1/2} \varepsilon,$$

where $\varepsilon$ is a vector of uncorrelated random variables with equal variance, unity.

Suppose that we want to find an approximation to the MES subsample of size $n$. One way to proceed is to start with the best spectral approximation of "length" $n$. This is

$$\Gamma_n = U_n \Lambda_n U_n^T,$$

where $\Lambda_n$ is the diagonal matrix of a set of $n$ largest eigenvalues: $\lambda_{(1)} \geq \ldots \geq \lambda_{(n)}$) (there may be a lack of uniqueness here, which we ignore) and $U_n$ is the $N \times n$ matrix of corresponding eigenvectors (as columns). The task, now, is to carry out MES on this problem. This corresponds to an MES problem with population design matrix:

$$X_S = U_n \Lambda_n^{1/2}.$$

Note also that $\prod_{i=1}^n \lambda_i = \det(X_S^T X_S)$.

The choice of the sample $Y_i, i \in s$ is simply a choice of $n$ rows of $X_S$, to form an $n \times n$ matrix $X_s$ to maximise $\det(X_s X_s^T)$. But $\det(X_s X_s^T) = \det(X_s^T X_s)$, because $X_s$ is square. Moreover $X_S^T X_S = X_s X_s^T + X_{\bar{s}}^T X_{\bar{s}}^T$.

Thus, this is exactly a saturated $D$-optimal design problem, for choosing $n$ points out of $N$ candidate points with $p = n$. We can even revert to the standard notation by defining:

$$f(x_i)^T = x_i$$

the $i$th row of $X_S$. If we are fortunate, the solution to the $n$-out-of-$N$ problem will also be solution of the global $D$-optimal design problem on the candidate set. At the very least we may obtain useful information about $s$.

We have a cheap lower bound for the MES entropy value. Applying the Binet-Cauchy theorem to $\det(X_S^T X_S)$ we have:

$$\binom{N}{n}^{-1} \prod_{i=1}^{n} \lambda_{(i)} \leq \max_s \det(X_s^T X_s) \leq \max_s \det(\Gamma_s) \qquad (9)$$

*Example 1.* Let

$$\Gamma = \begin{pmatrix} 2\,1\,0\,0 \\ 1\,2\,1\,0 \\ 0\,1\,2\,1 \\ 0\,0\,1\,2 \end{pmatrix}.$$

Considering the independent variable $x$ to be time this can be generated by the moving average process $Y_x = \varepsilon_x + \varepsilon_{x-1}$ with a dummy time factor $x = \{0, 1, 2, 3\}$. Let $n = 2$, then any pair $x_1, x_2$ with $|x_1 - x_2| > 1$ is MES.

The two largest eigenvalues are $\{\frac{5}{2} + \frac{\sqrt{5}}{2}, \frac{3}{2} + \frac{\sqrt{5}}{2}\}$ with (un-normalised) eigenvectors $\{(1, \frac{1}{2} + \frac{\sqrt{5}}{2}, \frac{1}{2} + \frac{\sqrt{5}}{2}, 1), (1, -\frac{1}{2} + \frac{\sqrt{5}}{2}, \frac{1}{2} - \frac{\sqrt{5}}{2}, -1)\}$, respectively. Since the problem is unaffected by scale multiples of the columns of $X_S$ we may take these vectors to be the columns.

This gives the $D$ optimality problem with candidate set (given by the rows) in 2-dimensions:

$$(1, -1), \left(\frac{1}{2} + \frac{\sqrt{5}}{2}, -\frac{1}{2} + \frac{\sqrt{5}}{2}\right), \left(\frac{1}{2} + \frac{\sqrt{5}}{2}, \frac{1}{2} - \frac{\sqrt{5}}{2}\right), (1, -1),$$

which is solved both exactly and globally by the design with equal mass at $\{0, 2\}$ or $\{1, 3\}$. Note that solution coincides with the MES solution. Interestingly the solution $\{0, 3\}$ does not arise in this case. The values in the inequality (9) are $\frac{5}{6} + \frac{\sqrt{5}}{3} < \frac{5}{4} + \frac{\sqrt{5}}{2} < 4$.

## References

W.F. Caseleton and J.V. Zidek. Optimal monitoring networks. *Statistics & Probability Letters*, 2:223–227, 1984.

M.V. Cover and J.V. Thomas. *Elements of Information Theory.* Wiley, New York, 1991.

V.V. Federov and P. Hackl. *Model-oriented design of experiments*, volume 125 of *Lecture notes in Statistics.* Springer Verlag, New York, 1997.

V.V. Federov and W.G. Müller. Comparison of two approaches in the optimal design of an observation network. *Statistics*, 20:339–351, 1989.

A. Hoffman, J. Lee, and J. Williams. New upper bounds for maximum entropy sampling. In A.C. Atkinson, P. Hackl, and W.G. Müller, editors, *mODa 6 - Advances in Model-Oriented Design and Analysis (Puchberg/Schneeberg, June 25-29, 2001)*, Contributions to Statistics, pages 143–153, Heidelberg, 2001. Physica-Verlag.

J. Kiefer. General equivalence theory for optimum designs (approximate theory). *Ann. Statist.*, 2:849–879, 1974.

J. Kiefer and J. Wolfowitz. The equivalence of two extremal problems. *Can. J. Math*, 2:363–366, 1960.

J. Lee. Constrained maximum entropy sampling. *Operations Research*, 43: 684–691, 1998.

W.G. Müller. *Collecting spatial data.* Contributions to Statistics. Physica-Verlag, Heidelberg, 1998.

P. Sebastiani and H.P. Wynn. Maximum entropy sampling and Bayesian experimental design. *J. Roy. Statist. Soc. B*, 62:145–157, 2000.

P. Shewry and H.P. Wynn. Maximum entropy sampling. *Applied Statistics*, 14:165–170, 1987.

H.P. Wynn. Optimum sub-measures with application to finite population sampling. In *Statistical decision theory and related topics, III, Vol. 2 (West Lafayette, Ind., 1981)*, pages 485–495. Academic Press, New York, 1982.

J.V. Zidek, W. Sun, and N.D. Le. Designing and integrating composite networks for monitoring multivariate Gaussian pollution fields. *Journal of Applied Statistics*, 49:63–79, 2000.

# Towards Identification of Patient Responses to Anesthesia Infusion in Real Time

G. Yin[1], H. Wang[2], L.Y. Wang[3], and H. Zheng[4]

[1] Department of Mathematics, Wayne State University, Detroit, Michigan 48202, U.S.A. gyin@math.wayne.edu

[2] Department of Anesthesiology, Harper University Hospital, Detroit Medical Center/Wayne State University, Detroit, Michigan 48201, U.S.A. howang@med.wayne.edu

[3] Department of Electrical and Computer Engineering, Wayne State University, Detroit, Michigan 48202, U.S.A. lywang@ece.eng.wayne.edu

[4] Department of Electrical and Computer Engineering, Wayne State University, Detroit, Michigan 48202, U.S.A. aj6823@wayne.edu

**Summary.** This paper is concerned with identification for patient responses to anesthesia infusion in real time. The model parameters are first classified by expert assessment. The initial model is then improved during the surgery by using real-time data via a two-step recursive stochastic approximation and projected least-squares estimation algorithm.

**Key words:** anesthesia infusion, identification, stochastic approximation.

## 1 Introduction

Anesthesia is an integral part of most surgical operations. Its objectives consist of achieving hypnosis (consciousness control), analgesia (pain control), and immobility (body movement control) simultaneously throughout surgical operations, while maintaining the vital functions of the body. These objectives are carefully balanced and maintained by a dedicated anesthetist using hypnotics, narcotics, and muscle relaxants. Anesthesia outcome is evaluated by anesthesia depth together with other physiological variables; see Gan, Glass, Windsor, Payne, Rosow, Sebel, and Manberg (1997), Mortier, Struys, De Smet, Versichelen, and Rolly (1998), Wang and Wang (2002b). To facilitate computer-assisted decision and control strategy development for improved anesthesia performance, it is highly desirable to develop representative models of a patient's BIS (bi-spectrum monitor for anesthesia depth) responses to drug infusion. Since patients differ dramatically in metabolism, pre-existing medical conditions, and surgical procedures, individualized models must be established for each patient with limited patient information and data points. Furthermore, the expert knowledge of an anesthesiologist plays critical roles in predetermining patient response characteristics. A knowledge-based and control-oriented modeling approach was introduced recently and applied to develop a feedback and predictive control strategy for anesthesia infusion Wang and Wang (2002a,b). The models have a Wiener structure and retain only the key characteristics of patient responses that are essential for control strategy development and can be determined by either data or expert knowledge.

This paper develops a real-time identification method that incorporates expert knowledge and real-time data to improve model accuracy. The model parameters are first classified by expert assessment. This initial model is then improved during the surgery by using real-time data via a two-step recursive stochastic approximation and projected least-squares estimation algorithm.

The overall system configuration is illustrated in Figure 1. Input signals include drug infusion rates from digital programmable infusion pumps. The patient response and conditions are measured by the BIS monitor (anesthesia depth), anesthesia monitor (patient vital signs), and also some variables from the anesthesia machine (ventilation conditions). Dynamic responses of the patient to changes in drug infusion rates can be measured by a data acquisition system that synchronizes signals from different monitors. Then, computer algorithms can be devised to execute real-time modeling and decision assistance functions. This paper will be limited to the aspect of real-time identification algorithms and their basic properties such as accuracy, computational burdens and convergence. Although the ultimate goal is to establish such properties for general multi-drug and multi-outcome responses, this paper will present results on single-drug-single-outcome identification results.

## 2 System

The system under consideration has a Wiener structure with input $u$ and output $y$: $x = Gu + \widetilde{w}$, $y = f(x) + \widetilde{v}$, where $G$ is a dynamic linear system, $f$ is a memoryless nonlinear function, and $\widetilde{w}$ and $\widetilde{v}$ are random noises. $G$ and $f$ represent the dynamic (or transient) and steady-state behavior of a patient's response to a drug infusion input, respectively. Simple examples of $G$ include $G(s) = \frac{1}{1+Ts}$ or $e^{-\tau s}\frac{1}{1+Ts}$. Here and henceforth, for a vector or matrix $z$, $z'$ denotes its transpose. For computerized data processing, $G$ will be discretized and represented by an ARMA (Auto-Regression and Moving Average) model $x_k = -a_1 x_{k-1} - \cdots - a_n x_{k-n} + b_0 u_k + b_1 u_{k-1} + \cdots + b_n u_{k-n} + \widetilde{w}_k = \varphi_k' \theta + \widetilde{w}_k$, where $\varphi_k' = [-x_{k-1}, \ldots, -x_{k-n}, u_k, u_{k-1}, \ldots, u_{k-n}]$ $\theta' = [a_1, \ldots, a_n, b_0, \ldots, b_n]$.

The prior information on $\theta$ is a bounded set $\Omega_1$, which is provided by prior expert knowledge. For example, suppose $G(s) = \frac{1}{1+Ts}$. The model parameter $T$ has a clear physical meaning of response speed and allows direct inputs from expert knowledge. A patient with fast drug response will have a small $T$. A linguistic assessment "fast response" can be translated by a pre-designed assessment mapping to a bounded set $[\underline{T}, \overline{T}]$. This information will then be mapped to a bounded set on $\theta$ after discretization.

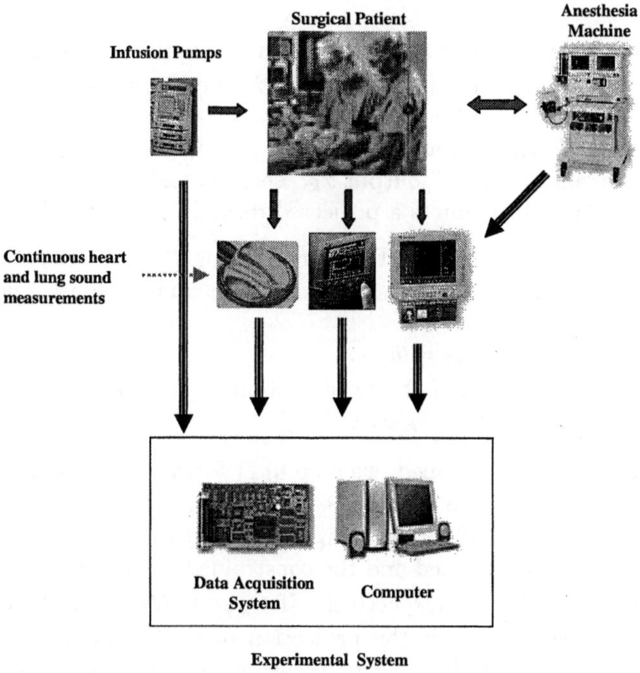


**Fig. 1.** Overall System Configuration

A physician's knowledge is often expressed by an approximate relationship between the input and outcome: Given typical input amounts $x_1, \ldots, x_m$, the approximate ranges of the predicted outcomes $[\underline{y}_1, \overline{y}_1], \ldots, [\underline{y}_m, \overline{y}_m]$. To accommodate this knowledge, we replace the usual bases $1, x, \ldots, x^{m-1}$ by a collection of appropriate functions $\Phi_i(x)$. Suppose that the prior information on $f$ is given by $m$ distinct mappings: $x_i \to [\underline{y}_i, \overline{y}_i]$, $i = 1, \ldots, m$. We parameterize $f$ by $y = f(x) = \sum_{i=1}^{m} C_i \Phi_i(x)$, where $\Phi_i(x)$ is the $m$th order polynomial $\Phi_i(x) = \frac{\prod_{j=1, j\neq i}^{m}(x-x_j)}{\prod_{j=1, j\neq i}^{m}(x_i-x_j)}$. $\Phi_i(x)$ satisfies $\Phi_i(x_j) = 0$, $j \neq i$; $\Phi_i(x_i) = 1$. In other words, $\Phi_i(x)$ are polynomial indicator functions on the set of points $\{x_1, \ldots, x_m\}$. The unknown parameters will be denoted by $\Theta = [C_1, \ldots, C_m]'$. Consequently, the prior expert information becomes precisely the prior interval information on the parameters $C_i \in [\underline{y}_i, \overline{y}_i]$, $i = 1, \ldots, m$. The expert knowledge will serve as the prior information and parameter bounds on the unknown parameters $\theta$ and $\Theta$. It will be generically expressed as $\theta \in \Omega_1$ and $\Theta \in \Omega_2$. Here, $\Omega_1$ and $\Omega_2$ are bounded sets.

### 2.1 Basic Two-Step Setup for Real-Time Learning Algorithms

Given bounded sets $\Omega_1$ and $\Omega_2$ and performance indices $e_1(\theta)$ and $e_2(\Theta)$, we start at $k = 0$ by selecting $\widehat{\theta}_0$ and $\widehat{\Theta}_0$ from $\Omega_1$ and $\Omega_2$ (such as their centers), respectively.
Step 1: At time $k$, the available $\widehat{\Theta}_k$ is used to map the output $y_k$ backwards to obtain the "intermediate" output $\widehat{x}_k$. Then $u_k$ and $\widehat{x}_k$ are used to update $\widehat{\theta}_{k+1}$. The update algorithm is a projected recursive identification scheme

1. $\theta^* = \arg\min_\theta e_1(\theta)$ subject to the constraint $\theta^* \in \Omega_1$. Denote the estimate at the $k$th step without constraint by $\check{\theta}_k$ and the constraint estimate by $\widehat{\theta}_k$.
2. If $\check{\theta}_{k+1} \in \Omega_1$, then $\widehat{\theta}_{k+1} = \check{\theta}_{k+1}$.
3. If $\check{\theta}_{k+1} \not\in \Omega_1$, then $\widehat{\theta}_{k+1}$ is the projection of $\check{\theta}_{k+1}$ onto $\Omega_1$, namely the closest element of $\Omega_1$ to $\check{\theta}_{k+1}$.

Step 2: From the $\widehat{\theta}_{k+1}$ obtained, we map $u_{k+1}$ forward to obtain $\widehat{x}_{k+1}$. Then $\widehat{x}_{k+1}$ and $y_{k+1}$ are used to update $\widehat{\Theta}_{k+1}$. The update algorithm is also the projected identification: $\Theta^* = \arg\min_\Theta e_2(\Theta)$ subject to the constraint $\Theta^* \in \Omega_2$. Denote the unconstrained and the constrained estimates at the $(k+1)$st step by $\check{\Theta}_{k+1}$ and $\widehat{\Theta}_{2k+1}$, respectively. If $\check{\Theta}_{k+1} \in \Omega_2$, then $\widehat{\Theta}_{k+1} = \check{\Theta}_{k+1}$. If $\check{\Theta}_{k+1} \not\in \Omega_2$, then $\widehat{\Theta}_{k+1}$ is the projection of $\check{\Theta}_{k+1}$ onto $\Omega_2$, namely the closest element of $\Omega_2$ to $\check{\Theta}_{k+1}$. This two-step iteration reduces the original optimization, which is nonlinear and non-convex and hence very difficult to implement numerically, to two consecutive convex optimization problems that are very easy to compute.

## 3 Recursive Algorithms

**Least Squares Algorithms.** Note that the statistics of $(x_k, \varphi_k)$ and $(y_k, \Phi_k)$ may not be known, but their measurements are available. We first construct the unconstrained algorithms and obtain its constrained counter part for $\widehat{\theta}_k$ and then write the corresponding part for $\widehat{\Theta}_k$. In what follows, to signify the dependence of the cost function on $k$, we write it as $e_{i,k}(\theta)$ for $i = 1, 2$. Consider the problem of minimizing $e_{1,k}(\theta) = \frac{1}{k}\sum_{i=0}^{k-1}(\widehat{x}_i - \theta'\varphi_i)^2$, where $\widehat{x}_k$ are obtained by a back mapping technique via $\Theta_k$. Suppose that the matrix $\Psi_k = \sum_{i=0}^{k-1}\varphi_i\varphi_i'$ is invertible with probability one for $k$ large enough. Then, the minimizing $\theta$ is given by $\widehat{\theta}_k = \Psi_k^{-1}\sum_{i=0}^{k-1}\widehat{x}_i\varphi_i$. The above equation can be put into a recursive form (known as the *recursive least squares*) by expanding $\widehat{\theta}_{k+1} = \Psi_{k+1}^{-1}\left[\sum_{i=0}^{k-1}\widehat{x}_i\varphi_i + \widehat{x}_k\varphi_k\right]$ and using matrix inversion lemma; see (Ljung and Söderström, 1983, pp. 18–20)). To further simplify the procedure, taking a first-order (in $\Psi_k^{-1}$) expansion yields a linearized least squares approximation

$$\widehat{\theta}_{k+1} = \theta_k + \Psi_k^{-1}\varphi_k\widehat{\theta}_k + \Psi_k^{-1}\varphi_k[\widehat{x}_k - \varphi_k'\widehat{\theta}_k],\ \Psi_{k+1}^{-1} = \left[I - \Psi_k^{-1}\varphi_k\varphi_k'\right]\Psi_k^{-1}. \tag{1}$$

The convergence of the recursive least squares algorithm can be analyzed by the stochastic approximation method using random and matrix step sizes. Similarly, using the cost function $e_{2,k}(\Theta) = \frac{1}{k}\sum_{i=0}^{k-1}(y_i - \Theta'\Phi_i)^2$, and assuming that $\Xi_k = \frac{1}{k}\sum_{i=0}^{k-1}\Phi_i\Phi_i'$ is invertible, we arrive at the recursive least squares procedure for estimating $\Theta^*$

$$\begin{aligned} \Xi_{k+1}^{-1} &= \Xi_k^{-1} - \frac{\Xi_k^{-1}\Phi_k\Phi_k'\Xi_k^{-1}}{1+\Phi_k'\Xi_k^{-1}\Phi_k}, \\ \widehat{\Theta}_{k+1} &= \widehat{\Theta}_k + \frac{\Xi_k^{-1}}{1+\Phi_k'\Xi_k^{-1}\Phi_k}\Phi_k\left[y_k - \Phi_k'\widehat{\Theta}_k\right], \end{aligned} \tag{2}$$

and the linearized least squares approximation

$$\begin{aligned} \widehat{\Theta}_{k+1} &= \Theta_k + \Xi_k^{-1}\Phi_k\widehat{\Theta}_k + \Xi_k^{-1}\Phi_k[y_k - \Phi_k'\widehat{\Theta}_k], \\ \Xi_{k+1}^{-1} &= \left[I - \Xi_k^{-1}\Phi_k\Phi_k'\right]\Xi_k^{-1}. \end{aligned} \tag{3}$$

**Adaptive-Filtering-Type Procedure.** Replacing the matrix gain $\Psi_k$ by a scalar sequence of step sizes leads to further simplification of (1). We obtain the following recursive algorithm

$$\widehat{\theta}_{k+1} = \widehat{\theta}_k + \varepsilon k\varphi_k[\widehat{x}_k - \varphi_k'\widehat{\theta}_k],\ \widehat{\Theta}_{k+1} = \widehat{\Theta}_k + \varepsilon k\Phi_k[y_k - \Phi_k'\widehat{\Theta}_k], \tag{4}$$

where $\varepsilon_k$ is a sequence of real numbers satisfying $\varepsilon_k \geq 0$, $\varepsilon_k \to 0$ as $k \to \infty$, and $\sum_k \varepsilon_k = \infty$. Algorithm (4) is a stochastic approximation type procedure, and belongs to the family of adaptive filtering algorithms; see Widrow and Stearns (1985) for references on adaptive filtering and Kushner and Yin (2003) for an updated account on stochastic approximation algorithms. Commonly used step-size sequences include $\varepsilon_k = O(1/k)$, $\varepsilon_k = O(1/k^\gamma)$ for some $0 < \gamma < 1$, among others.

**Constrained Algorithms.** For the constrained algorithms, we use projection operators $\Pi_1$ and $\Pi_2$ to denote the projections onto $\Omega_1$ and $\Omega_2$, respectively. Recall that $\Omega_1 = \{\theta : \underline{\theta}_i \leq \theta^i \leq \overline{\theta}_i,\ i = 1,\ldots,2n+1\}$ and $\Omega_2 = \{\Theta : \underline{\Theta}_i \leq \Theta^i \leq \overline{\Theta}_i,\ i = 1,\ldots,m\}$, respectively. That is, we let the $i$th component of the estimates $\widehat{\theta}$ and $\widehat{\Theta}$ be confined to the interval $[\underline{\theta}_i, \overline{\theta}_i]$ and $[\underline{\Theta}_i, \overline{\Theta}_i]$, respectively. The constrained algorithms can be written as

$$\widehat{\theta}_{k+1} = \Pi_1\left[\widehat{\theta}_k + \varepsilon k\left[\widehat{x}_k - \varphi_k'\widehat{\theta}_k\right]\right],\ \widehat{\Theta}_{k+1} = \Pi_2\left[\widehat{\Theta}_k + \varepsilon k\left[y_k - \Phi_k'\widehat{\Theta}_k\right]\right]. \tag{5}$$

Variants of the algorithms discussed such as constant-step-size algorithms Kushner and Yin (2003), and algorithms with averaging Bather (1989), Kushner and Yang (1993), Kushner and Yin (2003), Polyak (1990), Ruppert (1991), Schwabe (1993), Yin (1995), leading to asymptotically optimal performance, can also be used.

# 4 Convergence and Rates of Convergence

In this section, we study asymptotic properties including convergence and rates of convergence of the recursive algorithms proposed in the last section. In fact, our attention will be devoted to the constrained algorithms using iterate averaging and/or averaging of both iterates and observations. Sufficient conditions will be provided, and existing results will be used whenever possible. To proceed, we need the following assumptions.

(A) The sequences $\{(\widehat{x}_k, \varphi_k)\}$ and $\{(y_k, \Phi_k)\}$ are stationary martingale difference processes.
  2. The following moment conditions hold. $E\varphi_k\varphi_k' = A_\varphi$, $E\Phi_k\Phi_k' = A_\Phi$, $E\varphi_k\widehat{x}_k = B_\varphi$, $E\Phi_k y_k = B_\Phi$, where $A_\varphi$ and $A_\Phi$ are symmetric and positive definite. Moreover, $E[|\widehat{x}_k|^2 + |\varphi_k|^2 + |y_k|^2 + |\Phi_k|^2] < \infty$.

More general correlated noise (such as $\varphi$-mixing processes), allowing diminishing correlations between the remote past and distant future, can be dealt with; the interested reader is referred to Kushner and Yin (2003) and the references therein. However, for ease of presentation, we confine our attention to the uncorrelated signals.

**Convergence.** The main ideas are outlined, but the verbatim proofs will be omitted. Note that $\widehat{\theta}_{k+1} = \widehat{\theta}_k + \varepsilon_k\varphi_k[\widehat{x}_k - \varphi_k'\widehat{\theta}_k] + \varepsilon_k R_{\varphi,k}$, where $\varepsilon_k R_{\varphi,k} = \widehat{\theta}_{k+1} - \widehat{\theta}_k - \varepsilon_k\varphi_k[\widehat{x}_k - \varphi_k'\widehat{\theta}_k]$ is the vector having shortest Euclidean length needed to bring $\varepsilon_k\varphi_k[\widehat{x}_k - \varphi_k'\widehat{\theta}_k]$ back to the constraint set $\Omega_1$. To prove the convergence of the recursive algorithms, in lieu of considering the discrete iterates directly, we take a continuous-time interpolation and work with appropriate function spaces. To be more specific, define $t_0 = 0$, $t_k = \sum_{i=0}^{k-1}\varepsilon_i$, $m(t) = k$, $0 \le t_k \le t < t_{k+1}$, $= 0$, $t < 0$, and $\theta^0(t) = \widehat{\theta}_k$ for $t \in [t_k, t_{k+1})$, $\theta^k(t) = \theta^0(t + t_k)$. That is, $\theta^0(t)$ is the piecewise constant interpolation of $\widehat{\theta}_k$ and $\theta^k(t)$ is a sequence obtained by shifting $\theta^0(t)$ sequence that enables us to bring the "tail" of the sequence to the foreground. For a more detailed account on the interpolations, see Kushner and Yin (2003). Similarly, define $R_\varphi^0(t) = 0$ for $t \le 0$, $R_\varphi^0(t) = \sum_{i=1}^{m(t)-1}\varepsilon_i R_{\varphi,i}$, for $t \ge 0$, $R_\varphi^k(t) = R_\varphi^0(t + t_k) - R_\varphi^0(t_k)$, for $t \ge 0$, $R_\varphi^k(t) = -\sum_{i=m(t+t_k)}^{k-1}\varepsilon_i R_{\varphi,i}$, for $t < 0$.

To proceed, define a set $C(\theta)$ as follows. For $\theta \in \Omega_1^0$, the interior of $\Omega_1$, $C(\theta)$ contains only the zero element. For $\theta \in \partial\Omega_1$, the boundary of $\Omega_1$, $C(\theta)$ is the convex cone generated by the outer normal at $\theta$ at which $\theta$ lies. Owing to the projection used, it is easily seen that $\{\widehat{\theta}_k\}$ is bounded uniformly, and as a result $\{\theta^k(\cdot)\}$ is a sequence of uniformly bounded functions. Moreover, it can be shown that $\{\theta^k(\cdot), R_\varphi^k(\cdot)\}$ is equicontinuous in the extended sense (Kushner and Yin (2003)). By virtue of an extended version of the Arzelà-Ascoli theorem, there is a subsequence that converges to some continuous limit uniformly on each bounded time interval. Thus, we arrive at the following result; the detailed proof may be found in (Kushner and Yin, 2003, Chapter 5).

Under condition (A), the sequences $\{\theta^k(\cdot), R^k_\varphi(\cdot)\}$ and $\{\Theta^k(\cdot), R^k_\Phi(\cdot)\}$ are uniformly bounded and equicontinuous in the extended sense. There exist convergent subsequences with limits denoted by $(\theta(\cdot), R_\varphi(\cdot))$ and $(\Theta(\cdot), R_\varphi(\cdot))$ satisfying the projected ordinary differential equations

$$\begin{aligned} \dot{\theta} &= B_\varphi - A_\varphi\theta + r_\varphi, \quad r_\varphi(t) \in -C(\theta(t)), \\ \dot{\Theta} &= B_\Phi - A_\Phi\Theta + r_\Phi, \quad r_\Phi(t) \in -C(\Theta(t)), \end{aligned} \tag{6}$$

where $R_\varphi(t) = \int_0^t r_\varphi(s)ds$, and $R_\Phi(t) = \int_0^t r_\Phi(s)ds$. The significance of (6) is that its stationary points are precisely the values of the parameter we are looking for. In view of the expression (6), $r_\varphi(\cdot)$ is the term needed to keep $\theta(\cdot)$ in $\Omega_1$. Interested reader is referred to (Kushner and Yin, 2003, Section 4.3) for further details. The significance of the $r_\varphi(\cdot)$ for the drug infusion process is that the expert knowledge allows us to keep the dynamics within a bounded region.

In addition, suppose that there are unique stationary points $\theta^*$ and $\Theta^*$ to the ODEs (6). Then $\widehat{\theta}_k$ and $\widehat{\Theta}_k$ converge to $\theta^*$ and $\Theta^*$, respectively, w.p.1. Moreover, $\overline{\theta}_k \to \theta^*$ and $\overline{\Theta}_k \to \Theta^*$ w.p.1. The proof of convergence of $\{\widehat{\theta}_k\}$ follows from that of the argument in (Kushner and Yin, 2003, Chapter 5), whereas the convergence of $\overline{\theta}_k$ is a consequence of a familiar fact in analysis. The uniqueness assumption on the stationary point is guaranteed by choosing the projection bounds to be large enough such that $\underline{\theta}_i < 0$ and $\overline{\theta}_i > 0$, $i \le 2n+1$. An argument using a form of the Kuhn-Tucker points and the ideas of active constraints to verify the assertion is in (Kushner and Yang, 1993, Chapter 9). When $\theta \in \Omega_1^0$, $r_\varphi = 0$. Note that interior to $\Omega_1$, the dynamics of the trajectories are determined by the ODE $\dot{\theta} = B_\varphi - A_\varphi\theta$. This is a linear ODE and hence has a unique solution for each initial condition. The stationary point of this ODE is precisely $\theta^* = A_\varphi^{-1}B_\varphi$. In the actual calculation using the recursive algorithm, if the iterates repeatedly hover the bounding surface, we would increase the size of the projection region. Similar discussion applies to the recursive estimates $\{\widehat{\Theta}_k\}$.

**Rate of Convergence.** For the study of rate of convergence, we concentrate on both iterate averaging algorithms and algorithms with averaging in both iterates and observations in what follows. Let us begin with (5), under the assumptions above, $\widehat{\theta}_k \to \theta^*$ w.p.1 and $\theta^* \in \Omega_1^0$. Thus effectively, the truncation or projection can be dropped in the rate of convergence analysis. Define $\xi_k = \varphi_k(\widehat{x}_k - \varphi_k'\theta^*)$, $\widetilde{\theta}_k - \theta^*$, $\eta_k = \Phi_k(y_k - \Phi_k'\Theta^*)$, and $\widetilde{\Theta}_k = \overline{\Theta}_k - \Theta^*$. Dropping the projection term and using $D_{k|j} = \prod_{\ell=j+1}^{k}(I - \varepsilon_\ell A_\varphi)$, if $j < k$, $= I$, if $j = k$, we have

$$\widetilde{\theta}_{k+1} = D_{k|0}\widetilde{\theta}_0 + \textstyle\sum_{j=0}^{k} \varepsilon_j D_{k|j}\xi_j + \sum_{j=0}^{k} \varepsilon_j D_{k|j}[A_\varphi - \varphi_j\varphi_j']\widetilde{\theta}_j. \tag{7}$$

Using (7), we will be able to obtain a distributional limit. The statement is in the following theorem. Its proof can be obtained as in (Kushner and Yin, 2003,

Chapter 10), where we dealt with stochastic processes aspect of the problem by taking continuous-time interpolations, worked with interpolated processes, and proved a limit stochastic differential equation holding for the centered and rescaled sequence. For the algorithm given in (5), $(\widehat{\theta}_k - \theta^*)/\sqrt{\varepsilon_k}$ converges in distribution to a normal random variable with mean 0 and covariance $\Sigma_\varphi = \mathrm{E}(\xi_1 \xi_1')$, and $(\widehat{\Theta}_k - \Theta^*)/\sqrt{\varepsilon_k}$ converges in distribution to a normal random variable with mean 0 and covariance $\Sigma_\Phi = \mathrm{E}(\eta_1 \eta_1')$. The scaling factor $1/\sqrt{\varepsilon_k}$ together with the asymptotic covariances $\Sigma_\varphi$ and $\Sigma_\Phi$ give the desired rates of convergence. Furthermore, it can be demonstrated that iterate averaging and averaging in both iterates and observations lead to improved convergence speed.

## References

J.A. Bather. Stochastic approximation: a generalisation of the Robbins-Monro procedure. In *Proceedings of the Fourth Prague Symposium on Asymptotic Statistics (Prague, 1988)*, pages 13–27, Prague, 1989. Charles University.

T.J. Gan, P.S. Glass, A. Windsor, F. Payne, C. Rosow, P. Sebel, and P. Manberg. Bispectral index monitoring allows faster emergence and improved recovery from propofol, alfentanil, and nitrous oxide anesthesia. *Anesthesiology*, 87(4):808–815, 1997.

H.J. Kushner and J. Yang. Stochastic approximation with averaging of the iterates: optimal asymptotic rate of convergence for general processes. *SIAM J. Control Optim.*, 31(4):1045–1062, 1993.

H.J. Kushner and G. Yin. *Stochastic approximation and recursive algorithms and applications*, volume 35 of *Applications of Mathematics (New York)*. Springer-Verlag, New York, second edition, 2003. Stochastic Modelling and Applied Probability.

L. Ljung and T. Söderström. *Theory and practice of recursive identification.* MIT Press Series in Signal Processing, Optimization, and Control, 4. MIT Press, Cambridge, Mass., 1983.

E. Mortier, M. Struys, T. De Smet, L. Versichelen, and G. Rolly. Closed-loop controlled administration of propofol using bispectral analysis. *Anesthesia*, 53:749–754, 1998.

B.T. Polyak. A new method of stochastic approximation type. *Avtomat. i Telemekh.*, (7):98–107, 1990.

D. Ruppert. Stochastic approximation. In *Handbook of sequential analysis*, volume 118 of *Statist. Textbooks Monogr.*, pages 503–529. Dekker, New York, 1991.

R. Schwabe. Stability results for smoothed stochastic approximation procedures. *Z. Angew. Math. Mech.*, 73(7-8):T639–T643, 1993. Bericht über die Wissenschaftliche Jahrestagung der GAMM (Leipzig, 1992).

L.Y. Wang and H. Wang. Control-oriented modeling of bis-based patient response to anesthesia infusion. In *2002 Internat. Conf. Math. Eng. Techniques in Medicine and Bio. Sci., (Las Vegas, 2002)*, 2002a.

L.Y. Wang and H. Wang. Feedback and predictive control of anesthesia infusion using control-oriented patient models. In *2002 Internat. Conf. Math. Eng. Techniques in Medicine and Bio. Sci., (Las Vegas, 2002)*, 2002b.

B. Widrow and S.D. Stearns. *Adaptive Signal Processing.* Prentice-Hall, Englewood Cliffs, New Jersey, 1985.

G. Yin. Adaptive filtering with averaging. In *Adaptive control, filtering, and signal processing (Minneapolis, MN, 1993)*, volume 74 of *IMA Vol. Math. Appl.*, pages 375–396. Springer, New York, 1995.

# List of Contributors

**K.M. Anstreicher**
Tippie College of Business
University of Iowa
Iowa City, IA. U.S.A.
kurt-anstreicher@uiowa.edu

**A.C. Atkinson**
London School of Economics
London WC2A 2AE, UK
a.c.atkinson@lse.ac.uk

**A. Baldi Antognini**
University of Bologna
via Belle Arti 41
40126, Bologna, Italy
baldi@stat.unibo.it

**D. Basso**
University of Padova
Department of Statistics
Str. C. Battisti 241
35121 Padova, Italy
dario.basso@unipd.it

**S. Biedermann**
Ruhr-Universität Bochum
Fakultät für Mathematik
44780 Bochum, Germany
stefanie.biedermann@ruhr-uni-bochum.de

**A. Biswas**
Applied Statistics Unit
Indian Statistical Institute
203 B. T. Road, Kolkata 700 108, India
atanu@isical.ac.in

**B. Bogacka**
School of Mathematical Sciences
Queen Mary, University of London
Mile End Road, London E1 4NS, U.K.
B.Bogacka@qmul.ac.uk

**H. Dette**
Ruhr-Universität Bochum
Fakultät für Mathematik
44780 Bochum, Germany
holger.dette@ruhr-uni-bochum.de

**H. Evangelaras**
National Technical University of Athens
Department of Mathematics
Zografou 15773, Athens, Greece
harris11@central.ntua.gr

**V.V. Fedorov**
GlaxoSmithKline,
1250 So Collegeville Rd
PO Box 5089, Collegeville
PA 19426-0989, U.S.A.
Valeri.V.Fedorov@gsk.com

**N. Flournoy**
University of Missouri - Columbia
Department of Statistics
146 Middlebush Building
Columbia, MO 65211, U.S.A.
flournoyn@missouri.edu

**J.P. Gauchi**
INRA-Jouy en Josas
Biometrics Unit
78352 Jouy-en-Josas cedex, France
gauchi@banian.jouy.inra.fr

**A. Giovagnoli**
University of Bologna
via Belle Arti 41
40126, Bologna, Italy
giovagno@stat.unibo.it

**R. Harman**
Comenius University
Faculty of Mathematics, Physics and Informatics
Department of Probability and Statistics
Bratislava, Slovakia
harman@fmph.uniba.sk

**F. Hu**
University of Virginia
Department of Statistics
P.O. Box 400135
Charlottesville, Virginia 22904, U.S.A.
fh6e@virginia.edu

**C.P. Kitsos**
Technological Educational Institute of Athens
Department of Mathematics
Ag. Spyridonos, GR-12210 Egaleo, Greece
xkitsos@teiath.gr

**C. Koukouvinos**
National Technical University of Athens
Department of Mathematics
Zografou 15773, Athens, Greece
ckoukouv@math.ntua.gr

**J. Lee**
IBM T.J. Watson Research Center,
P.O. Box 218, Yorktown Heights,
New York, U.S.A.
jonlee@us.ibm.com

**S. Leonov**
GlaxoSmithKline
1250 So Collegeville Rd
PO Box 5089, Collegeville
PA 19426-0989, U.S.A.
Sergei.2.Leonov@gsk.com

**J. López-Fidalgo**
University of Salamanca
Departamento de Estadistica
Edificio de la Merced, Plza. de la Merced, 1-4, Fac. Ciencias,
Salamanca, Spain
fidalgo@usal.es

**S. Mandal**
Department of Statistics
University of Manitoba
Winnipeg, MB, R3T 2N2, Canada
saumen_mandal@umanitoba.ca

**I. Martínez**
University of Almería
Dpt. Statistics and Applied Mathematics
La Cañada de San Urbano s/n
04120 Almería, Spain
ijmartin@ual.es

**V.B. Melas**
St.-Petersburg State University
University av., 28
St.Petersburg, 198904, Russia
v.melas@pobox.spbu.ru

**C.H. Müller**
Carl von Ossietzky University Oldenburg
Institute for Mathematics
Postfach 2503
D-26111 Oldenburg, Germany
mueller@math.uni-oldenburg.de

**I. Ortiz**
University of Almería
Department of Statistics and Applied Mathematics
La Cañada de San Urbano s/n
04120 Almería, Spain
iortiz@ual.es

**A. Pázman**
Comenius University
Department of Probability and Statistics
Mlyns-ká Dolina
84215 Bratislava, Slovakia
pazman@center.fmph.uniba.sk

**A. Pepelyshev**
St.-Petersburg State University
Department of Mathematics
University av., 28
St.Petersburg, 198504, Russia
andrey@ap7236.spb.edu

**A. Ponce de Leon**
Universidade do Estado do Rio de Janeiro
Institute of Social Medicine
Rua São Francisco Xavier 524
Maracanã, Rio de Janeiro, Brasil
ponce@ims.uerj.br

**L. Pronzato**
Laboratoire I3S
Université de Nice Sophia-Antipolis/CNRS
Bat. Euclide, Les Algorithmes, BP 121
2000 route des Lucioles
06903 Sophia Antipolis Cedex, France
pronzato@i3s.unice.fr

**H.S. Rabie**
University of Missouri-Columbia
Department of Statistics
146 Middlebush Building
Columbia, MO 65211-4100, U.S.A.
hsr2mf@mizzou.edu

**E. Rodrigues Pinto**
Universidade Federal do Rio de Janeiro
COPPE/Department of Product Engineering
68504, Ilha do Fundao, Rio de Janeiro, Rj 21945-970, Brasil
edmilsonr@hotmail.com

**C. Rodríguez**
University of Almería
Department of Statistics and Applied Mathematics
La Cañada de San Urbano s/n
04120 Almería, Spain
crt@ual.es

**W.F. Rosenberger**
University of Maryland
Department of Mathematics and Statistics
Baltimore County, 1000 Hilltop Circle, Baltimore, Maryland 21250, U.S.A.
billr@math.umbc.edu

**L. Salmaso**
University of Padova
Department of Management and Engineering
Stradella San Nicola 3
36100 Vicenza, Italy
salmaso@stat.unipd.it

**É. Thierry**
Laboratoire I3S
Université de Nice Sophia-Antipolis/CNRS
Bat. Euclide, Les Algorithmes, BP 121

2000 route des Lucioles
06903 Sophia Antipolis Cedex, France
et@i3s.unice.fr

**B. Torsney**
Department of Statistics
University of Glasgow
Glasgow, G12 8QW, United Kingdom
b.torsney@stats.gla.ac.uk

**C. Trandafir**
University of Salamanca
Departamento de Estadistica
Edificio de la Merced, Plza. de la Merced, 1-4, Fac. Ciencias, Salamanca, Spain
a110619@usal.es

**Y. Tymofyeyev**
University of Maryland
Department of Mathematics and Statistics
Baltimore County, 1000 Hilltop Circle, Baltimore, Maryland 21250, U.S.A.
ytymof1@math.umbc.edu

**D. Uciński**
Institute of Control and Computation Engineering
University of Zielona Góra
ul. Podgórna 50
65–246 Zielona Góra, Poland
D.Ucinski@issi.uz.zgora.pl

**I.N. Vuchkov**
University of Chemical Technology and Metallurgy
1756 Sofia, Bulgaria
vuchkov@uctm.edu

**H. Wang**
Medical Center/Wayne State University, Detroit
Harper University Hospital, Detroit
Department of Anesthesiology
Michigan 48201, U.S.A.
howang@med.wayne.edu

**L.Y. Wang**
Wayne State University
Department of Electrical and Computer Engineering
Detroit, Michigan 48202, U.S.A.
lywang@ece.eng.wayne.edu

**É. Wolsztynski**
Laboratoire I3S
Université de Nice Sophia-Antipolis/CNRS
Bat. Euclide, Les Algorithmes, BP 121
2000 route des Lucioles
06903 Sophia Antipolis Cedex, France
wolsztyn@i3s.unice.fr

**H.P. Wynn**
London School of Economics
Houghton Street
London, WC2A 2AE, UK
h.wynn@lse.ac.uk

**G. Yin**
Wayne State University
Department of Mathematics
Detroit, Michigan 48202, U.S.A.
gyin@math.wayne.edu

**H. Zheng**
Wayne State University
Department of Electrical and Computer Engineering
Detroit, Michigan 48202, U.S.A.
aj6823@wayne.edu

# List of Referees

**A.C. Atkinson**
London School of Economics
Houghton Street, London WC2A 2AE, U.K.
a.c.atkinson@lse.ac.uk

**W. Bergsma**
EURANDOM,
P.O. Box 513
5600 MB, Eindhoven, The Netherlands
bergsma@eurandom.tue.nl

**A. Di Bucchianico**
EURANDOM and
Eindhoven University of Technology
Department of Mathematics and Computer Science
P.O. Box 513
5600 MB Eindhoven, The Netherlands
a.d.bucchianico@tue.nl

**V.V. Fedorov**
GlaxoSmithKline
1250 So Collegeville Rd
PO Box 5089, Collegeville
PA 19426-0989, U.S.A.
Valeri.V.Fedorov@gsk.com

**N. Flournoy**
University of Missouri - Columbia
Department of Statistics
146 Middlebush Building
Columbia, MO 65211
flournoyn@missouri.edu

**P. Hackl**
Wirtschaftuniversität Wien
Abteilung fr Wirtschaftsstatistik
Augasse 2-6,
A-1090 Wien, Österreich
peter.hackl@wu-wien.ac.at

**C.P. Kitsos**
Technological Educational Institute of Athens
Department of Mathematics
Ag. Spyridonos, GR-12210 Egaleo, Greece
xkitsos@teiath.gr

**J. López-Fidalgo**
University of Salamanca
Departamento de Estadistica
Edificio de la Merced, Plza. de la Merced, 1-4, Fac. Ciencias,
Salamanca, Spain
fidalgo@usal.es

**W. Müller**
Wirtschaftuniversität Wien
Abteilung fr Wirtschaftsstatistik
Augasse 2-6, A-1090 Wien, Österreich
werner.mueller@wu-wien.ac.at

**A. Pázman**
Comenius University
Department of Probability and Statistics
Mlyns-ká Dolina
84215 Bratislava, Slovakia
pazman@center.fmph.uniba.sk

**L. Pronzato**
Laboratoire I3S
Université de Nice Sophia-Antipolis/CNRS
Bat. Euclide, Les Algorithmes, BP 121
2000 route des Lucioles
06903 Sophia Antipolis Cedex, France
pronzato@i3s.unice.fr

**E. Riccomagno**
Department of Statistics
University of Warwick
Gibbet Hill Road,
Coventry, CV4 7AL, UK
emr@stats.warwick.ac.uk

**W.F. Rosenberger**
University of Maryland
Department of Mathematics and Statistics
Baltimore County, 1000 Hilltop Circle, Baltimore, Maryland 21250, U.S.A.
billr@math.umbc.edu

**R. Schwabe**
Otto von Guericke University Magdeburg
Institute for Mathematical Stochastics
PF 4120
D-39016 Magdeburg, Germany
rainer.schwabe@mathematik.uni-magdeburg.de

**B. Torsney**
Dept of Statistics,
University of Glasgow,
Glasgow, G12 8QW, United Kingdom
b.torsney@stats.gla.ac.uk

**I.N. Vuchkov**
University of Chemical Technology and Metallurgy,
1756 Sofia, Bulgaria
vuchkov@uctm.edu

**H.P. Wynn**
London School of Economics
Houghton Street,
London, WC2A 2AE,UK
h.wynn@lse.ac.uk

**A.A. Zhigljavsky**
School of Mathematics
Cardiff University
Senghennydd Road
Cardiff, CF24 4YH, UK
ZhigljavskyAA@cardiff.ac.uk

# List of Figures


| | | |
|---|---|---|
| Fig. 1 | Decrease in masked spectral bound for problem with $n = 63$, $s = 31$. | 6 |
| Fig. 2 | Comparison of bounds on problems with $n = 63$. | 7 |
| Fig. 3 | Comparison of bounds on problems with $n = 124$. | 8 |
| Fig. 1 | Two consecutive first-order reactions: locally-optimum designs when [A], [B] and [C] are measured. Left panel, times; right-panel, design weights. | 18 |
| Fig. 2 | Two consecutive first-order reactions when [A], [B] and [C] are measured. Continuous line, efficiency of Bayesian optimum design for $\lambda_0 = (0.2, 0.4, 0.6, 0.8, 1.0)^T$; dashed lines, efficiencies of the optimum designs for individual members of $\lambda_0$. | 19 |
| Fig. 3 | Two consecutive first-order reactions when [A], [B] and [C] are measured. Continuous line, efficiency of Bayesian optimum design for $\lambda_0 = (0.1, 0.2, 0.3, 0.9, 1.0)^T$; dashed lines, efficiencies of the optimum designs for individual values 0.2, 0.4, 0.6, 0.8, 1.0 for $\lambda$. | 20 |
| Fig. 1 | Sensitivity functions of locally $D$-optimal designs, unknown $\gamma$ and $\Lambda$; $\Sigma = I_2$. Top two curves: model (9), $\Lambda = \Lambda_{1F}$ (dashed, $\lambda_{ii} = 1, \lambda_{ij} = 0.6$) and $\Lambda_{0F} = 0.01 I_3$ (dash-dotted). Bottom two curves: reduced model (11), $\Lambda = \Lambda_{1R}$ (solid, $\lambda_{11} = \lambda_{22} = 1, \lambda_{12} = 0.6$) and $\Lambda_{0R} = 0.01 I_2$ (dotted). | 67 |
| Fig. 1 | The dependence of the support points $x_1^*$ and $x_2^*$ on $z$. | 103 |
| Fig. 1 | Evolution of $\rho(\kappa, \alpha)$ as a function of $\alpha$ in Example 1: $\kappa = 3$ and $\hat{\theta}^N_{TSLS}$ should be preferred to $\hat{\theta}^N_{PWLS}$ for $\alpha < 1/4$. | 123 |
| Fig. 2 | Evolution of $\log\det[\mathbf{M}_1^{-1}(\xi^*_{PWLS}, \theta)\mathbf{M}_2(\xi^*_{PWLS}, \theta)\mathbf{M}_1^{-1}(\xi^*_{PWLS}, \theta)]$ as a function of $\kappa$ (dashed line) and $\log\det\mathbf{M}^{-1}(\xi^*_{TSLS}, \theta)$ (full line) in Example 2. | 124 |
| Fig. 1 | $(q_a, q_b, q_c)$ such that $\mathrm{Re}(\lambda_1) < 1/2$. | 186 |
| Fig. 1 | (a) Responses vs. time for the $T$-optimal temperature (continuous and dashed lines denote the responses of the true and alternative models, respectively), and (b) the $T$-optimal temperature profile. | 197 |

Fig. 1 Predicted response. . . . . . . . . . 208
Fig. 2 Biased response. . . . . . . . . . 208
Fig. 3 Contour plot for variance. . . . . . . . . . 209
Fig. 4 Contour plot for $L$. . . . . . . . . . 209
Fig. 1 Overall System Configuration . . . . . . . . . . 221

# List of Tables


Table 1 Bayesian optimum designs for three five-point priors on $\lambda$. . . . 19

Table 1 Behaviour of the IMP test under $H_0$ with standard Normal and Cauchy errors. . . . . 38

Table 2 A comparison between IMP test, Lenth's test and the parametric $F$-test under $H_1$ with standard Normal and Cauchy errors. . . . . 39

Table 1 The support points $x_1^*, x_2^*$ of the locally $D$-optimal designs $\xi^* = \{x_1^*, x_2^*; 1/2, 1/2\}$ with respect to the parameter $\theta = (1, \theta_2)^T$. . . . . 43

Table 2 Locally $D$-optimal designs with respect to several restricted design spaces $[0, x_{max}]$ and different values of $\theta$. . . . . 44

Table 3 Standardized maximin $D$-optimal designs $\xi^*$ for model (2) with unrestricted design space with respect to various rectangular parameter spaces. . . . . 46

Table 4 Standardized maximin $D$-optimal designs $\xi^*$ for model (2) with unrestricted design space with respect to various rectangular parameter spaces. . . . . 47

Table 5 Standardized maximin $D$-optimal designs for model (2) with unrestricted design space with respect to various triangular parameter regions. . . . . 47

Table 6 Standardized maximin $D$-optimal designs for model (2) with respect to various rectangular parameter regions on restricted design spaces. . . . . 48

Table 7 Standardized maximin $D$-optimal designs for model (2) with respect to various triangular parameter regions on restricted design spaces. . . . . 48

Table 1 Exact proportion of allocations (and the SD's) and limiting proportion of allocations for exponential responses. Here $m = 5$, $c = 2$, $\lambda_2 = 1$ and 1000 simulations are taken. . . . . 57

Table 2 Exact proportion of allocations (and the SD's) and limiting proportion of allocations for normal responses. Here $m = 5$, $c = 0$, $\mu_2 = 1$, $\sigma_1 = 1$, $\sigma_2 = 1.2$ and 1000 simulations are taken. 57

Table 3 Exact proportion of allocations (and the SD's) and limiting proportion of allocations for normal responses. Here $m = 5$, $c = 0$, $\mu_2 = 1$, common $\sigma = 1$, only one covariate $x \sim N(1, 1)$, $\beta = 1$ and 1000 simulations are taken. .......... 57
Table 1 Coefficients of the Taylor expansions for $x_1^*$ and $x_2^*$ in a neighbourhood of $z = 1$ in terms of powers of $(z - 1)$. ........ 103
Table 2 Coefficients of the Taylor expansions for $x_1^*$ and $x_2^*$ in a neighbourhood of $z = 1$ in powers of $1/(z - 1)$. .............. 103
Table 1 Value of $\hat{C}_n$ in the location model for the LS, Minimum Hellinger Distance (Beran, 1978) and Minimum Entropy estimators ((3) for $ME_1$ and (4) for $ME_2$). The errors are standard normal, bi-exponential $(f(x) = (1/\sqrt{2}) \exp(-\sqrt{2}|x|))$, and Student's $t_\nu$ with $\nu = 3, 5$ and 10 degrees of freedom. ................................ 130
Table 2 Values $(T_n, D_n)$ of the trace $(\times 10^{-3})$ and determinant $(\times 10^{-2})$ of $\hat{C}_n$ in Example 2 for different $f$ and estimators (with $q$ outliers for the last three lines). ................... 131
Table 1 Efficiencies contained in cells of Table 2. ................... 160
Table 2 $A$-efficiencies in multi-factor models. ....................... 160
Table 1 Initial iteration results: Number of iterations needed to achieve $\max_{1 \leq j \leq J} \{F_j\} \leq 10^{-n}$ for $n = 1, 2, 3, 4$. ................ 168
Table 2 Iteration results with the proposed approach : number of iterations needed to achieve $\max_{1 \leq j \leq J} \{F_j\} \leq 10^{-n}$ for $n = 1$, 2, 3, 4. ................................................ 170
Table 1 Efficiencies for estimating the parameters $V$, $K$ and $F$ when $b = 5$. ................................................ 177
Table 2 Quasi-optimal compound designs and their efficiencies for estimating the parameters $K$, $V$ and $F$ for different values of $\lambda$ and $\mu$. ............................................. 179
Table 1 Simulated vs. theoretical covariance matrices for $\mathbf{Y}_n / \sum_{j=1}^{K} Y_{jn}$. ...................................... 188
Table 2 Simulated vs. theoretical covariances matrices for $\mathbf{N}/n$. ..... 190
Table 1 Second order simplex design for rubber mixture experiment. . 207

# Index

$D$-optimality, 90
$E$-optimality, 90, 91
$L$-optimality, 153
$TA$-optimality, 112
$TD$-optimality, 112
$TE$-optimality, 112
$TU$-optimality, 112
$\Phi$-optimality, 90
$\Phi_p$-optimality, 90
$\beta$-expectation tolerance region, 108
$\sqrt{n}$-consistency, 121, 128

adaptive estimation, 126, 128
Adjustable Biased Coin Design, 24
algorithm
– multiplicative, 164
almost exact density, 73
anesthesia, 219
asymptotic normality, 120, 121
asymptotically locally sufficient, 127

Bayesian
– methods, 213
– optimum design, 17
– tolerance region, 108
Box-Cox transformation, 14
Bradley-Terry model, 168

change of optimality criterion, 89
chemical kinetics, 15
compartmental model, 41
conditional entropy, 131
consecutive first-order reactions, 17
consistency, 120
continuation ratio model, 135
correlated errors, 21

decay
– first-order, 15
density
– almost exact, 73
derivative
– directional, 164
– vertex directional, 164
design
– Bayesian optimum, 17
– Ehrenfest-Brillouin, 27
– locally $D$-optimum, 15
– optimal regression, 166
– robust optimal, 19
– urn, 25
design measure, 130
differential equations, 21
directional derivative, 164
distribution
– optimizing, 165

efficiency, 17, 90, 126
Ehrenfest
– urn design, 23
– urn design, 25
Ehrenfest design
– asymmetric, 26
– symmetric, 26
Ehrenfest-Brillouin design, 27
entropy, 126, 212
errors
– correlated, 21

Fabian process, 75
first-order decay, 15
first-order reaction
– consecutive, 17

general equivalence theorem, 164
generalized $A$-optimality, 72
generalized $D$-optimality, 72

geometric mean, 14

Hellinger distance, 127
heteroscedastic model, 153

identifiability, 130
identification, 219
iteratively re-weighted least squares, 121

kernel estimation, 126, 128
Kullback-Leibler divergence, 127
kurtosis, 117, 120

locally $D$-optimum design, 15, 42
location model, 125

maximum entropy sampling, 1, 212
maximum likelihood, 117, 125
mean coverage tolerance region, 108
measure space, 214
minimum entropy estimation, 126
model
– Bradley-Terry, 168
– heteroscedastic, 153
– multi-factor, 153
multi-factor model, 153
multiplicative algorithm, 164
multivariate response, 15

nonlinear regression, 117, 125, 130

OIC, 90
optimal design
– $TA$, 112
– $TD$, 112
– $TE$, 112
– $TU$, 112
– regression, 166
– robust, 19
optimality
– $D$, 90
– $E$, 90, 91
– $\Phi$, 90
– $\Phi_p$, 90
– generalized $A$, 72
– generalized $D$, 72
– locally $D$, 15
optimality criterion
– change of, 89
optimization
– stochastic, 75
optimizing distribution, 165
orthogonally invariant criterion, 90

parameter sensitivities, 16
penalized weighted least squares, 120
penalty, 74
pharmacokinetics, 15
process
– Fabian, 75

randomized design, 117, 118
resolution
– spectral, 216
response
– multivariate, 15
robust optimal design, 19, 42

sampling
– maximum entropy, 1, 212
score function, 127
semi-parametric model, 125
spectral resolution, 216
spring balance weighing, 93
standardized maximin $D$-optimality, 42
stochastic optimization, 75
support points, 130

Taguchi method, 143
theorem
– general equivalence, 164
tolerance region
– $\beta$-expectation, 108
– Bayesian, 108
– mean coverage, 108
transform both sides, 15
transformation
– Box-Cox, 14

uniform convergence, 118
uniform Strong Laws of Large Numbers, 118

variance function, 118
vertex directional derivative, 164

weighing
– spring balance, 93
weighted least squares, 117

GPSR Compliance

*The European Union's (EU) General Product Safety Regulation (GPSR) is a set of rules that requires consumer products to be safe and our obligations to ensure this.*

*If you have any concerns about our products, you can contact us on ProductSafety@springernature.com*

In case Publisher is established outside the EU, the EU authorized representative is:

Springer Nature Customer Service Center GmbH
Europaplatz 3
69115 Heidelberg, Germany

**Batch number: 10577876**

Printed by Printforce, the Netherlands